EVERY BIT COUNTS
Posit Computing

Written by one of the foremost experts in high-performance computing and the inventor of Gustafson's law, Every Bit Counts: Posit Computing explains the foundations of a new way for computers to calculate that saves time, storage, energy, and power by packing more information into every bit than do legacy approaches. Both the AI and HPC communities are increasingly using the posit approach that Gustafson introduced in 2017, which may be the future of technical computing.

What may seem like a dry subject is made engaging by including the human and historical side of the struggle to represent numbers on machines. The book is richly illustrated in full color throughout, with every effort made to make the material as clear and accessible as possible, and even humorous.

Starting with the simplest form of the idea, the chapters gradually add concepts according to stated mathematical and engineering design principles, building a robust tool kit for creating application-specific number systems. There is also a thorough explanation of the Posit™ Standard (2022), with motivations and examples that expand on that terse 12-page document.

Prof. John L. Gustafson is a pioneer in high-performance computing, known for a breakthrough in parallel processing (Gustafson's law) for which he received the inaugural Gordon Bell Prize. He is a former Director at Intel Labs and a former Senior Fellow and Chief Product Architect at AMD. He now leads efforts to replace legacy computer arithmetic with high-efficiency next-generation approaches.

Chapman & Hall/CRC
Computational Science Series

Series Editor: Horst Simon and Doug Kothe

For more information about this series please visit:
https://www.crcpress.com/Chapman--HallCRC-Computational-Science/book-series/
CHCOMPUTSCI

EVERY BIT COUNTS

Posit Computing

John L. Gustafson

CRC Press
Taylor & Francis Group
Boca Raton London New York

CRC Press is an imprint of the
Taylor & Francis Group, an **informa** business

A CHAPMAN & HALL BOOK

Designed cover image: Sumei Chew

MATLAB® and Simulink® are trademarks of The MathWorks, Inc. and are used with permission. The MathWorks does not warrant the accuracy of the text or exercises in this book. This book's use or discussion of MATLAB® or Simulink® software or related products does not constitute endorsement or sponsorship by The MathWorks of a particular pedagogical approach or particular use of the MATLAB® and Simulink® software.

First edition published 2025
by CRC Press
2385 NW Executive Center Drive, Suite 320, Boca Raton FL 33431

and by CRC Press
4 Park Square, Milton Park, Abingdon, Oxon, OX14 4RN

CRC Press is an imprint of Taylor & Francis Group, LLC

Library of Congress Cataloging-in-Publication Data
Names: Gustafson, John L., 1955- author.
Title: Every bit counts : posit computing / John L. Gustafson.
Description: First edition. | Boca Raton, FL : CRC Press, 2025. | Series: Chapman & Hall/CRC computational science | Includes bibliographical references and index. | Summary: "Written by one of the foremost experts in high-performance computing and the inventor of Gustafson's law, Every Bit Counts: Posit Computing explains the foundations of a new way for computers to calculate that saves time, storage, energy, and power by packing more information into every bit than do legacy approaches. Both the AI and HPC communities are increasingly using the posit approach that Gustafson introduced in 2017, which may be the future of technical computing. What may seem like a dry subject is made engaging by including the human and historical side of the struggle to represent numbers on machines. The book is richly illustrated in full color throughout, with every effort made to make the material as clear and accessible as possible, and even humorous. Starting with the simplest form of the idea, the chapters gradually add concepts according to stated mathematical and engineering design principles, building a robust tool kit for creating application-specific number systems. There is also a thorough explanation of the Posit Standard (2022), with motivations and examples that expand on that terse 12-page document"-- Provided by publisher.
Identifiers: LCCN 2024036235 (print) | LCCN 2024036236 (ebook) | ISBN 9781032738062 (hbk) | ISBN 9781032738055 (pbk) | ISBN 9781003466024 (ebk)
Subjects: LCSH: Computer science--Mathematics.
Classification: LCC QA76.9.M35 G84 2025 (print) | LCC QA76.9.M35 (ebook) | DDC 004.01/51--dc23/eng/20240910
LC record available at https://lccn.loc.gov/2024036235
LC ebook record available at https://lccn.loc.gov/2024036236

ISBN: 978-1-032-73806-2 (hbk)
ISBN: 978-1-032-73805-5 (pbk)
ISBN: 978-1-003-46602-4 (ebk)

DOI: 10.1201/9781003466024

Typeset in Source Sans Pro
by KnowledgeWorks Global Ltd.

Publisher's note: This book has been prepared from camera-ready copy provided by the authors.

For Nan Ripley,
who is definitely not a number.

Contents

Preface

"I'D FIND MATH EVEN MORE INTERESTING IF WE DIDN'T HAVE TO DO ALL THAT MATH."

In 1920, a chemist might write a scientific paper with language like this:

> We dissolved borax in water.

By around 1960, the wording would be more like this:

> **Sodium borate was dissolved in water.**

In the present day, I would not be surprised to see it written like this:

> Boration was achieved in aqueous solution.

Technical writing seems to have acquired a style designed not so much to communicate to you as to impress you with how intelligent the author is. This is not that kind of book. Einstein once said, "You do not really understand something unless you can explain it to your grandmother." So I wrote this book with that in mind and tested it. I had a grandmother (not mine) with no math training read it, and it passed. She totally gets it. (Thank you, Ms. Levitt!)

Even though this is a book about computer arithmetic, it is aimed at a general audience. It presents the human side of things using the historical record along with my own personal experiences. You do not need calculus or anything more advanced than a K-12 math background to read this book. I have done everything in my power to make this very dry subject as clear and interesting as possible. And sometimes funny. Borderline outrageous.

The publisher has generously consented to the use of four-color printing on every page. Perhaps the hardest thing is all the binary numbers, which even computer science majors struggle to read. Color-coding the 0 and 1 bits in long bit strings makes them much easier to figure out. Most people are visual learners, and I spent more than half the time creating this book in crafting the illustrations that explain the concepts. Text backgrounds also are color-coded as guideposts to inform you of when there is math, computer code, important points, pitfalls, and so on.

There is at least one Exercise for the Reader in every chapter, for those who are ambitious. If you do not attempt them, you still might want to read the Answers to Exercises section because there is new material there that is not found elsewhere in the book.

There are no footnotes, end notes, or even References. Modern web search has made it so easy to find a source from the author's name and subject that a References section seems obsolete. There is also no "Theorem-Proof" structure. The opposite approach has been used, that of building up to an idea one step at a time, even allowing you to see mistakes made along the way. I speak from experience when I say that mathematicians will hate this book and its informal style. "Where are the Greek letters? Where is the cryptic notation? I feel like you are talking down to me." They are certainly entitled to their opinion.

> This is a book designed to be understood.

Acknowledgments

The governments of the US and Singapore, my two favorite nations, made this book possible.

Three people stand out as having enabled the research that led to the discovery of posit format and thus the contents of this book:

- Dr. Marek Michalewicz, former Director of Singapore's A*STAR Computational Resource Centre (A*CRC),
- Professor Tin Wee Tan, Director of the Singapore National Supercomputing Centre (NSCC), and
- Professor Weng-Fai Wong of the National University of Singapore's School of Computing.

After seeing the potential in my earlier efforts to find something better than floating-point arithmetic, they set up an incredible arrangement for me to move to Singapore and continue the research. Professor Wong also let me co-advise his two brilliant graduate students, Elavarasi Manogaran and Himeshi De Silva, who did hundreds of hours of experimental computational studies that steered the choices made in the design of posit arithmetic.

I am also at a loss to express my gratitude for those who reviewed my drafts and provided detailed (and much-needed) feedback. Hauke Rehr, besides being a crack mathematician with a sharp sense of correctness, has a seemingly superhuman ability to tell when a comma is in a font one point size too small, that sort of thing. Laslo Hunhold not only provided great feedback, but he advanced the state of the art with his own "takum" format in time for it to be included in the final chapter. My daughter, Janice Marquardt, was of immense help spotting things no other reviewer noticed.

The creation of posit format was triggered by a request from MIT's Kurt Keville, who told me in November 2016 that the RISC-V committee wanted a real number format that wasn't IEEE Std 754, and asked me to design one for them. After three weeks of struggling with modifications to Type II unums, I made the hardware-friendly break-through on December 3. I am also grateful to Kurt for detailed feedback for the draft of this book. Like my Singapore supporters, he is another *raison d'être* for the posit arithmetic format.

Dr. Siew Hoon (Cerlane) Leong of NSCC led the development of the SoftPosit library, the establishment of the posithub.org website, and the establishment of the Conference on Next-Generation Arithmetic (CoNGA) series. She also helped make many key decisions in the *Posit Standard (2022)*, including how the quire should handle exceptions. I do not know how I could have established a beachhead for the posit format without her boundless energy and keen eye for what needs to be done.

The cover artwork was done by a remarkable and multitalented Singapore artist, Sumei Chew. Without being a mathematician, she managed to capture the sense of wonder and genuine fun that exists in mathematics and that I hope readers will find in this book. She was a great friend to me during my years in Singapore and made my experience even better than it already was.

To my surprise and delight, cartoonist Sidney Harris allowed me to use his excellent cartoons that you will find scattered throughout the book. It is remarkable how perfectly many of his punchlines line up with the topic of the section where they appear. For a cartoonist, he somehow seems to know an awful lot about how scien-tists act when they're next to blackboards. Thank you, Sidney.

How to Read this Book

When jargon is introduced (some of it Industry-wide and some of it coined in this book), the first use is explained in a light-cyan **Definition** box, like this:

> **Definition:** The *Golden Zone* of a real number format is the dynamic range where its relative accuracy is as good as or better than that for IEEE Std 754 floats of the same precision.

If you forget a definition, check the Glossary (pages 353–361).

When someone else's writing or historical work is being cited, it is displayed in a parchment-colored box, like this:

> There are two fields whose total area is 1800 square yards. One produces grain at the rate of two-thirds of a bushel per square yard while the other produces grain at the rate of half a bushel per square yard. If the total yield is 1100 bushels, what is the size of each field?

A light red background color is used to signal the reader that something is mathematical, like

$$x = ((1 - 3s) + f) \times 2^{(1-2s) \times (r+s)}.$$

It is usually possible to skip over those formula displays without losing the flow of the main body text.

Similarly, you can usually skip over computer code, unless you're interested. It has a light green background. There are two kinds of computer code; one is something like C or C++ to show how something might be programmed in general, like this:

```
unsigned reverse (unsigned x) {
    x = ((x & 0 x55555555) << 1) | ((x >> 1) & 0 x55555555);
    x = ((x & 0 x33333333) << 2) | ((x >> 2) & 0 x33333333);
    x = ((x & 0 x0F0F0F0F) << 4) | ((x >> 4) & 0 x0F0F0F0F);
    x = (x << 24) | ((x & 0 xFF00) << 8) |
        ((x >> 8) & 0 xFF00) | (x >> 24);
    return x;
}
```

The other kind is live code that actually executes. This book was not created with a word processing program or LAT$_E$X; this entire book is a computer program written in *Mathematica*. For example, this code

```
morris8 = Union[morris8];
Print["morris8 has ", Length[morris8], " unique numbers."]
```

morris8 has 144 unique numbers.

actually executes; it generates what follows in the text. The input commands to *Mathematica* are in boldface **Courier**, a typewriter-like typeface, and the text output is in regular Courier, not bold. Sometimes the output is graphical.

Single-letter variables are concise, but sometimes longer names are better reminders of what is being represented. Words or shortened words in phrases are concatenated, with capital letters informing you about where a new word begins, a system called CamelCase. There is also lower-case camelCase where the first letter is always lower-case, and that is what is used in coining variable names like *maxPos* (for maximum posit value) and *eS* (for exponent size). Reserved words in *Mathematica* use CamelCase, so by using lower-case camelCase there is no chance of colliding with a reserved word.

While the purpose of this book is certainly not to teach *Mathematica*, it is helpful to understand that everything is computed with symbolic mathematics until you ask for it to be turned into a number with the `N[...]` command, like this:

```
Print[π]
Print[N[π]]
Print[N[π, 16]]
```

π

3.14159 2654

3.14159 26535 89793

Boldfaced `Courier` font is used to display binary numbers like **0101101**. Decimal numbers are left in the same font as the main text, like 1000.

Boxed text with this light blue background is important, not to be skipped:

> If you round to a lower precision, and then to a still lower precision, you can get an incorrect rounding compared to rounding in a single step.

In contrast, boxed text with a light yellow background is a warning that the text expresses a bad or fallacious idea, even if the idea seems reasonable at first glance:

> You cannot represent all the real numbers on a computer because there are an infinite number of them, and computers only have a finite amount of memory.

Finally, text highlighted with a gray background is an **Exercise for the Reader**, like this one:

> **Exercise for the Reader**: Convert the decimal number 2300 to an 8-bit posit with $eS = 1$. Caution: This number is in the Twilight Zone.

If you choose not to try the Exercises, Answers to Exercises is still worth reading since it contains ideas that are not in the main text.

1 The Posit Number Format

1.1 The Essence of the Idea

Whatever *x* turns out to be, there is a way to represent it on a computer using only three bits.

Most books about mathematical ideas build up a foundation of definitions and make the reader wait to see where it is all going. Instead, this first chapter will get you to the punchline in as few pages as possible. Later, we can look at the underlying details and reasons for why the thing works the way it does.

Here is a common misconception:

You cannot represent all the real numbers on a computer because there are an infinite number of them, and computers only have a finite amount of memory.

Suppose a computer uses three bits to represent numbers. There are eight possible bit patterns, and if we treat them as normal base-2 *unsigned* integers, their meanings on any conventional modern computer would be as follows:

```
000    zero
001    one
010    two
011    three
100    four
101    five
110    six
111    seven
  overflow
```

That is all you get. You can count from zero to seven with three bits, adding 1 to the last bit and propagating any carry bits, just like in grade school addition but with base-two (binary) instead of base-ten (decimal). If you try to add 1 to the representation of seven, **111**, you get overflow because you cannot store the carry bit **1000** so it wraps back to zero, **000**.

When computers need *signed* integers, the most commonly-used system is that the numbers flip from positive to negative as soon as the leftmost bit hits **1**, and then they count upward back to zero, like this:

```
000            zero
001            one
010            two
011            three          overflow
100        negative four
101        negative three
110        negative two
111        negative one
```

The leftmost bit (the most significant bit, or MSB) quickly informs us if a number is negative, so it is often called the *sign bit*, and the convention here is to color it **red** since accountants have used that color to indicate negative amounts of money for so long that "being in the red" is a synonym for being in debt.

The unsigned integers overflow if you count up from **111**, but the signed integers overflow if you count up from **011**, so it is a matter of where you put the discontinuity. If you count up from negative one, **111**, you ignore the carry bit overflow and get **000**, so the system automatically does the right thing.

However: Suppose you instead assigned the following meanings to the bit patterns:

000	0
001	Real x such that $0 < x < 1$
010	1
011	Real x such that $1 < x < \infty$
100	Anything that is not a real
101	Real x such that $-\infty < x < -1$
110	-1
111	Real x such that $-1 < x < 0$

That covers the entire real number line, $-\infty < x < \infty$, and if x is not even a real number, it provides a way to represent that as well. We abbreviate that as NaR for "Not a Real" and it includes forms of infinity, indeterminate forms like $\frac{0}{0}$, complex numbers like $\sqrt{-1}$, and even things that are not numbers at all, like if a computer programmer tries to convert the character string "zebra" into this number format. It is *not a real*.

It is not a very precise system, but it *does* provide a unique representation for every x.

The rightmost bit (least-significant bit, or LSB) is color-coded **magenta** and is called the *ubit* (short for **u**ncertainty **bit**, which rhymes with cubit). Exact numbers are represented with bit strings that end in 0. Bit strings that end in a 1 mean the *open interval* between exact values (or between an exact value and ±∞). The open interval between a and b is written (a, b). Like, $\sqrt{2}$ and π are both represented by 011 since they satisfy $1 < x < \infty$. The ubit is the fundamental idea underlying *unum* arithmetic, where *unum* is a shortening of "**u**niversal **num**ber." For that reason, it is pronounced "yoo-num" and not Latin "oo-num" as in *E Pluribus Unum*.

Including the ubit is exactly like writing $\sqrt{2} = 1.414 \cdots$ where the "$\cdots$" means, "There are more nonzero digits after the last one shown: $1.414 < \sqrt{2} < 1.415$." That is mathematically true. If you say $\sqrt{2} = 1.414$, you are making an error just the same as you would be making an error to say $2 + 2 = 5$. The presence or absence of the "$\cdots$" takes only one bit to represent, and that is the ubit, just like the sign of a number takes only one bit to represent.

Notice that we do not color-code the sign bit for 000, as a reminder that zero has no sign. We also do not code it for the 100 = NaR exception since exceptions are also unsigned. Nor do we color-code the ubit of NaR since it makes no sense to regard "Not a Real" as an exact quantity.

Notice that the unums map to a ring, because just as with the signed integer format, `111` being right next to `000` makes perfect sense. The real values and NaR are shown in **violet** on the outside of the ring, with the bit strings shown on the inside.

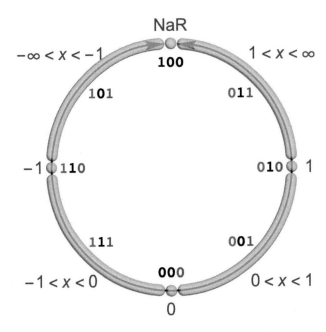

This unum *ring plot* shows the entire real number line mapped to only three bits.

There is a concept called the *projective reals*, which means thinking of the real number line as a projection of a circle onto a line, with the source of light at the top of the circle:

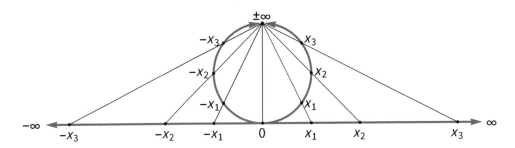

The *projective reals* concept maps the infinite line to a finite circle.

I have added two things to this very old concept. First is the observation that signed integers map elegantly to the circle in the same way, flipping sign at the top of the circle. The second is that placing –1 and 1 at the east and west points on the circle allows symmetry about the horizontal axis when we take the *reciprocal* of a number.

To save space, (a, b) means the open interval $a < x < b$. The three-bit system has a *vocabulary* of eight items: $\{NaR, (-\infty, -1), -1, (-1, 0), 0, (0, 1), 1, (1, \infty)\}$. As an ordered set, we treat NaR as if it were $-\infty$, less than all the other items, though it also serves as the dumping ground for anything else that is not a real number.

We can do arithmetic on both the exact numbers and the ranges. For instance, the square of the interval $(0, 1)$ is exactly the same set of values, $(0, 1)$. That may not seem like a big deal, but it actually means that this system is more powerful than what any calculator can do. Try entering a number between 0 and 1 into a calculator app (or a real calculator, which at this point is something of an antique) and square it. Square the result. Keep doing that over and over, and eventually one of two things will happen: The display will erroneously say the result is exactly 0, or it will say it is "Not a number." I tried it just now on my laptop, and got the second case.

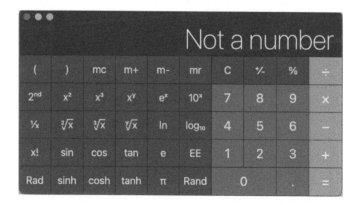

The software for the calculator app is certainly using a lot more than three bits. Yet it cannot produce a mathematically correct answer to "What is the square of 1e–50?" because the power-of-ten notation only has two digits so it goes as small as 1e–99 (that is, 10^{-99}). The calculator is saying that 10^{-100} is "Not a number"; I beg to differ.

Here is another example that does not involve underflow or overflow. Enter 2 and take the square root. Apply the square root button over and over and you get closer and closer to 1.

After entering 2 and pressing the square root 49 times, I see this result:

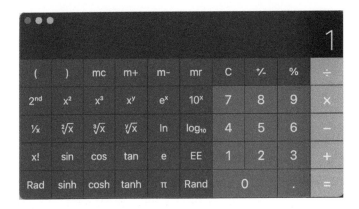

I took the square root one more time, and it did not change, which is interesting. It has some "guard digits" and knows more about the answer than it shows on the display. But take the square root yet one more time, and I get this:

The display shows exactly 1. But that *is not true.*

What would the three-bit system do? The square root of $(1, \infty)$ is the same set of real numbers, $(1, \infty)$. You can take the square root of $(1, \infty)$ as many times as you like and it will never produce the mathematically incorrect result 1.

1.2 Raising the Precision

If we use four bits instead of three, we can now insert exact values between the ones we had before. There are many possible value choices, but here is a logical one:

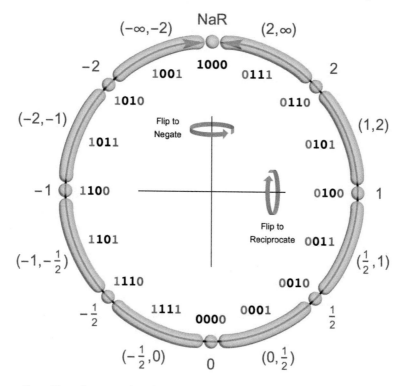

Appending a bit to the unum ring plot shows how specificity increases for the open intervals.

Insert ½ between 0 and 1, and reflect that to the negative side. But we can also reflect about the *horizontal* axis by taking the reciprocal, so 2 goes between 1 and ∞. Notice that reciprocation also works for the open intervals, too. The one exception is that while the reciprocal of 0 is NaR, the reciprocal of NaR should stay NaR. Once you land on an invalid or indeterminate result, it is vital that NaR is preserved under every operation, so that it shows up in the final result to alert that things went awry.

In most number systems, negation is perfect, so once you have a way to add a and b, you also have a way to subtract b from a, by computing $a + (-b)$. With the system shown above, once you have a way to multiply a and b, you also have a way to divide a by b, by computing $a \times \left(\frac{1}{b}\right)$. That does the right thing if b is NaR, because $\frac{1}{b}$ is also NaR and anything multiplied by NaR is NaR, preserving the information that something went wrong in the computation.

How do you negate a signed binary integer? The rule is this: *Flip all the bits and add 1, and ignore any overflow.* For example, if the integers are of size eight bits, the integer 5 in binary is

00000101.

Flip all the bits to get

11111010.

Then add 1 and we are done. For 8-bit signed integers, this is the bit pattern for –5:

11111011

The only case where you get carry overflow is if you try to negate zero. For example, **00000000** flips to **11111111**, and adding 1 to that should be **100000000** (nine bits), but you only have eight bits so the leftmost bit is discarded. This is why zero has no sign; if you try to negate zero this way, you get back zero.

NaR has the same lack of sign, and the same charming property: In the ring, treat the four-bit values as integers so NaR is **1000**. Flip the bits to **0111**, add 1 to get **1000**, and you are back to NaR.

To "flip about the vertical axis" to turn x into $-x$, treat the bit string as a signed integer and negate it. To "flip about the horizontal axis" to turn x into $\frac{1}{x}$, ignore the sign bit and negate the other bits (that is, flip them and add 1). If there is a carry, do a logical OR with the sign bit. That way, zero = **0000** reciprocates to NaR = **1000**. The rule also turns NaR back into NaR, so there are no exception cases.

The reason we should care about minimizing exception cases is that exception cases make hardware for computer arithmetic wasteful and messy (and slower), and harder to prove correct in all cases.

1.3 Rigor or Guesswork? The Tough Decision

"Perfect is the enemy of good." — Voltaire

This brings us to a difficult choice. For a system that lets us work with real numbers on computers, we can continue refining this *perfect* system of exact numbers and open ranges between exact numbers, or we can settle for a *good* system where there are only exact numbers and we round every real number to one of them, creating and accumulating mathematical errors in doing so.

This is a battle between two schools of thought, one that sometimes resembles a religious war. The "good enough" system came first in computing, which is to represent real numbers in imitation of scientific notation, like 6.022×10^{23} where we have some fixed amount of storage for a digit followed by a fraction, times a number base raised to a power. The amount of storage for the power is also fixed. The digit-fraction part can be negative, zero, or positive, and so can the exponent. A number represented by such a format is called a *floating-point number*, or "float," for short. Unlike integer computations, which are exact so long as you stay within the representable range, float computations are almost always inexact. But users find them "good enough" for many things we want to do with computers. Usually.

The term *interval arithmetic* is used for an approach introduced some sixty years ago where all quantities are treated as closed sets of real numbers $[a, b]$, that is, real x such that $a \le x \le b$. Exact values are expressed as $[a, a]$. The endpoints are expressed with floats, and in doing arithmetic, the rule is always to rigorously *bound* the result and hope that the bound is tight enough to have practical value.

If you had a choice between using a computer to get mathematically correct answers as opposed to guessing its way through a calculation and *probably* being good enough, who would ever pick the latter? The answer is: *Almost everybody*. In the marketplace of ideas, approximate arithmetic with rounding errors has beaten out interval arithmetic. The main reason is that unless you *really* know what you are doing, using intervals in a computation may produce something like [−∞, ∞] for the result. The set of experts who know how to stop interval bounds from becoming ridiculously conservative (much wider than they should be) is not a large group, but it is a zealous one that sometimes argues for rigorous computing with the passion of a tent revivalist minister. In all the technical computing that gets done in the world, I would be surprised if the amount done with interval arithmetic is more than one percent.

For that reason, most of this book is going to do something that resembles what floats do and assume that we *round* after performing arithmetic on the quantities. But they will use the projective reals framework described above and *not* be floats. What should we call them? My first thought was to call them a *guess*, since rounded quantities are a way of guessing your way through a computation. However, "guess" can be pejorative, so let's use a more optimistic term: *posit*. When used as a noun, it is defined as follows:

---

posit | ˈpäzət |

noun *Philosophy*
A statement that is made on the assumption that it will prove to be true.

1.4 The Posit Format

With that justification, we now replace the open intervals in the ring plot with exact values, so that the last bit is no longer a ubit. No more magenta 0 or 1 bits at the end, but we introduce two new bit colors, **gold** and **brown** for reasons explained below. We also go back to color-coding the sign bit of 0 and NaR, because it turns out we can eliminate regarding them as "exceptions."

Here is a ring plot for 4-bit posits:

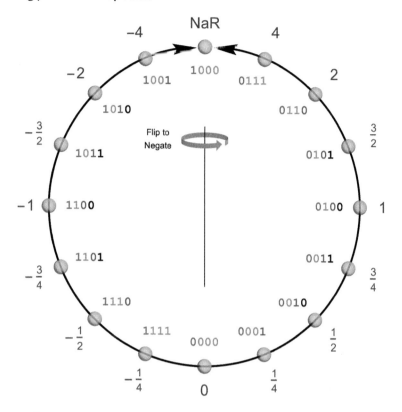

The posit ring plot replaces open intervals with a single point in the interval.

The "in-between" rules are as follows when the last bit is a **1**, where we used to have an open interval:

- Between NaR and a real number, insert twice the real number.
- Between 0 and a real number, insert half the real number.
- Otherwise, between x and y, insert their average, $(x + y)/2$.

For example, between 2 and NaR, we insert 4, by the first rule. Between $-\frac{1}{2}$ and 0, we insert $-\frac{1}{4}$, by the second rule. And between $\frac{1}{2}$ and 1, we insert $\frac{3}{4}$, by the last rule. Notice how this set of rules keeps the largest magnitude numbers the exact reciprocals of the smallest magnitude numbers.

The "Flip to Reciprocate" part of the diagram is removed because it is not true when the number is not a power of 2. The reciprocal of $\frac{3}{4} = 0.75$ is $\frac{4}{3} = 1.33\cdots$, but the mirror image in the ring is $\frac{3}{2} = 1.5$ instead. There is a way to fix that, but it can make the computer hardware more complicated, and that idea will be shown later.

1.5 The Regime

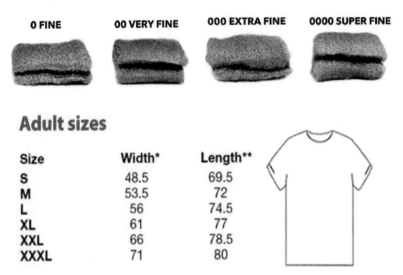

Steel wool grades getting finer and shirt sizes getting larger resemble *regime* notation.

The *regime* of a posit is the bits that follow the sign bit that are identical, shown in gold, until they hit the opposite bit, shown in **brown**.

For the four-bit posits, their meaning as an integer r is as follows (sign bit not shown):

regime	r
000	$-\infty$
001	-2
01 ·	-1
10 ·	0
110	1
111	2

The regime value r is determined by the number of identical bits in a row, k, before hitting an opposite bit, if it occurs in the stored bits. You see, all bits after the last bit in a posit are implied **0** bits. Just like when you say a number is 1.53, with no other information, you have to treat it as meaning 1.5300000000⋯. So, if the regime bits are 111 which means they go all the way to the last bit, the implied bits that lie beyond are all **0** bits; call them *ghost bits*. The very next ghost bit is a **0** that serves as the stop bit for the run of 1 bits.

What if the regime is 000, going all the way to the end without an opposite bit to end the run? Then it represents an *infinitely small integer*, $-\infty$, because it never reaches its opposite bit and is an infinite string of 0 bits.

It there are k identical bits in a row, the value of r is

$$r = \begin{cases} -k & \text{if the first regime bit is } 0, \\ k-1 & \text{if the first regime bit is } 1. \end{cases}$$

Remember the rule for negating an integer, where you flip all the bits and add one to the LSB? It works for the regime notation as well, but in a completely different way. Like, if you negate regime 110 = 1, flip bits to 001, add 1 but read the result as a regime 01· where the "·" bit is not part of the regime since it is after the opposite bit. Or try negating regime 10· = 0. Flip the regime bits to 01· and add 1 to the last bit of the regime (which is the second bit, not the third), and you get back 10·, that is, 0 in regime notation. Similarly, if you try to negate the regime for $-\infty$, you get back the same bit pattern, so $-\infty$ acts exactly like NaR. It is "the point at infinity."

Regime notation works a lot like clothing sizes: L, XL, XXL, XXXL for large, extra-large, and so on. Or for small sizes, sizes 0, 00, and 000 indicate smaller and smaller clothing sizes.

> **Definition**: The *radix point* is like a decimal point, but for any number base (including binary).

The "·" bits after the opposite bit are *fraction bits* representing the fraction f. There is an implied radix point "·" after the last bit of the regime. So far, we only have the possibility of one fraction bit, and in binary, .0 means $f = 0$ and .1 in binary means $f = \frac{1}{2}$. "But what if there aren't any fraction bits?" you might ask. The answer is, "There are *always* fraction bits. If they are beyond the end of the number, they all have value 0." And there is nothing wrong with having a fraction value of $f = 0$.

The value of the sign bit is s, that is, 0 if the sign bit is 0 and 1 if the sign bit is 1. We can now give the formula for the value x represented by a posit from its bit string:

$$x = ((1 - 3s) + f) \times 2^{(1-2s) \times (r+s)}.$$

Let's practice by decoding 1011. The sign bit is 1, so $s = 1$. The regime is 01, so k is one bit in a row and $r = -1$. The fraction has one bit showing (a .1 followed by an implied infinite string of 0 bits) so $f = \frac{1}{2}$. Apply the formula:

$$x = ((1 - (3 \times 1)) + 1/2) \times 2^{(1-(2\times 1))\times(-1+1)}$$
$$= (-3/2) \times 2^0 = -3/2$$

But does the formula work for the case of $x = 0$? Surprisingly, it does. The bit string is 0000, so $s = 0$, $r = -\infty$, and $f = 0$. The formula simplifies to $1 \times 2^{-\infty}$, which is 0.

What happens for NaR, though? Its bit string 1000 means $s = 1$, $r = -\infty$, and $f = 0$. The formula simplifies to $(-1) \times 2^{\infty} = -\infty$. That is certainly not a real. We simply make the rule that $-\infty$ is our stand-in for anything that is not a real, not just $-\infty$.

That really is the punchline of this chapter, and it works for any posit precision. A 1-bit posit is possible, but all it can represent is 0 or NaR.

As a taste of what is to come, let's take the format out for a spin, with 16-bit posits. What is the largest and smallest magnitude posit values we can represent with the same number of bits? Call those extremes *maxPos* and *minPos*. *maxPos* is 0111111111111111 with k = fifteen 1 bits in a row, so $r = 14$; *maxPos* $= 2^{14} \approx 1.6 \times 10^4$. Reciprocation is perfect; *minPos* is 0000000000000001 with k = fourteen 0 bits in a row, so $r = -14$ and *minPos* $= 2^{-14} \approx 6 \times 10^{-5}$. More than 8 orders of magnitude.

If you decode 0110100100100010 you get the rational number $\frac{3217}{1024} \approx 3.1416$, that is, π accurate to about five decimal places. To put this in perspective, the closest that IEEE Standard 16-bit floats can get to π is $\frac{201}{64} = 3.140625$, with only about three decimals of accuracy. This is a small sampling of the advantages of the posit format.

> **Exercise for the Reader**: The standard 16-bit float closest to $\sqrt{17}$ is $\frac{33}{8} = 4.125$. Find the rational number represented by 16-bit posit 0111000000111111. Compare the closeness of those approximations to $\sqrt{17} = 4.1231056\cdots$

If we are going to be examining and comparing *error* and *accuracy*, we had better do so with careful thinking and clarity. That is the subject of Chapter 2.

2 | Inaccuracy Done Right

This did not actually happen. It is a computer-generated image.

2.1 Absolute versus Relative Error

Like so many things in numerical analysis, we have become accustomed to some concepts that are widely accepted but are actually rather sloppy in their logic. The way we measure *error* of various types is one of those concepts. For instance, here is a widely-used way of measuring the error of a result computed with real numbers:

$$absolute\ error = \left| x_{computed} - x_{exact} \right|$$

There is nothing wrong with this; if the values are identical, their difference is 0 so there is no error. Absolute error is what to use when someone lies about their age, say, or how fast they were driving, or how many glasses of wine they have had. But if a calculation produces 315 instead of 314, doesn't the error look very similar to returning 3.15 instead of 3.14? A number system designed for *real* numbers usually spans a large dynamic range. Simply subtracting numbers may be the best measure when the values being compared are within an order of magnitude of one another.

For situations with many orders of magnitude, the *relative error* of a computed value is traditionally defined this way:

$$traditional\ relative\ error\ \text{is defined as}\ \left| \frac{x_{computed} - x_{exact}}{x_{exact}} \right|.$$

There are some reasons you should be dissatisfied with this definition. If you compute −1 when the correct answer is 100, then the above relative error formula would only be 1.01. If you do not even know the sign of a number, then you know essentially nothing about the answer since the sign is the most significant part of a number. An error formula should return "indeterminate" in such cases. Even declaring the relative error to be infinite is too flattering.

If you did an experimental measurement that produced two different values x_1 and x_2, you might want to talk about the relative error in the measurement. But you do not know the x_{exact} value. Depending on whether you put x_1 or x_2 in the denominator, you get a different relative error. That does not seem right.

For another, the formula is quite different for numbers and their inverses. Suppose $x_{computed} = 0.001$ but $x_{exact} = 0.0001$. Then the relative error is 9. But if we were instead computing the reciprocals of the numbers, the relative error would take $x_{computed} = 1000$ and $x_{exact} = 10\,000$ as inputs, and the above formula would reassure us that the relative error is only 0.9. It makes no sense that an answer can be made to look ten times more accurate simply by taking reciprocals. If you knew a wavelength with a relative error of 0.1, would it not bother you to know the frequency (inversely related to wavelength) of the same component with a relative error of 0.2? These problems go away if instead of taking the differences between two numbers, we take the difference of the *logarithm* of the two numbers.

2.2 Three Types of Logarithms

Engineers who work with signals that have a wide dynamic range (like sound) have long had a solution when comparing numbers with ratios, which is to use *decibels*. A ratio of 10 is 10 decibels (10 dB), also known as a *bel*, though that word is not used very often. A ratio of 1 dB means the ratio is $10^{1/10} = 1.2589\cdots$. Decibels measure the difference of the logarithms (base 10) of two (positive) quantities x_1 and x_2, which is the same as the logarithm of their ratio:

$$dB = \left| \log_{10}(x_1) - \log_{10}(x_2) \right| = \left| \log_{10}\left(\frac{x_1}{x_2}\right) \right|.$$

The absolute value makes x_1 and x_2 interchangeable. Hence, it can be used to measure experimental errors. Also, it produces the same result whether you use x_1 and x_2 as inputs, or $1/x_1$ and $1/x_2$. So *that* looks like a mathematically sound definition. But if x_{exact} is 100, say, and we compute $x = 110$, the error is about 4 dB, which looks nothing like the traditional relative error value of 0.1. Can we have it both ways, with the elegance of logarithms of ratios but also something that resembles traditional relative error?

There are three types of logarithms we will be using here. Base 10 logarithms are good for human understandability since we tend to think in powers of 10 like hundreds, thousands, billions, etc. Computer scientists tend to use base 2 logarithms more than any other type, so much so that a base 2 logarithm has its own abbreviation: $\log_2(x)$ is often written as $\lg(x)$. For example, if told that "The memory address space is 4096 items," then a computer scientist immediately infers that there are $\lg(4096) = \lg(2^{12}) = 12$ bits in the address.

A third kind, *natural logarithms*, uses $e = 2.71828 \cdots$ as the base because that base simplifies expressions that arise in higher math. It also has its own abbreviation: $\log_e(x) = \ln(x)$. Here are graphs of those three kinds of logarithm:

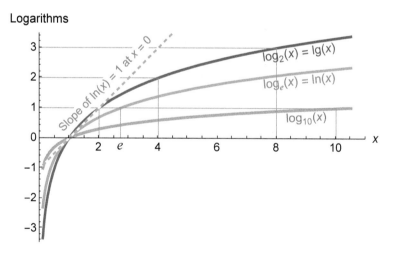

log$_2$ is computer-friendly, log$_e$ is math-friendly, and log$_{10}$ is human-friendly.

The decibel-type approach would work with any base logarithm, because $\log(a) - \log(b) = \log(a/b)$ for any kind of logarithm. There is a subtle reason for choosing the *natural logarithm* base $e = 2.718 \cdots$ instead of base 10 to measure relative error: as an error becomes small, natural log error becomes close to traditional relative error. As the figure shows, the slope of $\ln(x)$ near $x = 1$ is 1.

The log definition meets the other requirements of being immune to measuring inverses or interchanging the two values. So, here is the definition we will use:

> **Definition**
>
> The *relative error* of $x_{computed}$ and x_{exact} is
>
> $\texttt{relErr}\left(x_{computed}, x_{exact}\right) =$
>
> $$\begin{cases} 0 & \text{if } x_{computed} = x_{exact}, \\ \text{NaR} & \text{if } x_{computed} \text{ and } x_{exact} \text{ are of different sign,} \\ \left| \ln\left(\frac{x_{computed}}{x_{exact}}\right) \right| & \text{for all other cases.} \end{cases}$$

Suppose the exact answer is 3.00 but we compute 3.03; the traditional relative error is $\frac{0.03}{3.00} = 0.01$ but the formula above is $\ln\left(\frac{3.03}{3}\right) = 0.00995\cdots$ which is very close, so this means traditionalists should not be too unhappy with the updated definition. We have Peter Lindstrom to thank for this perceptive observation.

What if x_{exact} is zero? Unless $x_{computed}$ is also zero, the relative error is ∞, which makes some people very uncomfortable because it certainly seems like a computed value like 0.000001 should get credit for *some* resemblance to zero. Sorry, but no. In the world of real numbers, the log of zero is infinitely far from the log of any nonzero value, just as the log of infinity is infinitely far from the log of any finite value. There is a human tendency to mentally flip back to wanting absolute error when thinking about the relative accuracy of very small values; resist that tendency, because it can land you in trouble.

What if $x_{computed}$ does not even have the right *sign*? Then the relative error is NaR. Literally. The formula takes the log of a negative number, which is a complex number and Not-a-Real. Otherwise, we have the desired symmetry and $\texttt{relErr}(x, y) = \texttt{relErr}(y, x) = \texttt{relErr}\left(\frac{1}{x}, \frac{1}{y}\right)$.

Usually, we care about the amount of error and not whether it is too low or too high. If we leave out the absolute value $|\ldots|$ in the expressions, we get *signed relative error*. The graph on the next page shows signed traditional relative error, traditional relative error of the reciprocals, and the natural log relative error used here:

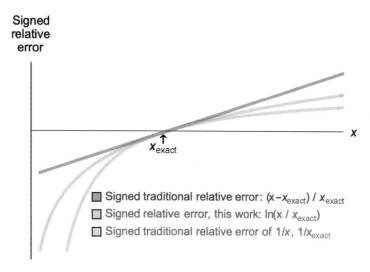

The $\ln(x/x_{exact})$ relative error is between the traditional and traditional reciprocal relative error.

The $\left|\ln(x/x_{exact})\right|$ case is "inaccuracy done right," a definition that has mathematical self-consistency but also bears a close resemblance to the traditional definition of relative error when the amount of error is small.

2.3 Accuracy as the Inverse of Error

Knowledge is the inverse of uncertainty.

That is, the more tightly we can provably bound what a result can be numerically, the more knowledge we have about its true value.

Similarly, *accuracy* is the inverse of relative error, and we already use this when we say, "The answer is accurate to five decimals." If the relative error is about 0.001, then we know the answer to about 3 decimals; if the relative error is 0.0000001, then we know the answer to 7 decimals. Mathematically:

Definition

The *decimal accuracy* of $x_{computed}$ and x_{exact} is

$$\textbf{decAcc}\left(x_{computed}, x_{exact}\right) =$$

$$\begin{cases} \infty & \text{if } x_{computed} = x_{exact}, \\ \text{NaR} & \text{if } x_{computed} \text{ and } x_{exact} \text{ are of different sign}, \\ -\log_{10}\left|\ln\left(\frac{x_{computed}}{x_{exact}}\right)\right| & \text{for all other cases.} \end{cases}$$

A computer scientist or numerical analyst might sometimes want to know, instead, the number of *bits* to which an answer is correct. That is easy. Just use logarithms base 2 instead of base 10 to find *binary accuracy*. Or compute the decimal accuracy and multiply by $\log_2(10) = 3.32\cdots$, to get the same thing.

It looks peculiar to have expressions that mix two different logarithm types, but the natural log makes traditional relative error resemble our ratio-based relative error. Then, the log of *that* shows the accuracy in a form suited for humans (log base 10) or for computers (log base 2).

Appendix A has the code for a *Mathematica* function, **decAcc**[*x*, *y*], based on the above definition of decimal accuracy. Let's try it on approximations to π:

```
decAcc[3.14, π]
decAcc[3.1416, π]
decAcc[π, π]
```

3.29491844

5.63107 5206

∞

That is π accurate to about three decimals, about five decimals, and infinitely accurate. *Mathematica* uses symbolic methods, so π can be written exactly.

2.4 The Accuracy of a *Number System*

Now that we have a definition of accuracy, we can also analyze the accuracy of a number system. Suppose we are using 8-bit posits, as described in Chapter 1. We only need to consider the accuracy of non-negative values, since accuracy of negative representations is the same. Here are all 128 non-negative 8-bit posits, shown on a ring plot. Call the values they represent the **posit8** set.

To make the figure easier to read, the radial lines indicate changes in the regime value, which occur at integer powers of 2. Notice that half the values are between 0.5 and 2, where the spacing is closest, so accuracy is highest.

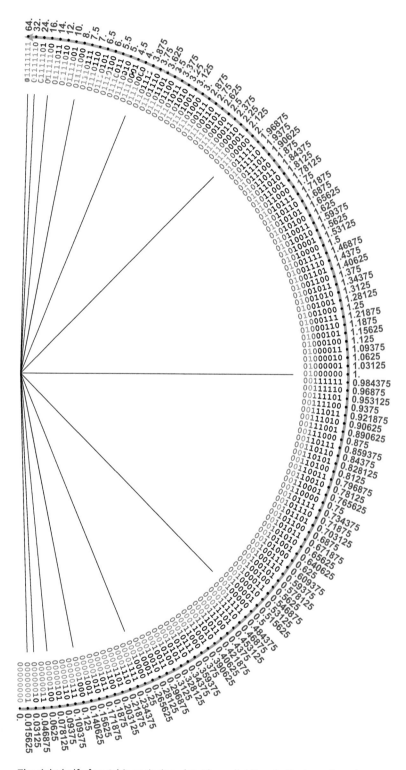

The right half of an 8-bit posit ring plot. The radial lines indicate regime changes.

The rule for rounding can be summarized as "round a real *x* to the nearest posit; if two posits are equally near, choose the one ending in a **0** bit." There are some important additional details, but it will serve for now. Let's first study three consecutive values in the **posit8** set: {7.5, 8, 10}. The tie point between 7.5 and 8 is 7.75. The tie point between 8 and 10 is 9.

If we happen to compute a value that is exactly 7.5, 8, or 10, then the accuracy is infinite. Between those exactly-representable values, accuracy dips to a minimum at the tie point.

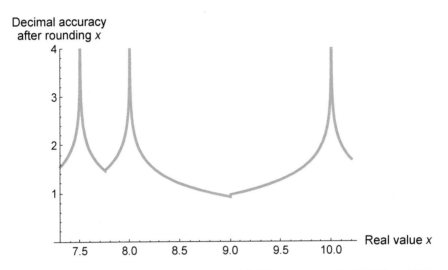

The peaks actually go to infinity, since accuracy is infinite at exact points 7.5, 8.0, and 10.0.

Notice the slight discontinuity at the two minima. That is caused by the difference between the *geometric* mean of two values *x* and *y*, $\sqrt{x \cdot y}$, and the *arithmetic* mean, $(x + y)/2$. The discontinuities are at the halfway points, but those are *not* the places where the decimal error (defined as a logarithmic distance) is minimized. The minimum error would happen at the geometric means, $\sqrt{7.5 \times 8} = 7.7459669\cdots$ and $\sqrt{8 \times 10} = 8.944\cdots$ instead of the arithmetic means, 7.75 and 9, which at this very low 8-bit precision is a large enough discontinuity to see in the graph, barely.

Hence, the accuracy of a number system depends on the rounding policy. By our definition of relative error, the worst-case decimal accuracy between adjacent values x_k and x_{k+1} occurs at their arithmetic mean, $(x_k + x_{k+1})/2$, relative to the smaller magnitude *x*.

At the arithmetic mean, the worst-case decimal accuracy is

$$-\log_{10}\left|\ln\left(\frac{(x_k+x_{k+1})/2}{\min(|x_k|,|x_{k+1}|)}\right)\right|$$

$$\approx \textbf{decAcc}(x_k,\,x_{k+1}) + \log_{10}(2).$$

In other words, the worst-case decimal accuracy of a numerical vocabulary is very close to the decimal accuracy of the adjacent points and adding $\log_{10}(2) \approx 0.3$.

Here is a decimal accuracy plot for a range of real values rounded to the `posit8` vocabulary:

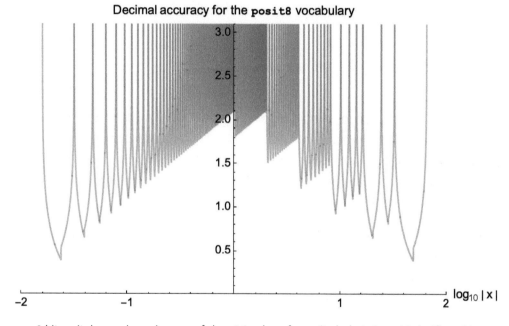

8-bit posits have a dynamic range of almost 4 orders of magnitude, but at most 6 significant bits.

The graph is thorough in showing the accuracy of the entire number system over its dynamic range, but the vertical spikes to infinity at the exact points make it difficult to read. Let's instead be pessimistic and look at the *worst case decimal accuracy* for each point in the number set.

The following graph shows just the bottom of each trough. If you ignore the jaggedness of the right half, it makes a symmetrical tent shape. The relative accuracy is tapered away from zero.

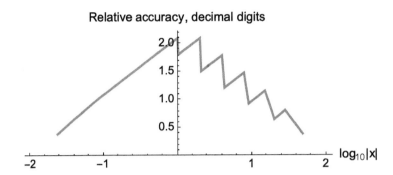

This is the usual form of an accuracy plot: Decimal relative accuracy as a function of magnitude.

For posits of higher precision, the jaggedness is less pronounced and the decimal accuracy looks even more like a symmetrical triangular shape that peaks in the center.

A sequence of calculations that depend on previous rounded results is generally only as accurate as the least-accurate calculation, like a chain being only as strong as its weakest link. That is why it makes sense to use the pessimistic minimum accuracies and assume a chain of calculations will stumble into something near the bottom of each trough.

There is another way to plot the decimal relative accuracy, and that is as a function of the *bit pattern* treated as a binary integer instead of as the real value it represents by posit format rules. This illustrates how half the bit patterns are in the highest-accuracy region, defined in the next Section.

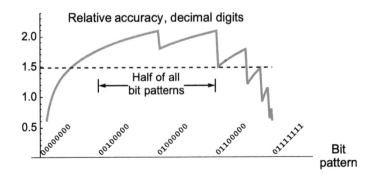

Accuracy can also be graphed as a function of the *bit pattern*, shown here for positive 8-bit posits.

2.5 The *Fovea*

The highest-accuracy region has for years been called the "sweet spot" by posit users, but Will Wray came up with a clever, short word for it: The *fovea*, like the part of the retina of an eye that has the highest resolution:

> **Definition:** The *fovea* is the set of posits for which the regime is only two bits long, either 01 or 10. This leaves the maximum number of significant bits for the fraction and thus the highest relative accuracy.

Half of all posits are in the fovea, as is evident from the ring plot. One-fourth of all posits have a regime length of three bits (001 or 110) and thus have one less bit of significance than the fovea posits. One-eighth of all posits have a regime length of four bits, and so on.

To get the most accurate results with posits, it helps to try to stay in the fovea or not too far away from it. Wandering into very large or very small magnitudes, even temporarily, causes loss of relative accuracy that might propagate to the final answer and make it less accurate. Fortunately, most modern numerical methods are already carefully designed to avoid overflow and underflow and thus stay near the center of the dynamic range. Those methods are already well-designed for posits.

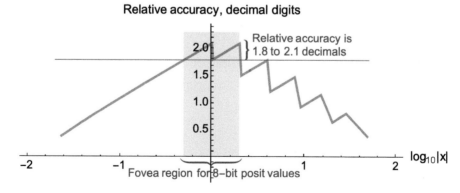

The *fovea* for 8-bit posits is where $\frac{1}{2} \le |x| \le 2$.

2.6 Fixed-Point Accuracy

Scaling a number by an integer power of two is like moving the radix point. What if you did not have to do that, but could simply assume a fixed location for the radix point? That would free up the bits that we use to indicate scaling, for more significand bits. Fixed-point formats are like dollars and cents, where there are two digits to the right of the radix point. You can treat every dollar amount as an integer multiple of the number of cents, $\frac{1}{100}$ of a dollar.

Suppose we have an 8-bit format regarded as a signed integer, representing multiples of $2^{-4} = \frac{1}{16}$. That is, it has four bits to the right of the radix point, and the 256 different values it could represent would be

$$
\begin{array}{ll}
\textbf{0000.0000} & 0 \\
\textbf{0000.0001} & \frac{1}{16} = 0.0625 \\
\textbf{0000.0010} & \frac{1}{8} = 0.125 \\
\quad \vdots & \quad \vdots \\
\textbf{0111.1111} & 7 + \frac{15}{16} = 7.9375 \\
\textbf{1000.0000} & -8 \\
\textbf{1000.0001} & -\left(7 + \frac{15}{16}\right) = -7.9375 \\
\quad \vdots & \quad \vdots \\
\textbf{1111.1111} & = -\frac{1}{16} = -0.0625
\end{array}
$$

Call the positive values the `fixed8` set. The points are equispaced. What would the decimal accuracy plot look like for the set of positive values, compared with the one for `posit8`? Here they are on a single plot:

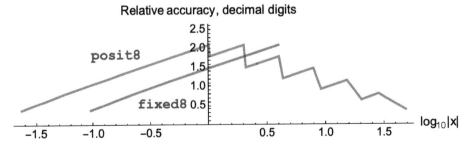

Accuracy plot of 8-bit posits versus 8-bit fixed point

The plot shows that fixed-point format does not support a wide dynamic range and represents small-magnitude values with much less accuracy than large-magnitude values. That is usually not what you want for technical computing. There are applications where fixed-point format is exactly what you want (like financial programs), which is also where absolute error might be more useful than relative error.

2.7 ULPs and Relative Accuracy "Wobble"

The term *ULP* is for Unit in the Last Place, and it has some uses for measuring inaccuracy. It is usually pronounced to rhyme with "gulp," not as "you-ell-P." Measuring ULPs assumes that as you count up in binary, the real number that is represented also increases in value. This is not true for all real number formats. It is one way to measure the distance between a computed result and a (correctly rounded) exact result: How far apart are their bit strings?

Suppose we use `fixed8` and the correct answer is $\frac{1}{8}$ = `0000.0010`, but the calculation misses and produces $\frac{3}{16}$ = `0000.0011` instead. That is an error of one ULP. As a relative error, though, it is $\left|\log\left(\frac{3}{16}/\frac{1}{8}\right)\right| \approx 0.41$, which is a rather poor result. The result has only about 0.38 decimals of relative accuracy.

At the other extreme, suppose the correct answer is $\frac{127}{16}$ = 7.9375 = `0111.1111` but the calculation is too low by just one ULP, so that we instead get $\frac{126}{16}$ = 7.875 = `0111.1110`. The resulting relative error is a much smaller number: $\left|\log\left(\frac{127}{16}/\frac{126}{16}\right)\right| \approx 0.008$ and we have about 2.1 decimals of relative accuracy. So, fixed-point format suffers a wide range of relative accuracy that becomes smaller for small magnitude results.

In scientific notation, an ULP is an error in the last decimal of the significand, assuming the exponent is correct. Like, 2.99×10^8 versus 3.00×10^8 for the speed of light in meters per second.

> **Definition**: A *decade* is the range of real values from $1.00\cdots0\times10^{n}$ to $9.99\cdots9\times10^{n}$ for some integer exponent n and some fixed number of decimal digits in the significand.

So "decade" in this context has nothing to do with a 10-year span of time. It is a numerical range that spans 10 dB. If a number format has a minimum positive value of 10^{-6} and a maximum positive value of 10^{6}, we might say, "the format covers twelve decades of dynamic range." Another common term is *order of magnitude*, so if something is off by a factor of a thousand, either too small or too large, it is off by "three orders of magnitude" (log base 10 of a thousand).

Back to the subject of ULPs: If we have three significant decimals in our scientific notation system, a 1-ULP error near the bottom of the decade is 1.01 versus 1.00, which has a relative error of about 0.01 and a decimal accuracy of about 2 decimals. But at the top of the decade, 9.98 versus 9.99 is instead a relative accuracy close to 0.001, ten times smaller, and the decimal accuracy is close to 3 decimals. The decimal accuracy ramps up from 2 to 3 decimals for every power of ten in the scale factor 10^{n}, producing a decimal accuracy that looks like this sawtooth shape:

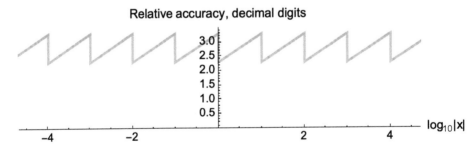

The "wobble" in relative decimal accuracy for scientific notation is a full decimal from worst to best.

> Some refer to this as "wobbling precision."

But that is a misnomer. The *precision* is not wobbling at all. The precision is a constant three decimals. This phenomenon should be called *wobbling relative accuracy* or *wobbling relative error*. Notice that the wobble is not from 2 to 3 decimals, but from about 2.3 to 3.3 decimals since when used as a number system, the worst error happens midway between two representable numbers, as explained in Section 2.4.

The less wobble, the better. Since calculations are usually only as accurate as their least accurate operation, it is usually best not to have deep "notches" of low accuracy that the calculation can stumble into.

One way to reduce wobble is to use base 2 for the scale factor, not base 10 or base 16. That reduces the wobble to $\log_{10}(2) \approx 0.3$ decimals.

Definition: A *binade* is the range of real values from $1.00\cdots0 \times 2^n$ to $1.11\cdots1 \times 2^n$ for some integer exponent n and some fixed number of bits in the significand.

With some number systems, including posits, the significand digits always represent a value between 1 and 2, ignoring the 2^n scaling factor and the sign.

Early IBM mainframes used a form of scientific notation with powers of 16 instead of powers of 10, to be more suited to binary computers. But base-16 exponents make the wobble even worse than base-10 exponents. A real number slightly greater than 1 would be expressed with 16 times less accuracy than a real number slightly less than 1. Base-2 has much less wobble, as this figure shows:

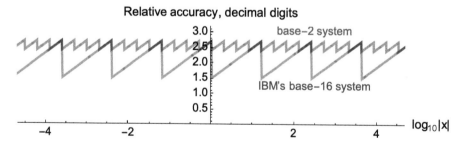

Using base 16 reduces relative accuracy and makes the wobble four times worse than using base 2.

Posits do not work like scientific notation, but they still represent binades times scale factors, so you see similar notches in their relative accuracy plots.

For example, the shape of the sawtooth in the fovea region of `posit8` is exactly like the sawtooth shape in the previous accuracy plots:

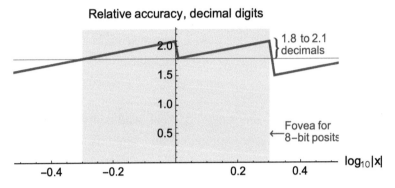

This is a close-up view of the fovea, overlayed by part of the usual accuracy plot.

2.8 Losing Something in Translation

Here is the picture to keep in mind when using a computer to do calculations on real numbers:

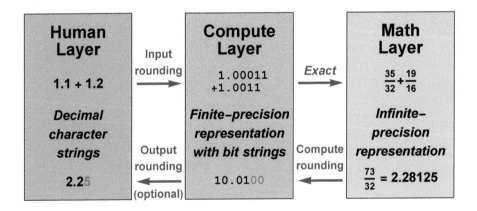

These are the three layers to remember when using computers to work with real numbers.

The diagram shows three layers: Human, Computer, and Mathematical. It also shows an example of how a simple-looking calculation like "1.1 + 1.2" can wind up pretty butchered by the three rounding steps. The example in the diagram uses the `posit8` set for the compute layer, but the same thing would happen with fixed-point or floating-point or any other representation with binary representation. The reason is that $1/10 = 0.1$ in decimal becomes an infinite repeating pattern in binary, `0.0001100110011`⋯ the same way $1/3$ becomes the infinite repeating pattern 0.333333⋯ when expressed as a decimal.

Input rounding occurs when decimals get turned into binary. Here, the closest binary to 1.1 is **1.00011** = $\frac{35}{32}$ = 1.09375. The closest binary to 1.2 is **1.0011** = $\frac{19}{16}$ = 1.1875. The only time there is no rounding is when a decimal just happens to be exactly expressible as a finite binary string, like 0.375 = **0.011**. It is not just fractions that suffer input rounding. If you try to add 9 and 11, for example, neither integer is in the vocabulary of **posit8**, and a computation will round those inputs. Any number system that has a decent dynamic range will have this problem: Large values *sample* the space of integers, skipping over swaths of them instead of including all of them in an interval by simply counting up.

The translation *to* the Math Layer is perfect. If you've got the time, God's got the integers. The computation in the Math Layer does not need to be perfect.

> In practice, a computer does *just enough* calculation in the Math Layer that the returned result will be indistinguishable from one done with infinite precision.

That is, the Math Layer takes into account that there will be rounding to go back to the Compute Layer and therefore it can perform the *finite* amount of work necessary to reach that result. In the example, the rational numbers are added correctly to be $\frac{73}{32}$ = 2.28125. The Math Layer is extra registers with more bits, hidden algorithms, and other techniques working with data that are not accessible to the user.

Compute rounding is what happens whenever the result of an operation requires more bits to express than are available in the Compute Layer. The closest number in our vocabulary to $\frac{73}{32}$ is $\frac{9}{4}$ = 2.25.

The final blow to correctness comes in *output rounding*, where a binary number is expressed in decimal. It is always possible to do that exactly, but it can be quite bulky, and doing so can give you the misimpression that you have far more significance in the answer than you actually have. For example, $\frac{1}{1024}$ = 0.00097 65625 exactly, but if the $\frac{1}{1024}$ was the result of a calculation involving rounding errors, then an output rounding of $\frac{1}{1024}$ ≈ 0.001 conveys the inaccuracy in the output. As a rule of thumb, the significance expressed in the output should not exceed (at least not by much) the significance expressed in the input, and "1.1 + 1.2" uses two-digit inputs. Hence, rounding the result 2.25 to a two-digit output would become 2.2 in the Human Layer if we use banker's rounding (ties go to the nearest even).

You may be thinking, "Why not have the computer do the extra work of computing in decimal, eliminating the input and output rounding errors?" It is a great question. Sometimes, that is indeed what should be done. There are computer number formats that support exact decimal representation. However, the argument for computing in binary format goes like this: Typically, you do vastly more compute rounding than input rounding or output rounding. Computing with decimal formats is at least twice as slow as computing in binary formats. If you enter a number, perform thousands of (rounded) calculations, and format the result for output, then using decimal format protected you for only two of all of those thousands of calculations. And decimal format slowed you down by at least a factor of two. Usually, that is not a good tradeoff.

Before leaving the topic of translation error, there is an important question to answer for every binary number format: What is the minimum number of decimals needed to express a binary result such that it converts back to that same binary result? For example, the smallest positive value in **posit8** is $\frac{1}{64} = 0.015625$, but do you need all those decimals to ensure that when it is converted back to a **posit8** value, it converts to $\frac{1}{64}$? Would 0.0156 suffice? How about 0.016?

It turns out that all you need is 0.02, just a single decimal of significance, to make sure it converts back to $\frac{1}{64}$. That is economical, *if* you are sure that "0.02" in the Human Layer will be converted back to that same low-precision format the next time that value is needed.

Near the middle of the range where **posit8** values have more significance, you need as many as three decimals to express them such that they convert back to the correct posit. For example, consecutive **posit8** values $\frac{33}{32}, \frac{17}{16}, \frac{35}{32}$ are expressible as exact decimals 1.03125, 1.0625, 1.09375, but all you need is 1.03, 1.06, and 1.09 to make sure they convert back to $\frac{33}{32}, \frac{17}{16}$, and $\frac{35}{32}$.

A small program lets us test every value to see how many decimals are needed. It is related to the decimal accuracy, but it is not the same. It is much more chaotic, and it is difficult to find a closed-form equation for how many decimals are needed. Here is an empirical plot for the **posit8** values:

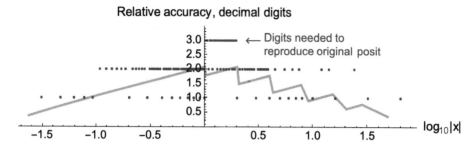

For each 8-bit posit, a dark magenta dot shows the number of decimals needed to specify it uniquely.

Perhaps 8 bits is somewhat too low of a precision to show the pattern. Here is the plot for the 16-bit posits mentioned at the end of Chapter 1:

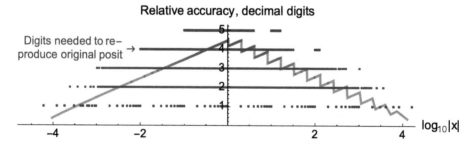

For 16-bit posits, a range of 1 to 5 decimals suffice to specify each one.

That shows that five decimals suffice to express a **posit16** in a way that re-encodes as the same posit. But it also shows one of the advantages of tapered accuracy: For large-magnitude and small-magnitude values, it is more economical to express numbers with scientific notation, like "−6e−5" (five characters) instead of "−0.00006" (eight characters). Even though the "e" notation takes extra characters, it saves space for extreme magnitudes. With tapered accuracy, the more you need scientific notation, the fewer decimals need to be displayed. The smallest positive **posit16** value is $\frac{1}{16\,384}$, which can be expressed as "+6e−5". For the higher-accuracy values like the ones in the fovea, five decimals are needed but not the "e" notation, like 1.0001, 1.0002, 1.0004 for the three **posit16** values that follow 1.0.

2.9 Defining *Overflow* and *Underflow*

2.9.1 The Overflow Fallacy

A common misconception is this one:

> Overflow occurs when a result is larger than the largest representable value.

That is not quite true, because a result can legitimately *round down* to the largest representable value and thus stay in range. So we offer this definition instead:

> **Definition**: *Overflow* occurs when a result is too large in magnitude to round it to a real number in the vocabulary of the format.

For an example with an 8-bit fixed-point format, the negative number of greatest magnitude is –8. Numbers are spaced $\frac{1}{16} = 0.0625$ apart, so tie points are $\frac{1}{32} = 0.03125$ from each value in the vocabulary. Any result x where x is in the interval (–8.03125, –8) rounds to –8 without being an "overflow." For positive values, the largest value in `fixed8` is 7.9375, and $\frac{1}{32}$ beyond that is 7.96875, so values in (7.9275, 7.95875) do not overflow even though they are larger than the maximum representable real number in the format.

There is a similar definition for underflow. Here we use "too small" to mean "too low in magnitude" (that is, too close to zero) and not "too close to –∞."

> **Definition**: *Underflow* occurs when a result is too small in magnitude to round it to a nonzero number in the vocabulary of the format.

Again using `fixed8`, we could multiply $\frac{1}{16}$ and $\frac{9}{16}$ which has the exact result $\frac{9}{256}$, but that has to be rounded to the smallest positive value $\frac{1}{16}$ or be classified as underflow. The tie point is $\frac{1}{32} = \frac{8}{256}$, so it rounds up to $\frac{1}{16}$ and does *not* underflow. The relative error is rather high, however, the same as the relative error of rounding 9 up to 16. That is a relative error of $\left| \ln\left(16/9\right) \right| \approx 0.58$.

The following diagram summarizes the careful definition of underflow and overflow that takes into account rounding. Relative widths are not to scale.

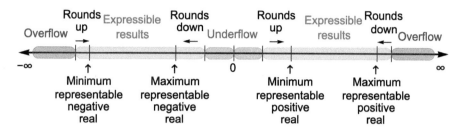

Underflow and overflow need to take rounding into account.

Notice that the result has to be signed. If the result of a calculation is correctly expressable as zero, that is not underflow. Whether the endpoints of the Expressible Results range are open or closed depends on the specific rounding rules, like "ties go to even" or "ties go towards zero" and so on.

2.9.2 Infinite Relative Error or Saturate?

In some legacy formats for real numbers, there are representations of ∞ and –∞, and overflow can return the appropriate signed infinity. Similarly, underflow returns zero. Notice that this creates an infinitely large relative error; if the correct answer is a signed value x and we instead return 0 or ±∞, the relative error is $\left| \ln(\pm\infty/x) \right| = \left| \ln(0/x) \right| = \infty$.

Legendary computer scientist Donald Knuth weighs in on the underflow issue in his *The Art of Computer Programming* series:

> It has unfortunately become customary in many instances to ignore exponent underflow and simply to set underflows to zero with no indication of error. This causes a serious loss of accuracy in most cases (indeed, it is the loss of all the significant digits), and the assumptions underlying floating point arithmetic have broken down, so the programmer really must be told when underflow has occurred. Setting the result to zero is appropriate only in certain cases when the result is later to be added to a significantly larger quantity.

Volume 2, *Seminumerical Algorithms*, 2nd edition, page 206.

Posit arithmetic instead *saturates* to the nearest signed posit. This preserves information about the sign of the result, and while the relative error can be large, at least it is known and it is *finite*. The "certain cases" Knuth mentioned can be handled in software for posit calculations.

2.10 The Inaccuracy of Operations

Another thing we need to be able to measure is the amount of error caused by rounding after an operation, like addition or squaring or any other function. When adding or multiplying *integers* on a computer, you either get the exact answer or you overflow/underflow the range you can represent. When working with *real* numbers, you have no such luxury.

A figure of merit for a real number format is the fraction of the cases it actually *is* possible to represent the result of an operation without rounding. And if you do have to round, the relative error of rounding should be as small as possible. We can look at the worst-case rounding error and the average rounding error. If we compute all the errors and sort them from smallest to largest, it helps us visualize the overall inaccuracy of an operation.

2.10.1 An Easy Example: The Square Root Function

The square root operation is easy to evaluate because there is no possibility of overflow or underflow. It reduces the set of representable (nonnegative) real numbers to a subset of those numbers. Let's apply the error assessment to the square root function for the **posit8** set and the **fixed8** set:

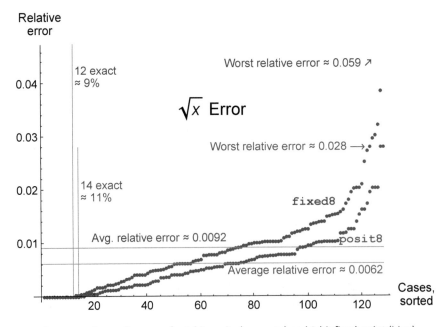

Square root rounding error for 8-bit posits (magenta) and 8-bit fixed-point (blue)

While this shows posits to be superior for square root operations for accuracy, remember that the hardware for fixed-point arithmetic is simpler and faster than that for posits, so there is an engineering tradeoff.

2.10.2 A Two-Input Example: Multiplication

For two-input functions (like $a + b$, $a - b$, $a \times b$, a/b, and a^b) and small sets of possible inputs, we can exhaustively tabulate the errors for all possible pairs of inputs. Let's try that with multiplication, again comparing `fixed8` and `posit8`. It is sufficient to test the nonnegative numbers since negating an input does not change the error. But unlike the square root, it is possible for multiplication to overflow or underflow. We can count the number of times that the result

- overflows
- underflows
- is exactly representable in the format
- is rounded to the nearest number in the format

In the last case, we compute the relative error from rounding, and find the worst-case relative error, and the average error including results with no error (exact). Start with `fixed8`.

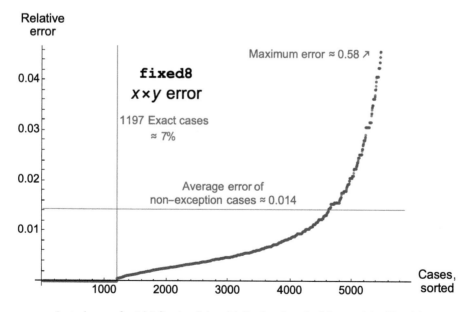

Sorted errors for 8-bit fixed-point multiplication (much of the graph is off-scale)

If we lay out the multiplication table graphically, using colors to indicate what happens to this imperfect representation of real values, it looks like this:

Exact	Rounded	Underflow	Overflow
1197	10 503	20	4664
7%	28%	0.12%	64%

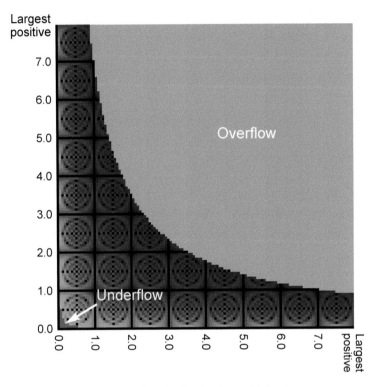

Closure plot for 8-bit fixed-point multiplication

The **black** pixels indicate an exact result, that is, cases where the product just happens to be exactly representable as a fixed-point number. Light Blue indicates overflow, which occurs for an alarmingly large fraction of the multiplication table. Red indicates underflow. Both use the careful definition that the number cannot legitimately round to the nearest representable real. The **green** pixels are where the result is rounded to a number that the format can represent; the brighter the color, the larger the relative error.

For the `posit8` set, the average relative error appears to be greater at first glance:

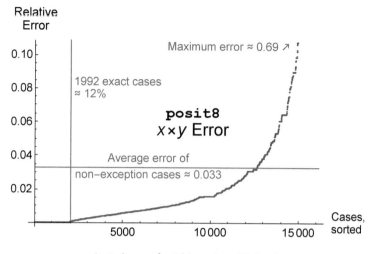

Sorted errors for 8-bit posit multiplication

However, that is a small price to pay for a massive increase in the resilience to overflow, as the graphical table of errors from multiplication shows:

Exact	Rounded	Underflow	Overflow
1992	14 151	119	122
12%	86%	0.73%	0.74%

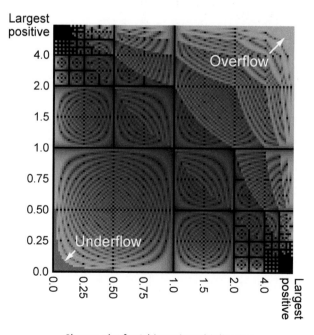

Closure plot for 8-bit posit multiplication

The area of overflow has been reduced to a tiny fillet in the table, very similar (but not identical) to the area of underflow.

> **Exercise for the Reader**: Why is there a slight asymmetry between the overflow cases and the underflow cases for `posit8` multiplication?

Now that we are armed with tools for measuring inaccuracy accurately, we can start comparing number formats and ways to use them for calculations.

"SMEDWICK, WE HAVE REASON TO BELIEVE YOU'VE BEEN COMMITTING SCIENCE FRAUD."

3 More Power!

Exponentially increasing power in the classic 1956 sci-fi movie *Forbidden Planet*

The first explanations of the full posit format were published in 2017, and in the intervening years much has improved about how to explain it more clearly. If you have already seen explanations of the posit format, do not be surprised if the following differs slightly from the early papers on the subject.

3.1 Increasing Posit Dynamic Range

Start with a three-bit posit ring, but instead of the regime bits representing a power of 2, let them represent a power of $2^2 = 4$. That doubles the dynamic range:

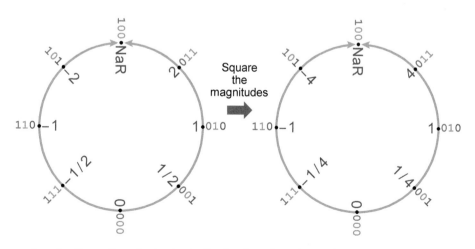

Squaring the northwest value (and adjusting the reflected values) doubles dynamic range.

It also creates a new rule for inserting values when we append bits. As usual, appending a **0** bit does not change the value; but when we append a **1** bit we now can insert an integer power of 2 between the values.

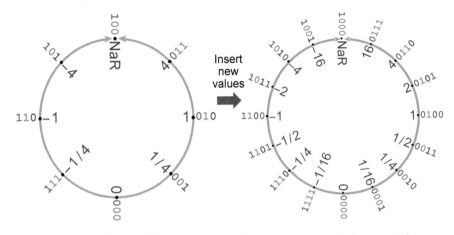

Between powers of 2 that differ by more than 1 in the exponent, insert the geometric mean.

When a bit is appended after a regime termination bit, we call that an *exponent bit* and color code it in **blue**. If it is a **1** bit, the flanking values are 2^k and 2^m and the inserted value is $2^{(k+m)/2}$. The regime bits now serve as a kind of super-exponent, where the value they represent is a power of 4 scaling, not a power of 2. The exponent bit then lets us fill in the gap between 1 and 4, or between 1/4 and 1. To make it clearer, we can append one more bit so that some fraction bits appear for magnitudes 1/4 to 3 (shown in **black** as usual). The exponent bit lets us fill the gap between 4 and $4^2 = 16$ with $4^2 \cdot 2^1 = 8$. Integer powers of 2 still have perfect reciprocals, their mirror image about the horizontal axis. The mapping to 2's complement integers is still monotone, as shown in this ring plot:

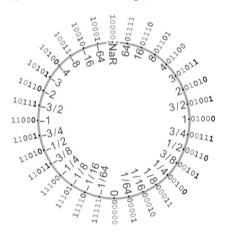

Ring plot of 5-bit posits with 1 exponent bit

Suppose we want still more dynamic range. We can square the meaning of each regime bit again, to $4^2 = 16$, so the value in the northeast position is 16. To fill in between $16^0 = 1$ and $16^1 = 16$ means scaling by 1, 2, 4, or 8, that is, 2^e where e ranges from 0 to 3. As an unsigned binary, we can do that with exponent bits `00`, `01`, `10`, and `11`. Here is the ring plot for a 5-bit posit with a 2-bit exponent field:

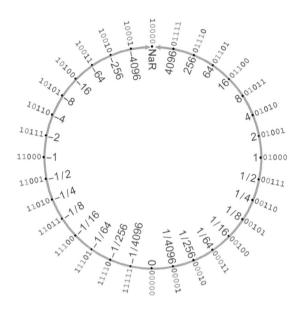

Ring plot of 5-bit posits with 2 exponent bits

The number of bits in the exponent field is *eS*, short camelCode for *exponentSize*. The value in the northeast position is called the *uSeed*, and it is equal to $2^{2^{eS}}$. Every time *eS* goes up by 1, the *uSeed* is squared and the dynamic range doubles in size, at the expense of taking away one bit from the significant bits in the fraction part. As *eS* increments starting at 0, *uSeed* is 2, 4, 16, 256, 65 536,… a *very* rapidly-increasing series. Be careful with it.

Chapter 1 introduced the simplest form of posit with no exponent bits (*eS* = 0), but by introducing exponent bits we have considerably more power. Here are the complete rules for introducing new values on the posit ring as the precision *n* grows:

- Appending a **0** bit to an existing posit does not change the represented value.
- Between NaR and a real number, insert the real number times *uSeed*.
- Between 0 and a real number, insert the real number divided by *uSeed*.
- Between 2^k and 2^m where k and m differ by more than one, insert $2^{(k+m)/2}$.
- Otherwise, between x and y, insert their average, $(x + y)/2$.

We also generalize the formula for a posit value. With $eS = 0$ in Chapter 1, we had

$$x = ((1 - 3s) + f) \times 2^{(1 - 2s) \times (r + s)}$$

where s is the sign bit value, f is the value of the fraction bits, and r is the value represented by the regime. Each increment in the r value makes the exponent part of the above formula larger by 2^{eS}. The exponent bits then represent an unsigned integer that fills in all the integers between those regime strides, from 0 to $2^{eS} - 1$. This seems weird at first, because it mixes two different ways to write a number. Imagine writing counting numbers with Roman numerals for the ten's place and 0–9 numerals for the one's place, so twenty-three would be "**XX**3" and thirty-nine would be "**XXX**9". Or we could use Roman numerals for the hundred's place and 00–99 numeral for the ten's and one's place, so two hundred seventy-one would be "**CC**71". That is similar to how the regime bits and exponent bits work together.

For example, if $eS = 2$, each increment of r raises the power of two by a stride of $2^{eS} = 4$, and the two exponent bits `00`, `01`, `10`, `11` represent $e = 0, 1, 2, 3$ to fill in the gaps between those multiples of 4. The formula for x generalizes to

$$x = ((1 - 3s) + f) \times 2^{(1 - 2s) \times (2^{eS} \times r + e + s)}.$$

The function above is the **pToX** function (for posit-to-X) and the code for it is in Appendix A. If the formula produces $-\infty$, it returns **Indeterminate** to accommodate all the cases where the result is not a real number, as the Not-a-Real (NaR) value.

> **Exercise for the reader**: In going from $eS = 1$ to $eS = 2$, the *uSeed* jumps from 4 to 16. What happens if we try to use 8 as the *uSeed* value?

The tent-shaped relative accuracy curves double in width with each increase in eS, but the top of the fovea drops by about 0.3 decimals because the fraction has one less bit.

Here is a plot of 16-bit posit decimal accuracy with *eS* set to 0, 1, 2, and 3.

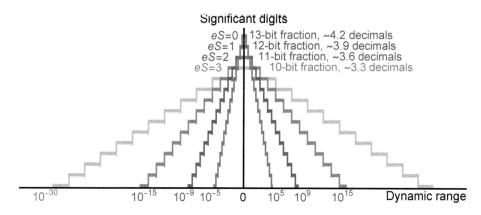

Accuracy plots for 16-bit posits as *eS* ranges from 0 to 3

In that figure, the wobble is simplified as a bar, and the annotated accuracy refers to the *minimum* accuracy.

3.2 *Ghost Bits* and the *Twilight Zone*

As an example, let's look at the posit ring for 6-bit precision and *eS* = 2:

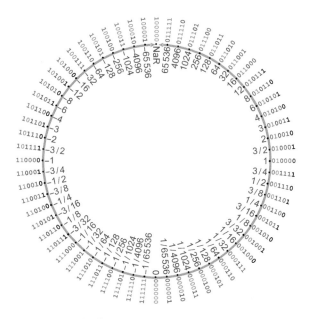

Ring plot for 6-bit posits with 2 exponent bits

We need a notation for the number of bits of precision *n* and the exponent size *eS*. Let's use angle brackets ⟨*n*, *eS*⟩, so the above ring shows "posit⟨6, 2⟩" format.

The ring plot shows 2 exponent bits, shown in **blue**. But wait, you say; some of the values do not have *any* exponent bits, and some have *only one* exponent bit. Not true; there are always 2 exponent bits, but any bits beyond the end of the number are **0** bits, just as we did for the fraction bits. Ghost bits can be exponent bits.

The area where the exponent bits are partly or entirely beyond the least significant bit (LSB) is called the *Twilight Zone* because years of experience has shown this slightly-weird region to be the source of most of the bugs in creating a posit environment.

Consider the value 1024 in the ring, with posit representation **011101**. The right way to think of that is as six stored (explicit) bits followed by an infinite string of ghost bits, and in this case, the first ghost bit is the second bit of the two-bit exponent field:

$$011101|000000\cdots$$

It is even possible for regime bits to be ghost bits, implied **0** bits beyond the stored bits. For example, look at the representation of $-1/65\,536$:

$$111111|000000\cdots$$

The opposite bit that ends the regime is in the trailing **0** bits, not shown explicitly.

As an example of the kind of mistake that is easy to make, suppose we want to encode the number 600 in the closest posit⟨6, 2⟩ form, properly rounded. We have a choice between 256 or 1024, both of which are in the Twilight Zone. The halfway point between the two numbers is $(256 + 1024)/2 = 640$, which might make it seem like 600 should round down to 256 as the closer point. However, that is not the way rounding works. You instead ask, "If we had posit⟨$n + 1$, eS⟩ format instead of posit⟨n, eS⟩ format, what would be the new halfway point and which side would the value be on?" Bit $n + 1$ might be an exponent bit, not a fraction bit.

If we had seven bits instead of six, we could represent **0111011**, which would decode as 512. Since 600 is greater than 512, the nearest representation as a posit is to round *up*, to 1024. This is consistent with our definition of relative error, since the relative error of choosing 1024 is $\left|\ln\!\left(1024\big/600\right)\right| \approx 0.5$ but the relative error of choosing 256 is $\left|\ln\!\left(256\big/600\right)\right| \approx 0.9$. On a logarithmic scale, 600 is closer to 1024. In the Twilight Zone, the tie point is also a point where relative accuracy is continuous instead of having a slight discontinuity (see Section 2.4).

3.3 Is the Opposite Bit in the Regime "Wasted"?

Another common initial reaction to the posit format is that the terminating bit of the regime is a "wasted" bit, and there ought to be a way to define it differently so that every bit counts. So, *is* it wasted?

In the posit⟨6, 2⟩ format, suppose we have $x = $ `010111` representing 12, and we add 2 to that. The result of 14 is midway between representable values 12 and 16 and the nearest even binary is 16 = `011000`. We need to round *up*. Notice what happens to the last four bits when we increment the posit representation for 12. Taking the color-coding away, `010111` increments to `011000`. The amazing thing is that the carry extends the regime by one more bit, the most-significant exponent bit turns into the opposite bit terminating the regime, and the exponent shifts over by one place to form the last two represented bits of the posit. All fraction bits become ghost bits. This happens *automatically*. The result, 16 = `011000`, is exactly what it should be. There is no need to check for carry overflow in each field. Everything just works.

> When you round, you round the *binary* (computed to more than the desired number of bits) and *not the real value* it represents.

The terminating bit of the regime is essential for this to work. It is definitely not wasted. Every bit counts.

A common mistake is to worry about rounding the fraction part, and once that is rounded, assemble the parts of the posit (sign bit, regime, exponent, and fraction). This will create more bugs than you find in a tropical rainforest. Always assemble the parts of the posit, computed with extra bits of precision, and *then* apply rounding.

3.4 Converting Real Numbers to Posits

Section 3.1 showed how to turn a posit bit string *p* into a real number *x* (or NaR). This section explains the algorithm that converts a real number *x* (in any format) into a posit string *p*. This routine is called **xToP**, and the code for it is in Appendix A. For clarity, this method takes the absolute value of the input, but it is also possible to directly convert a real value without doing that.

Assume the following quantities have been defined or computed as global variables based on the requested precision n and exponent size eS:

- *uSeed* is $2^{2^{eS}}$.
- *minPos* is the smallest positive posit value, $uSeed^{2-n}$.
- *maxPos* is the largest positive posit value, $uSeed^{n-2}$.

In the following, assume $n > 2$. The algorithm for $n = 2$ can be expressed in a single sentence: "Any non-real becomes NaR, 10; zero becomes 00; positive reals become 01; and negative reals become 11."

3.4.1 Check for Non-Real Inputs

The first step is to determine if the input x represents a real number. If it is stored in integer format, that works. Integers are real numbers. It could also be a complex number $a + bi$ where b happens to be zero. It could be an input string like "1.6e-19" that follows the rules of a computer language for expressing a real number in scientific notation. If it fails the test of being a real number, return the n-bit string for NaR, 1000⋯000.

3.4.2 Take the Absolute Value and Check for Special Cases

Setting $y = |x|$ makes some steps simpler. That does not lose the information about whether x is positive or negative (or zero), since we do not alter the input value x and can always refer back to it.

If $y = 0$, return the n-bit string consisting of all 0 bits, 0000⋯000. It is important to take care of possible NaR or 0 cases before trying to find the regime bits. This also takes care of the cases where $y \geq maxPos$ or $y \leq minPos$:

If $y \geq maxPos$,

    Return the posit for *maxPos* times the sign of x,

Else if $y \leq minPos$,

    Return the posit for *minPos* times the sign of x,

End if.

This also takes care of one of the rounding rules, that magnitudes outside the dynamic range saturate to the (signed) largest or smallest value. Also, handling these special cases assures that the opposite bit ending the regime will fit in the n-bit string built in the next subsection.

3.4.3 Encode the Regime Bits

Since computer environments usually support unsigned integer types, it is easy to treat the posit bit string as an unsigned n-bit integer. Having already dealt with the NaR and 0 cases, the regime can be at most $n - 1$ bits long. With an n-bit unsigned integer p we can store the regime starting in the Most Significant Bit (MSB) since we do not need a sign bit, and that will leave an extra bit at the end that we can use to do proper rounding.

> If $y \geq 1$, the regime will be a string of **1** bits:
> > Set $p = 1$ and $i = 2$. The number of posit bits found so far is i.
> > While $y \geq uSeed$ and $i < n$,
> > > shift p left by one place and set the LSB to **1** (lengthen the regime).
> > > Divide y by $uSeed$ and increment i.
> > End while.
> > Shift p left by one place to prepare for the next phase, and increment i.
> Else $y < 1$, so the regime will be a string of **0** bits:
> > Set $p = 0$ and $i = 1$ (off by 1 from the $y \geq 1$ case, since we seek $1 \leq y < uSeed$).
> > While $y < 1$ and $i \leq n$,
> > > multiply y by $uSeed$ and increment i.
> > End while.
> > Set $p = 1$ (the opposite bit ending the regime) and increment i.
> End if.

The "Shift p left by one place" creates a **0** as the LSB, which is the opposite bit ending the regime in the case where the regime is a string of **1** bits.

3.4.4 Encode the Exponent Bits

Converting x to the posit format has proceeded from left to right, and finding the exponent bits is no different. We know that $1 \leq y < uSeed$. If the format uses $eS = 0$ then there is nothing to do, so skip to finding the fraction bits. Otherwise, continue bisecting to find the exponent bits from most significant to least significant. Computing $uSeed = 2^{2^{eS}}$ is actually an easy thing for a computer to do, because a **1** bit shifted left by eS places is the unsigned integer for 2^{eS}, and then a **1** bit shifted left by that unsigned integer is $2^{2^{eS}}$. In the C computer language, it would be concisely expressed as `1 << (1 << eS)`.

So the algorithm for extracting the exponent bits looks like this:

Set $e = eS - 1$.
While $i < n$ and $e > 0$,
     Shift p left by one place to make a place for an exponent bit.
     Set $scale = 1 << (1 << e)$.
     If $y \geq scale$, set the LSB of p to 1 and set $y = y / scale$.
     Increment i, since we have used another posit bit for the exponent.
     Set $e = e - 1$.
End while.
Shift p left by one place to prepare for the next phase, and increment i.

If the bit counter i actually reaches n before e decrements to 0, the exponent bits will be clipped off and the posit will be in the "Twilight Zone." Otherwise, we will be left with the significand y, with $1 \leq y < 2$.

3.4.5 Encode the Fraction Bits

Since y is now a fixed-point number between 1 and 2, its binary bit string always begins with a 1 bit, so there is no need to store that. The 1 is implied. Hence, we first find the fraction by subtracting 1 from y. Then each bit of the fraction is found by doubling the fraction and testing if it is larger than 1; if it is, install a 1 bit in the posit bit string, subtract 1 from y, and repeat until there is no more room for fraction bits.

Once all $n + 1$ bits are populated, the test of whether the conversion was exact is very simple: Is $y > 0$? If it is, then the ubit is 1 because there are more bits after the last one. But if $y = 0$, then the conversion landed exactly on a posit value, so the ubit is 0. Remember, the ubit is needed for the banker's rounding.

Set $y = y - 1$.
While $i < n$,
     Shift p left by one place. This puts a 0 in the LSB position.
     Set $y = 2 \times y$.
     If $y \geq 1$, set the LSB of p to 1 and set $y = y - 1$.
     Increment i, since we've used another posit bit for the fraction.
End while.

After this step, $i = n$, so we have a bit string with n bits. But it does not have the sign bit, so we have one more bit than will be in the rounded value. Call it b_{-1}.

3.4.6 Round and Apply the Sign of x

Banker's rounding is rounding a value x to the nearest representable number p, and if there are two possibilities for p equally close to x, the rule is to choose the one that has *even* representation. In binary, even numbers end in **0** and odd numbers end in **1**. The reason for the "ties go to even" rule is that it makes the rounding *unbiased*. Imagine a bank that needs to calculate interest to the penny on billions of dollars of transactions; even the slightest bias up or down could quickly amount to a fortune. There have been schemes to do this, using rounding to get something for nothing.

Banker's rounding is the rounding mode for the posit format, and it is the *only* rounding mode. No matter whether you are rounding integer format, fixed-point, floats, or posits, the principle is the same. The algorithm for banker's rounding needs just *three bits*, shown in the following figure for the case of an 8-bit format:

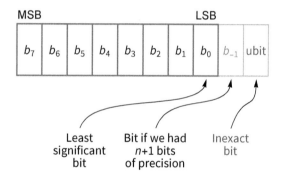

The three bits needed for banker's rounding

For an n-bit format, our convention is to label bits with subscripts from 0 for the LSB up to $n - 1$ for the MSB. The three bits we need are the LSB, a "b_{-1}" bit we can call the *rounding bit*, and a bit we saw in Chapter 1, the ubit. If ubit $= 1$, that indicates there are more significant digits after b_{-1}, like writing "$\cdots$" after the last digit of a number.

If b_{-1} is **1** and the ubit is **0**, the value is at the tie point; it is the value that would be represented if there were one more bit. To make the rounded value even, if b_0 is 0, do nothing (that is, round down by ignoring b_{-1}); but if b_0 is 1, add **1** to the LSB and propagate the carry to the left. With posit format, there is no possibility of overflow.

If b_{-1} is **1** and the ubit is **1**, meaning there are some **1** bits after bit b_{-1}, then the value is above the tie point and needs to be rounded up. The four cases for rounding a value x to a posit are as follows:

- If *x* is 0 or NaR, return the respective posit representing that value.
- If |*x*| ≥ *maxPos,* return the posit for *maxPos* with the same sign as *x*.
- If |*x*| ≤ *minPos*, return the posit for *minPos* with the same sign as *x*.
- If *minPos* < |*x*| < *maxPos*, encode *x* to *n* + 1 bits; if that representation is exact, set the ubit to 0, else set the ubit to 1. If b_{-1} is 1 and (b_0 is 1 or the ubit is 1), add 1 to the b_0 position and propagate any carry.

If you like digital hardware design and can read circuit diagrams, the figure below shows what banker's rounding looks like for a number with the extra b_{-1} bit and some trailing bits in temporary storage.

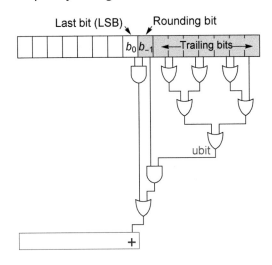

Logic circuit for banker's rounding

The structure under the trailing bits is an "OR tree" that produces a ubit of 1 if any trailing bits are nonzero, or 0 if all trailing bits are zero. Even if you do not read circuits, you can see that banker's rounding does take logic gates and logic delays. If you did not care about bias in rounding, you could simply drop all the bits after b_0. Some processors opt to do that with legacy formats, which can lead to an insidious problem that goes beyond biased errors accumulating in the result: Lack of *portability* and *reproducibility*. When a programmer discovers that a program produces different results on different computers, it leads to many unproductive hours spent trying to figure out why, culminating in the discovery that rules were bent in an invisible way. That is why there is only one rounding mode for the posit format.

If b_{-1} (the LSB of *p*) is 1,

   If b_0 (the next-to-last bit of *p*) is 1 or *y* > 0, set *p* = *p* + 1.

Shift *p* right by one place to discard the b_{-1} bit and make room for the sign bit.

If *x* < 0, return –*p*, else return *p*.

3.4.7 An Example

Code-like descriptions can be tedious to read; an example should make it easier to see how to turn a real number into posit representation. Let's convert the decimal number $x = -0.003$ into an 8-bit posit with $eS = 2$. The *uSeed* is $2^{2^2} = 16$.

Since x is a real number and not zero, we do not have to worry about those special cases with an infinitely long regime of **0** bits. Its magnitude is also between *minPos* and *maxPos*, so there is no saturation.

Set $y = |x| = 0.003$. Begin by noticing that the regime will start with a **0** bit because y is less than 1. In other words, the posit for y is in the southeast quadrant of the ring. That consumes our first bit, so set $i = 1$ to keep track of the number of bits used in p.

$$p \qquad\qquad\qquad y$$
$$\bullet \bullet \bullet \bullet \bullet \bullet \bullet \bullet 0 \qquad\qquad 0.003$$
$$\uparrow$$
$$i = 1$$

Shift p left to make room for the next bit. Scale y by multiplying by *uSeed* = 16 and check if it is still less than 1. It is, so we have another **0** regime bit. Increment i.

$$p \qquad\qquad\qquad y$$
$$\bullet \bullet \bullet \bullet \bullet \bullet \bullet 00 \qquad\qquad 0.048$$
$$\uparrow$$
$$i = 2$$

Again shift p and scale y by *uSeed*; 0.048 times 16 is 0.768, which is still less than 1:

$$p \qquad\qquad\qquad y$$
$$\bullet \bullet \bullet \bullet \bullet 000 \qquad\qquad 0.768$$
$$\uparrow$$
$$i = 3$$

Shift p left to make room for the next regime bit; this time, when we scale y by *uSeed*, the result $0.768 \times 16 = 12.288$ is larger than 1, so terminate the regime with the opposite bit and move on to finding the exponent bits.

$$p \qquad\qquad\qquad y$$
$$\bullet \bullet \bullet \bullet 0001 \qquad\qquad 12.288$$
$$\uparrow$$
$$i = 4$$

The first exponent bit is a scale factor of $2^{2^{eS-1}} = 2^{2^1} = 4$, and since y is larger than 4, the most significant exponent bit will be a 1. Shift p left and OR in the 1 bit. Divide y by the scale factor of 4; $12.288/4 = 3.072$.

$$p \qquad\qquad y$$
$$\cdots 00011 \qquad\qquad 3.072$$
$$\uparrow$$
$$i = 5$$

The next (and last) exponent bit is a scale factor of $2^{2^{eS-2}} = 2^{2^0} = 2$. Since $y > 2$, the next exponent bit will again be a 1. Shift p left and OR in the 1 bit. Divide y by the scale factor of 2; $3.072/2 = 1.536$, and now we know the significand: $y = 1.536$. We are done with the regime and exponent bits that determine the power-of-2 scaling.

$$p \qquad\qquad y$$
$$\cdot\cdot 000111 \qquad\qquad 1.536$$
$$\uparrow$$
$$i = 6$$

To find the fraction bits, first subtract the implicit 1 from the significand $y = 1.536$ to get just the fraction part, 0.536. Double the fraction to get 1.072; since that is greater than 1, the first fraction bit is a 1:

$$p \qquad\qquad y$$
$$\cdot 0001111 \qquad\qquad 1.072$$
$$\uparrow$$
$$i = 7$$

That last bit will be in the b_0 position in the final posit bit string, but we need one more bit for banker's rounding, the b_{-1} "rounding bit." Subtract 1 from y and double it to get $y = 0.144$. That is not greater than 1, so the next fraction bit is 0, which is in gray to remind us that it is a temporary bit beyond the last bit of the posit. And that is the nth bit, so now we are ready to round.

$$p \qquad\qquad y$$
$$00011110 \qquad\qquad 0.144$$
$$\uparrow$$
$$i = 8$$

The b_{-1} is 0, so there is no need to round up. Shift p one place to the right, which does two necessary things at once: It discards the b_{-1} and makes b_0 the LSB, and it shifts in a sign bit 0 on the left indicating a positive number.

$$p$$
$$00001111$$

The final step is to apply the sign of x. If x were positive, we would be done, but since $x = -0.003$, we need to negate p. Flip all the bits to `11110000` and add 1 to the LSB:

$$p$$
$$\texttt{11110001}$$

That is the closest posit representation to $x = -0.003$. The posit decodes as $-\frac{3}{1024} = -0.0029296875$ which is -0.003 to one significant digit.

> **Exercise for the Reader**: Convert the decimal number 2300 to an 8-bit posit with $eS = 1$. Caution: This number is in the Twilight Zone.

The thing to notice about encoding a posit is that it progresses from MSB to LSB, one bit at a time. That seems like the logical and obvious thing to do, but be aware that it is *not* the way reals are commonly encoded using legacy formats.

3.4.8 Some Advice for Anyone Building a Posit Environment

If you presently have a way to work with real values (such as IEEE floats, which will be described in the next chapter), make sure they have enough precision to do exact posit calculations. For example, here is a common pitfall:

> "I want to experiment with ⟨32, 2⟩ posit format and I have fast 64-bit floating-point arithmetic handy. I will write a routine **pToF** to convert 32-bit posits to 64-bit floats so I can then do the arithmetic operation on the floats, and then use a routine **fToP** to convert the result back to the posit format."

The problem, and this is perhaps surprising, is that 64-bit floats have insufficient precision. The 32-bit posits have significands with as many as 28 bits. When multiplied, the product can have 56 significant bits. But a 64-bit IEEE float has only 53 significant bits. There are situations where the final conversion back to a posit will be off by one in the last bit and that will create incorrect experimental results.

A much safer approach is to use integer arithmetic. After checking for NaR, convert a posit to a number of the form $k \times 2^m$ where k and m are signed integers with enough precision to cover the necessary dynamic range, so that you can stay in the Math Layer long enough to go back to the Compute Layer with the correct rounding.

Another piece of advice is to test the round-trip conversion **pToX** and then **xToP** exhaustively and plot **pToX**(**xToP**(x)) as a function of x with graphics tools. Start with very low precision, like $n = 4$ bits, debug, and work up with eS values from 0 to, say, 3. Also plot the discrete function **xToP**(**pToX**(p)) for all possible p up to a precision n, beyond which exhaustive testing becomes impractical. Experience shows that this is a quick and effective way to find bugs.

The **pToX**(p) function is the easiest to code and to get right; there are 2^n possible bit patterns for an n-bit posit, so for small n you can generate and display the real x value and check that the routine is decoding every p correctly. You do not have to worry about rounding. But the **xToP**(x) function is more challenging since there are many more possible inputs x than the number of possible outputs p, and there are many opportunities to create off-by-one programming errors. Here are example plots of a posit⟨6, 1⟩ environment using the **pToX** and **xToP** functions in Appendix A.

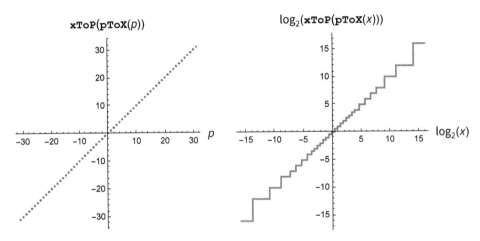

The plot on the left should always look like a discrete version of $y = x$, where the input and output posit bit strings are treated as signed integers. The plot on the right is a log-log plot; it uses the $\log_2$ of x values from *minPos* to *maxPos* (in this case, $2^{-16} = \frac{1}{65536}$ to $2^{16} = 65536$). It should look like a staircase wobbling about the line $y = x$, with increased wobble at the smallest and largest magnitudes. If there is a bug in either **xToP** or **pToX**, it usually shows up instantly in these "round-trip" plots.

3.5 A Simple Notation for Rounded Arithmetic

When FORTRAN was introduced in the 1950s, it was touted as "FORmula TRANslation," hence the name. It allowed you to write algorithms with a notation very similar to algebraic notation, albeit with an asterisk "$*$" for the multiplication operator, and a few other tweaks. You could declare variables a, b, and c to be of REAL type and then write a line of code like

```
c = a + b
```

and it looked for all the world like you were computing with real-valued variables. But that was a deception. If the variables are *integers*, then it is a fairly honest thing to do because you really will get the correct integer unless there is underflow or overflow. But with *real* values? Not likely.

For most of the many possible input values a and b, you will not get a REAL that exactly expresses "a + b" nor will the computer tell you anything about how far off it is from the correct value. FORTRAN has lost its all-caps look and become Fortran, and it is now fading as a computer language. But it made more than one generation of programmers highly vulnerable to the surprise that what looks like real-number math on a computer is not exact math at all, but instead is an approximation that can sometimes go horribly wrong without warning.

One can imagine a rigorous programming language where you have to write

```
c = ROUND(a + b)
```

or else you get an error message from the language compiler. But that would make code look pretty ugly.

Here is an idea that might at least find its way into Integrated Development Environments (IDEs) if not actual source code for a language. IDEs can color-code your program with different colors for comments, keywords, and so on, even if that is not part of the official language you are programming in. An IDE that understands rounding error and makes the hazards visible could be a huge help to the programmer.

Suppose we compute a value with exact math, convert that to an expressible (exact) posit using correct rounding, then convert the posit back to its mathematical value? In other words,

$$\text{ROUND}\,(\textbf{a}\; op\; \textbf{b}) = \text{pToX}(\text{xToP}(\textbf{a}\; op\; \textbf{b})).$$

This assumes that the **a** *op* **b** on the right hand side is computed in the Math Layer, that is, exactly. Throughout this book, the overbar operator is assigned to that composite function **pToX**(**xToP**(…)). So

$$c = \overline{a + b}$$

means that exact inputs *a* and *b* will be added exactly, rounded to the nearest real that posits can represent, and then presented back to the programmer as a value that can be expressed without rounding error.

That is much less intrusive than having to type "**ROUND**" everywhere. Some papers on rounding error put a circle around the + − × / operators for their rounded versions, like "⊕" for "rounded addition." But how would you do that for square root, or cosine? The overbar works for every operator. And you can use it on constants and single-argument operators as well. For example,

$$x = \overline{\log_{10}(\overline{3.1})}$$

instructs a computer to round the real value 3.1 to the nearest representable posit value, then compute the log base 10 of that to enough accuracy that the nearest representable posit is known. It still looks a lot like $\log_{10}(3.1)$, but the overbar decorations remind you that *you need to be suspicious*. This definition of the overbar as an operator is part of the code in Appendix A, and you will see a lot of it in this book.

3.6 As Simple as Possible, But No Simpler

"The supreme goal of all theory is to make the irreducible basic elements as simple as possible without having to surrender the adequate representation of a single datum"

—*Albert Einstein*, 1933

The folk paraphrasing of Einstein's words is usually something like, "A theory should be as simple as possible, but no simpler." Which is more succinct, though not his exact words.

Posits are designed with Einstein's viewpoint. It would be wonderful if we could do an entire calculation in the Math Layer and never have to worry about the finite nature of physical computers, but that is usually either impossible or is so complicated to do (by using symbolic methods, not just numerical methods) that it would run very slowly. Given that we *must* use finite precision in the Compute Layer and the Human Layer, and that we want the distribution of representable numbers to resemble the distribution of values that would occur if all computations are exact, posits are about as simple as possible. But no simpler.

3.7 Not the Current State of Things

So far, we have presented the positive message that *it is possible* to represent real numbers, either as exact values or ranges between exact values (using the ubit), and it is also possible to round after operations and sacrifice provable error bounds for speed. The unum and posit approach is derived through logical steps starting from low precision and working up, resulting in an exception-free definition.

But at the time of publication, this is not what we see offered to us in processor hardware for working with real numbers. Not even close. Which means we need the backstory on how we got to where we are now. In the next chapter, I can almost guarantee the reader a few surprises, and some information you will not find anywhere else.

Please keep this in mind as you read Chapter 4.

4 How We Got Into This Mess

A robot vacuum as envisioned in the 1960s, and what robot vacuums actually look like, lately. We similarly make the mistake of trying to get machines to mimic humans in our number format designs.

There is a strong human tendency to design technology to behave the way people do, instead of figuring out what is *optimal for the technology*. That is why some of the earliest digital computers wasted huge amounts of hardware to operate on decimal numerals instead of in binary, for example, when the only reason humans use base-10 is the physiological accident that our hands have ten fingers.

A few cultures used base 20 (fingers and toes?) and numerals in Maharashtra were base 5, suggesting one-handed counting. Bases 60 and 12 are still in modern use for time and degrees.

This chapter examines a sampling of the vast history of how humans have recorded and used numbers, and how those methods have proved to be far from optimal for *machines*. It also will tell the story of where the present-day IEEE Std 754 definition of floating-point formats actually came from; it's not what most people think.

4.1 The Amazing Babylonians

The ancient Babylonians were no slouches at mathematics, even going back as far as 1800 BC. They recognized the merits of using 60 as a number base, which persists to the present day in the form of 60 seconds per minute and 60 minutes per hour, as well as 360° per circle. The Egyptians may have been just as capable, but they used ink on papyrus instead of clay tablets left to bake in the sun, so we have much less surviving evidence of their skills. Here is a particularly impressive Babylonian relic, redrawn for clarity and annotated with modern numbers in white:

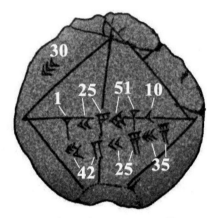

Ancient Babylonians knew the square root of 2 to seven decimals.

First, notice how logical the Babylonian number notation is. At the lowest level, it is unary in that it repeats the 𒌋 symbol for ten and the 𒁹 symbol for one to notate two-digit positional numbers between 1 and 59. If they needed a zero, they left that position blank. At the level above that, the system uses positional notation from left to right like our modern decimals, but base 60 instead of base ten. The side of a square is labeled 30 in the upper left. Across the diagonal of the square is their notation for $1 + \frac{25}{60} + \frac{51}{60^2} + \frac{10}{60^3}$, which in decimal works out to $1.414213\cdots$, which is the square root of 2 correct to the first seven decimals! The side of the square is labeled as 30, so the length of the diagonal is $30 \times \sqrt{2} \approx 42 + \frac{25}{60} + \frac{35}{60^2} \approx 42.4264$, again uncanny accuracy to six decimal places.

The relic is not a "tablet" but someone's scratch work from almost four thousand years ago. They knew how to find the diagonal of a right triangle, so maybe what we call the Pythagorean Theorem should be called the Babylonian Theorem. The backside of that block of clay has been partly erased but seems to show a 3-4-5 right triangle, the simplest case where the diagonal is an integer, $\sqrt{3^2 + 4^2} = 5$.

The Babylonians also knew how to solve the quadratic equation and compute compound interest rates (exponentials), and they could approximate the area under a curve by using close-fitting trapezoids (integral calculus). They knew how to solve systems of linear equations. I could go on. Somehow, though, their knowledge and well-designed number system was lost instead of spreading to the empires that followed, like that of the Romans.

4.2 Roman Numerals

"NOW WITH THE _NEW_ MATH..."

It is amazing that a civilization as successful as the Roman Empire was able to thrive with one of the most cumbersome number systems ever devised. Since we still see Roman numerals everywhere, from the hour labels on analog clocks to their use in numbering sections of publications, perhaps they do not seem that painful. But ask yourself this: How would you *multiply* **XIV** by **LXV** by hand, without first translating to some other number system?

Here is a question: Are Roman numerals positional notation, or are they more like tally marks and numeral systems that do not use position? I argue that they are *slightly* positional, but not all the way there. For example, suppose we write the integer 1984 in Roman numerals. The conventional way would be **MCMLXXXIV**, with big numbers mostly on the left and small numbers mostly on the right.

However, if we break that up as **M CM L X X X IV**, they could be written in any pattern and it would still mean 1984:

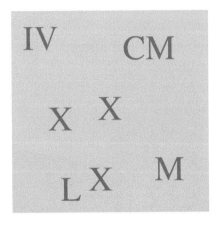

Roman numerals are not positional notation.

The only positional thing about Roman numerals is the subtractive pairs **IV, IX, XL, XC**, and so on where the value of the first letter is subtracted from the value of the second. Otherwise, they might as well be coins thrown on a table and added up. Hence, *slightly* positional. There are examples, say of the twenty-second Roman Legion writing their number as **IIXX**, since the way they described themselves in Latin translates as "the two and twentieth". Sometimes a scribe would write **VXL** for 45 instead of **XLV**, probably an indication that scribes of the era found the rules just as confusing and error-prone as we do today.

It's not commonly known, since in modern times we only use Roman numerals to write positive integers, but the Romans *did* have a way to write fractions.

$\frac{1}{12}$	$\frac{1}{6}$	$\frac{1}{4}$	$\frac{1}{3}$	$\frac{5}{12}$	$\frac{1}{2}$	$\frac{7}{12}$	$\frac{2}{3}$	$\frac{3}{4}$	$\frac{5}{6}$	$\frac{11}{12}$
·	:	∴	::	⁙	S	S·	S:	S∴	S::	S⁙

The system actually almost makes sense, like the pips on dice or dominos. However, besides integers divided by 12, they had special fractions with their own symbols, like Ɔ for $\frac{1}{288}$ and Ƨ for $\frac{1}{72}$. This type of thing is what you get when number systems form organically like human language, with people making it up as they go along, instead of being *designed* from first principles.

So how *did* the ancient Romans perform such common tasks as adding up the cost of items in a transaction? They converted Roman numerals to an abacus-type representation. Not necessarily beads on wires the way we usually picture an abacus, but as simple as pebbles lying in a groove that can be slid back and forth. And that's definitely positional and fairly well-suited to adding and subtracting.

We have no record of their pebble-in-groove type of abacus, but the British Museum has an excellent example of a portable Roman mechanical abacus:

Roman abacus

The results could then be converted back to Roman numerals. The abacus indicates symbols for a thousand, ten thousand, one hundred thousand, and a million. The rightmost two columns show the fraction counter as the number of twelfths, and a way to indicate how many thirds of $\frac{1}{12}$ after that. *That* is positional notation, and it is even ordered the same way as our positional numeral system today.

The design of Roman numerals suggests they originated in a medium where it took a lot of work just to make a line. Carving **IV** in stone takes three chiseled lines, instead of **IIII**. Writing **X** takes only two chisel lines, instead of four lines for **VV**. Obviously an **L** is far easier to carve than **XXXXX**. Considering how difficult it is to chisel a curve, the notation **C** for one hundred and **D** for five hundred suggests they did not expect to use such large-magnitude integers very often. For a thousand, Romans eventually used **M** (the most common form today) but earlier used this symbol: ↀ. That symbol is clearly not chisel-friendly, so it is easy to see how **M** (short for *mille*, Latin for *thousand*) came to be substituted when carving stone. There is also a symbol Ð for five thousand, and ↂ for ten thousand (a ↂ is visible on the abacus above) which appear in the present-day Unicode character set.

Sometimes it is said that the Indians invented the concept of zero. They may have been the first to use zero as a numeral, but the above abacus shows a zero: all beads down. Any culture that uses an abacus has a representation for zero, and it is the starting point for any calculation. The Romans could certainly communicate zero as *nullus*, the Latin adjective for "none."

When we use modern computers, the computer converts our decimal input into binary, performs the operations, then converts the result back to decimal so we can more easily read and understand it. That seems to be how Europe worked for hundreds of years; they would *communicate* numbers with Roman numerals, but if they had to do any actual *arithmetic*, they converted to abacus format with positional notation, did the operation, then converted back to Roman numerals. The Romans were not as arithmetically challenged as their numeral system suggests. They obviously had all the powers of 10 they needed to engineer the Colosseum.

4.3 Almost Binary Encoding: Cistercian Numerals

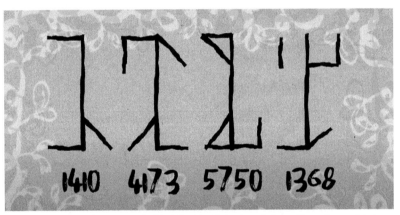

Cistercian Numerals were probably killed by Gutenberg's invention of movable type.

If tally marks are one of the bulkiest ways to express numbers, *Cistercian numerals* have to be one of the most compact ways humans have ever developed to write numbers. They were introduced around 1200 AD by the Cistercian monastic order as an alternative to Roman numerals, and were used for a few hundred years. They originated in a form of shorthand used by English shopkeepers, because they were economical not only for area but also for pen strokes.

Their definition varied from monastery to monastery, but the most common original form for digits 1 to 9 looked like this:

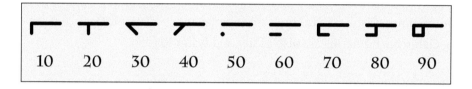

For the number of tens, reflect about the horizontal axis:

For the number of hundreds, flip the unit symbols about the *vertical* axis:

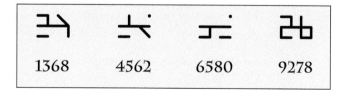

There is one corner left, and they used it for the thousands place:

But instead of treating these symbols as numerals, they combined them into a single glyph, like this:

If you want a zero in any place, you simply leave that corner unadorned. While a plain bar would represent zero, there is no evidence that a plain, unadorned bar was ever used. Cistercian numerals were used for numbering things, not for arithmetic, and it did not occur to them to start numbering with 0 instead of starting with 1.

For early hand-copied books, paper was precious and they made the most use they could of the space it offered. Imagine writing a page number like 1368 above in Roman numerals: **MCCCLXVIII**, ten characters. In Cistercian, you can write 1368 with four or five pen strokes, picking the pen off the paper three times, but if you write it in Roman numerals, you need eleven pen strokes, picking the pen off the paper ten times. Cistercian is about three times more efficient. Once you get used to them, Cistercian numerals are also easier and faster to read.

As they evolved, their users started writing them vertically, and the 5 was often written as a combined 1 and 4 instead of as a dot:

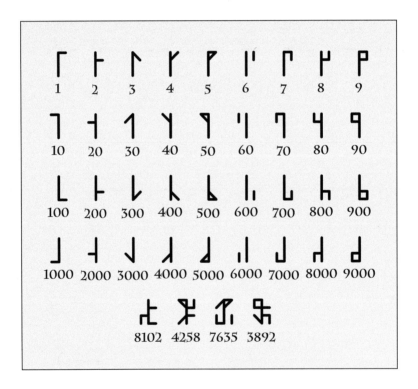

Later form of Cistercian numerals, with four examples

It is a positional notation, with up to four places. In fact, it is *two-dimensional* positional, not the one-dimensional left-to-right positional notation we are used to.

There are elements of binary-coded decimal (BCD) in Cistercian numerals. They were *so close*. Like, the symbols for 7, 8, and 9 are just the symbol for 6 combined with the symbol for 1, 2, and both 1 and 2. They were tantalizingly close to discovering a *logical binary encoding*. Suppose the inventors had instead decided to define 1 through 9 like this, where the new line segments are shown in red and the line segments that are dropped are shown in gray:

Exercise for the Reader: What would Cistercian numerals look like for base-16 (hexadecimal) notation, if we use the "revised Cistercian" shown above? What would be the range of integers that could be encoded that way?

So, what happened to Cistercian numerals? They are very easy to write, and with a little practice, easy to read (certainly easier than reading Roman numerals for integers that large). What happened was this man and his invention, around 1450 AD:

Gutenberg and his invention

Johannes Gutenberg's world-changing invention of a printing press with movable type for characters cast in lead. How many bins would be needed to hold every Cistercian numeral? *Ten thousand*.

Unlike the hand-copied Bibles produced by the Cistercian monks, none of the Gutenberg Bibles have page numbers. When Hindu-Arabic numerals finally spread throughout Europe, printing with just ten characters 0–9 was a thousand times easier than printing Cistercian numerals with movable type. Cistercian numerals fell into extreme obscurity.

They had an unusual resurgence among French Freemasons, around 1780. Tight secrecy of special knowledge is a hallmark of Freemasonry, and knowledge of how to write numbers in a concise way that was utterly baffling to non-members fitted perfectly into a group with secret handshakes, secret passwords, and secret oaths.

Another resurgence almost happened in the 1930s, in Nazi Germany. Perhaps Nazi scholars noticed that the 0–9 system came from India via Arab traders and felt that system therefore went against German nationalism, whereas Cistercian numerals are entirely European in origin. The resurgence did not get far. Besides being very impractical to print, just try to imagine how difficult it would be to invent a *typewriter* for them. Even with backspacing and over-typing, you would need an additional 36 keys to type each of the four corner digits plus a key for the central bar.

Amusingly, Cistercian on a modern display made with light segments would fit the (revised) Cistercian system beautifully. Imagine going to an airport and seeing flight numbers shown with a single character on a display with just 16 on-off segments (in addition to the always-on center stem), yet able to represent any of ten thousand flight numbers.

What 4-digit flight numbers would look like had Cistercian survived

Who knows; perhaps someday Cistercian will be added to Unicode using a set of *composable* characters, which would take up just one Unicode slot (64 characters).

4.4 Hindu-Arabic Numerals Spread to Europe

In the 1st century AD, Indian notation for numbers 1–9 looked like this:

Like so many number systems, it starts out as tally marks and switches to something easier to write when it reaches four tally marks. It was a positional system like we use today. We do not have a record of them using a circle for zero that long ago, so they probably used a blank to indicate a zero in a position like the ten's place or the hundred's place.

Imagine trying to write those numerals as quickly as possible; that means switching to cursive where you avoid taking the marker off the writing medium while writing a single numeral. Also, there seems to be some human preference for making characters similar in height, so the first three numerals need some vertical padding.

That would explain how the numerals evolved in India to look like this by the 9th century, with every numeral taking only a single squiggle to write:

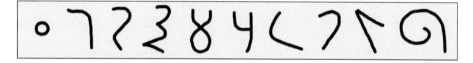

By then there was definitely a small circle to indicate a zero, and it is easy to see the tally marks for one, two, and three turned into cursive. It is confusing to us that the symbol for five looks exactly like a modern 4, but no, that really does mean five.

Arab traders picked up the Indian system, and as it spread from east to west, by the 11th century the numerals evolved into this:

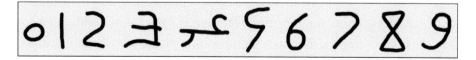

Perhaps because these are recognizable to us now (though the 4 is a little peculiar), the numerals became known as "Arabic" despite being an Indian invention. "Hindu" used to refer to the region and culture, and only later meant the religion.

The first European to realize and publicize the power of positional notation with 0–9 numerals was an Italian named Leonardo Bigollo Pisano ("Leonardo, traveler from Pisa"), an accomplished mathematician who published *Liber Abaci* (Book of Calculation) in about the year 1200 AD, very close to the time the Cistercian monks created their numeral system. Never heard of him, you say?

Leonardo Bigollo Pisano

He was the son (fi, in Italian) of a man named Bonacci, so he is today better known by his patronymic name:

Fibonacci.

Yes, *that* Fibonacci. Besides bringing us the 0–9 positional system, he introduced Europe to the "Fibonacci series" 1, 1, 2, 3, 5, 8, 13,... which was known to the Indians, where each number is the sum of the two previous numbers.

The techniques we learn in elementary school for pencil-and-paper arithmetic spread quickly, and those using those techniques were called *algorists*, as opposed to the *abacists* who still used an abacus to calculate. It was as if a cult of warriors trained in Asian martial arts had arisen among a population that only knew how to fist-fight, and they knew how much advantage it gave them. Being able to use a writing instrument to add, subtract, multiply, and divide very large numbers much faster than can be done with an abacus felt like a superpower to those who took the time to learn the algorist art. The algorists understood the secret power of being able to notate zero in particular, for which the word was one that sounds like "cipher"; to this day that word conveys a clandestine way of doing things.

Finally, Western civilization had a number system with mathematical underpinnings and one that was easy to write by hand, no pebbles or beads needed. A string of numerals d_i that could be any symbol from 0 to 9 was a shorthand for a polynomial of the number ten:

$$d_n\, d_{n-1} \cdots d_0 \text{ means}$$
$$d_n \cdot 10^n + d_{n-1} \cdot 10^{n-1} + \ldots + d_1 \cdot 10 + d_0.$$

Isaac Newton and Leibniz share credit for the invention of calculus in the 1600s, but Leibniz also did something no one else did. He noticed that positional notation also worked for bases other than ten, and in particular he found great elegance in using just the symbols 0 and 1 to represent any number. Binary arithmetic. Leibniz invented it in 1689, though he gave ample credit to the Chinese for ancient notations that used a similar idea. An English translation of his introduction to the idea follows:

> The ordinary reckoning of arithmetic is done according to the progression of tens. Ten characters are used, which are 0, 1, 2, 3, 4, 5, 6, 7, 8, 9, which signify zero, one, and through successive numbers up to nine inclusively. And then, when reaching ten, one starts again, writing ten by "10", ten times ten, or a hundred, by "100", ten times a hundred, or a thousand, by "1000", ten times a thousand by "10000", and so on.
>
> But instead of the progression of tens, I have for many years used the simplest progression of all, which proceeds by twos, having found that it is useful for the perfection of the science of numbers. Thus I use no other characters in it bar 0 and 1, and when reaching two, I start again. This is why two is here expressed by "10", and two times two, or four, by "100", two times four, or eight, by "1000", two times eight, or sixteen, by "10000", and so on.

Explication de l'Arithmétique Binaire, Die Mathematische Schriften, volume 7, page 223.

In seeing the concept of binary, I think Leibniz was the first person who "got it" and realized it was a game-changer. It took a few hundred years for the rest of the world to catch up with him. But again, binary notation is *not* human-friendly, nor is it economical to set in type. To write a number like 72 in decimal is **1001000** in binary, seven characters instead of just two characters. The human eye is better at pattern-matching alphabet-like symbols than it is at traipsing over a long sequence of just two symbols and keeping track of which **0** or **1** digit in the list is being viewed. Binary is not as bulky as using tally marks, but still bulky compared to decimal.

Cistercian numerals are super-compact, but were once impractical for typesetting. So the Hindu 0–9 positional system won out as the practical middle ground. Perhaps, hundreds of years from now, humans will have evolved to use hexadecimal as their primary digit system, eliminating the errors in human-computer number conversion.

4.5 Scientific Notation and Automatic Computing

The word "anthropomorphize" describes a recurring theme in this chapter. Here is the usual dictionary definition:

anthropomorphize | ˌan-thrə-pə-ˈmȯr-fīz |

verb
to attribute human characteristics to nonhuman things

It could also mean "to design human characteristics into nonhuman things," like when robots are made to look as much like humans as possible (androids) instead of being optimized for their intended purpose. We can talk about doing so as *anthropomorphization*. A seven-syllable mouthful, but a useful word.

Charles Babbage proposed the Analytical Engine in 1837, and it was to have an ingenious (and ambitious) system of mechanical gears that could perform programmable computations with *decimal* positional notation. He anthropomorphized the human 0–9 numeral system into the machine. That made the machine so ambitious that Babbage's plan to actually build it never got off the ground. It was to have had a memory of 1000 decimal numbers, with 50 digits each. Why so many digits? Because that gives you *dynamic range* of from 1 to 10^{50}, using simple integer odometer-like hardware. The design predated scientific notation or the concept of a floating decimal point, and it greatly complicated the gear-based design.

It was commonplace to avoid the use of large numbers of zeros in a number using phrases like "One followed by twenty zeros," but what we now call *scientific notation* was first specified carefully by Leonardo Torres Quevedo in 1914. He wrote a remarkable paper in French, the title of which translates as "Essays on Automatics," about the use of machines to imitate not just the physical actions of humans but their reasoning powers as well (that is, he was anthropomorphizing).

Here is a translated excerpt that laid out a potent idea:

Very large numbers are as embarrassing in mechanical calculations as they are in conventional calculations (Babbage planned 50 wheels to represent each variable, and even then they would not be sufficient if one does not have recourse to means which I shall describe below or to something similar). In the latter, they are usually avoided by representing each quantity by a small number of significant figures (six to eight at the most, except in exceptional cases) and by indicating by a point and zeros, if necessary, the order of size of the units represented by each digit.

Sometimes also, so as not to have to write a lot of zeros, we write the quantities in the form $n \times 10^m$. We can greatly simplify this notation by arbitrarily establishing three simple rules:

I.   n will always have the same number of digits (six, for example).
II. The first digit of n will be tenths, the second hundredths, etc.
III. We will write each quantity in this form, n;m.

Thus, instead of 2435.27 and 0.00000341862, they will be respectively, 243527;4 and 341862;–5.

I have not indicated a limit for the value of the exponent, but it is obvious that, in all the usual calculations, it will be much smaller than one hundred, so that, in this system, one will write all the quantities involved in the calculations with only eight or ten digits.

Reprinted in *The Origins of Digital Computers*, Brian Randell ed., page 102.

There it is. The invention of floating-point format, and also the invention of scientific notation of the form $n \times 10^m$ where the significant digits are n, and m is the exponent. He perceptively said the number of significant digits should be six, or eight at most, matching what we now call "single precision" floats. What he writes as "341862;–5" is what a modern program would print as "`0.341862e-5`", very close to current notation. He did not propose a method of representing *negative* values, notice; his "n" is a fraction ranging from 0.100000 to 0.999999, not a signed value. Some people will stop at nothing to avoid dealing with negative numbers.

The "quantities involved in the calculations" correspond to the temporary use of higher precision to minimize rounding error, but not as high as double precision. He could not have anticipated that programmers would want to over-insure against rounding error by using double-precision floats with 15-decimal accuracy for *every* real variable.

Notice that there is no provision for storing the number zero, nor does he consider any exceptions that arise, such as dividing by zero.

Quevedo also accurately estimated the dynamic range needed for conventional calculations as well within 10^{-99} to 10^{99}. He knew that 10^{-9} to 10^9 would not be sufficient, so he elected a two-digit (decimal) exponent representation, and then it became overkill, but sufficient to do the job. There is nothing to indicate that he considered number bases other than the number of fingers that humans have.

In the late 1930s, Konrad Zuse conceived the Z-1 computer as a *mechanical* binary computer that stored its **0** or **1** states using sheet-metal and pin linkages. He built it on the family dining room table. Remarkably, it implemented the Quevedo idea of a separate significand (15 bits) and exponent (7 bits). Zuse could not get the representation of zero working, though he knew it was needed; he fixed that in a later version, the Z-3, that used electromechanical relays instead of linkages. Both the Z-1 and Z-3 computers worked, so they were history's first binary computers capable of floating-point calculations. The only thing they lacked was electronic switching speed.

John Atanasoff, in late 1937 (independently of Zuse) realized the need for binary representation and Boolean arithmetic, but he wanted an all-electronic machine so he used vacuum tubes. The machine used 50-bit (equivalent to 15-decimal) integers, not floating-point, to solve systems of linear equations. Atanasoff conceived of the idea of a data store separate from whatever device performs arithmetic, and coined the term "memory" for that part of a computing device.

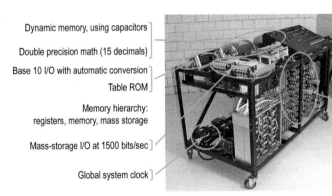

Dynamic memory, using capacitors
Double precision math (15 decimals)
Base 10 I/O with automatic conversion
Table ROM
Memory hierarchy: registers, memory, mass storage
Mass-storage I/O at 1500 bits/sec
Global system clock

Fully-electronic calculation
Base-2 internal arithmetic
Parallel processing (30 PEs, SIMD)
Modular, interchangeable components
Boolean logic for arithmetic
+ − × ÷ calculation from each PE

The reconstructed 1942 Atanasoff-Berry Computer with its unprecedented ideas

A scaled-down prototype was finished in 1938, and the full-scale design for solving up to 29 equations in 29 unknowns was finished in 1942. World War II effectively ended the efforts of both Atanasoff and Zuse.

In 1940, John Mauchly and J. Presper Eckert were planning an *analog* device at University of Pennsylvania, but when Atanasoff communicated to Mauchly the idea of "true" computing, as he called it (what we now call *digital* computing), Mauchly and Eckert pivoted from building an analog Electronic Numerical Integrator (ENI) to building a digital Electronic Numerical Integrator And Computer (ENIAC), finished in February 1946. Instead of true binary with Boolean logic, they wanted the ENIAC to do arithmetic the way people do, so they used the on-off states of vacuum tubes to count from 0 to 9 and then do a carry operation, imitative of the classic algorithms for elementary school arithmetic that Fibonacci had conveyed from the Indians. Constants were ten-digit decimal integers, set by mechanical rotary dials; it held twenty such constants.

4.6 The Emergence of Bit-Field Mentality

Bit-field mentality is a line of reasoning that goes something like this:

> We have *n* bits in the computer word size. Let's use *these* bits to represent *this* part of the number, *those* bits to represent *that* part of the number, …

That is what Quevedo and Zuse did. Had Quevedo actually built his design, he would have realized that the devil lies in the details. Zuse noticed that representing zero was hard, creating an exception to the rules of scientific notation. Bit field design stems from the urge to anthropomorphize, and it leads to inefficiency and non-mathematical behavior.

In the early 1950s, when IBM was just introducing electronic (vacuum tube) computing, they used bit-field mentality to design their *integers*. "Let's use the first bit to represent the sign of the integer, and the rest of the bits to represent the magnitude." At the time, it seemed as good a way as any. Engineers could look at bit patterns and (almost) read the binary, with a little practice, since the sign-magnitude representation mimicked the human way of writing integers with a "+" or "−" in front of positional-notation digits. Like, in six-bit precision, 000101 means +5 and 100101 means −5.

But what about two ways to write zero, as "positive zero" 0**00000** and "negative zero" 1**00000**? Do they represent the same real number, zero? Or is there something else going on? If I test 1**00000** < 0**00000** , will the computer return True or False? The workaround is, *if the magnitude bits are zero, the sign bit does not count.* That means the hardware has to test for a special case, which slows things down; it also means "every bit counts" is not true, at least not always. It was one of the first cases of a redundant bit pattern: more than one bit pattern to represent the same number.

That was just the beginning of problems with the humanlike sign-magnitude representation. Think back to grade school, probably around 5th grade, when you first learned about negative numbers and how to add and subtract signed integers. You had to master, in your mind, a flowchart that looks like this:

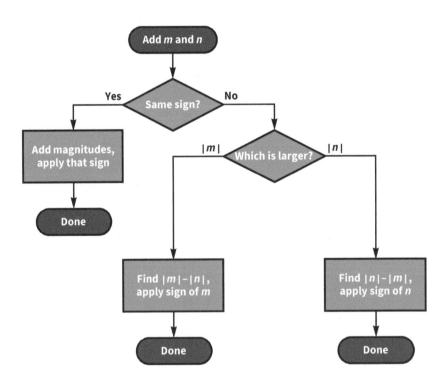

How to add signed integers, elementary school style

It was not long before more careful thinkers realized that using 2's complement arithmetic (described in Chapter 1) makes this entire flowchart go away and avoids "negative zero." The circuits for adding unsigned integers and signed integers become one and the same, except for defining where they overflow. It was a less human-friendly way to encode integers, but much more computer-friendly, and every computer now uses 2's complement format to store integers.

When you combine bit-field mentality with anthropomorphized scientific notation, you get *floating-point format*. This is quite different from an engineering approach of first setting out desirable properties of a format as design goals. It is also different from the mathematical approach of asking what is a logical mapping of a set of bits to a vocabulary of real numbers.

The most successful scientific computers of the 1950s were from IBM and a "word" of data was 36 bits, not 32 bits. The IBM 704 was the first commercial system to support floating-point arithmetic, with vacuum tubes as switching elements.

The IBM 704 was a 1950s vacuum tube computer that supported floating-point arithmetic. Photo from Unità Centrale-Museo Scienza Tecnologia Milano

It could perform a floating-point add or subtract in 84 microseconds, which was quite a respectable speed for that era; to do a multiply or divide took about five times as long.

That model and its more reliable transistor-based successors used this 36-bit format for single-precision floating-point numbers:

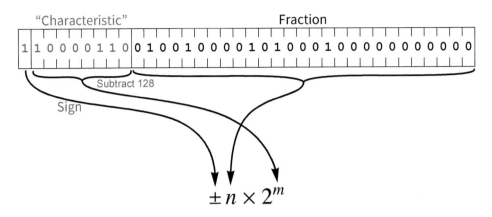

Early IBM floating-point format

The word "Characteristic" is in quotation marks because it was IBM's term for the exponent, but it is not the correct mathematical term. In the days when logarithm tables were used to multiply and divide numbers, the fraction part of log(x) was called the *mantissa*, and if you subtracted the fraction from log(x), the integer part that was left was the *characteristic*. So if the log(x) was 2.30, the mantissa was 0.30 (the part you looked up in the tables) and the characteristic was 2 (for which you were expected to be able to add or subtract with pencil-and-paper). But if the log(x) was –2.30, the mantissa was 0.70 and the characteristic was –3, always keeping the mantissa as a value between 0 and 1. As the above figure shows, that is *not* how IBM's floating-point worked. The "Characteristic" bit field should have been called "Biased Exponent" where the bias is 128. To make matters worse, Arthur Burks started calling the fraction part the "mantissa," again confusing floats with logarithms, and the confusion over terminology persists to this day.

In the above example, the exponent bits represent $m = 6$ and the fraction bits represent $n = \frac{18\,513}{2^{16}} \approx 0.282486$. The sign bit being 1 means the value is negative (sign-magnitude representation, like IBM used for signed integers in the 1950s). So the value is $-n \times 2^m = -2^6 \times \frac{18\,513}{2^{16}} = -\frac{18\,513}{1024} \approx -18.079101$ (the accuracy was slightly more than 8 decimals). The documentation for the computer uses the above example and states that the value represented is –18.0790, which suggests that its author worked it out manually and made a transcription error or a rounding error.

Because IBM also introduced Fortran in 1956, the first commercial language of higher level than Assembler and one that recognized the floating-point instructions, IBM began its path to dominance of scientific computing, with Sperry (the UNIVAC series) and Burroughs as its main rivals.

IBM's format looks much like the one proposed by Quevedo, with two exceptions:

- There is a way to represent negative values.
- The base of the scale factor is 2, not 10.

Like Quevedo's proposal, there are a lot of different ways to represent the same value, and that is never a good thing. Every bit should count, but if the fraction bits are all zero, then the exponent bits do not count. Nor does the sign bit.

To make matters worse, you can divide the fraction by two (shift the bits right by one place) and increase the exponent by one, and it is *exactly the same number* (if the rightmost bit of the fraction was a **0**). Similarly with shifting left and decreasing the exponent. So, instead of having 2^{36} different real values, there are far fewer. The "information per bit" is reduced by redundant ways to represent the same value.

> **Exercise for the Reader**: An eight-bit format should be able to represent as many as $2^8 = 256$ different values. Suppose we use a format like that of the IBM 704, but with a sign bit, three exponent bits (biased by 4), and four fraction bits; how many distinct real values can be represented?

4.7 The Rise and Fall of IEEE Std 754

There were dozens of small companies (or small divisions of large companies like General Electric) trying to enter and succeed in the computer business in the 1960s and 1970s, with names that are mostly forgotten. They had wildly different floating-point formats. The number of bits for exponent and fraction (and where they were placed) were different; the way of encoding the signs of exponent and fraction were different; some were base-2, some base-8, and some base-10.

One base, that used by the biggest player in the industry, was base-16. IBM switched from the 704 format described above to "Hexadecimal Floating Point" (HFP) when it introduced its highly successful System/360 family of mainframes in 1964, and the word size was changed from 36-bit to 32-bit; the wobbling accuracy caused by using 16 as the number base (see Section 2.7) caused their format to wobble between about 6.3 and 7.5 decimals of relative accuracy, effectively making the accuracy only 6.3 decimals in calculations of any length. The IBM 704's single precision at least produced over 8 decimals of relative accuracy as the outputs of plus-minus-times-divide, which was frequently sufficient. Many scientific applications produced unacceptably inaccurate results when ported to IBM's new design.

The solution was obvious: Use double-precision (64-bit) float format, which was supported on the System/360 from its inception. That was back in the days when fetching and storing data to memory was *much* faster than doing an arithmetic operation on a floating-point number, so switching to double-precision was a no-brainer. It became routine for programmers of technical codes to make all the real variables double precision, without thinking about which variables actually needed to have around 15 decimals of accuracy.

Arithmetic operations got dramatically faster over time, but memory access over physical distances improved slowly. "The Memory Wall," where it takes hundreds of times longer to fetch or store a float than to do arithmetic with it, did not become obvious until the late 1970s. Rewriting software is extremely expensive, so we are now trapped in our use of overkill precision for many applications involving real numbers, largely because of that one decision by IBM to use HFP in the design of the System/360.

Why did some kind of *de facto* standard not form to coalesce around a single way to represent real numbers? Having a unique format means it is very difficult to switch vendors, and vendors liked that. The data files computed with the existing machine might contain a mix of integer and character and float data, and while you might be able to port integer and character data to a different computer brand, extracting the float data and converting it could be a nightmare. If the destination format had a smaller dynamic range, that would cause overflow and underflow. If it had fewer bits of fraction, that could cause accuracy to drop below what was acceptable. It was what marketing people call a "lock"; once a customer, always a customer.

4.7.1 The Computing Industry Gets Emancipated

Chip designer Intel was on a roll in the 1970s, having pivoted from making the first commercial DRAM memory chips to making the first microprocessors. Their first microprocessor, the Intel 4004, operated on four bits of data at a time, and the chip had about 2300 transistors; it took considerable innovation to squeeze the design into a single chip. It became practical for the first time to put some smarts into microwave ovens and traffic light controllers, and perhaps build four-function calculators. Intel co-founder Gordon Moore speculated that the optimal density of transistors would double every year; he later revised that to every *two* years, and chip designer Carver Mead dubbed that prediction "Moore's law." Engineers were designing much more capable microprocessors like the Intel 8080, which could make computers so cheap that they could one day become *personal computers*.

But there was another plot arc going on, between the heavy-hitters in the computing industry: Sperry, IBM, Honeywell, and other makers of the large-scale "mainframe" computers of the 1960s. Mauchly and Eckert had used their experience with the government-funded ENIAC to develop computers practical enough to sell commercially through Sperry. Sperry had patented the idea of an electronic digital computer and was charging royalties from other computer makers. IBM brokered a patent-sharing agreement with Sperry, so IBM actually benefited from the barrier to entry into the market created by the royalty charges much the way Sperry did.

At Honeywell, someone stumbled on the fact that Atanasoff's 1937–1942 electronic digital computer was shown to Mauchly well before Mauchly and Eckert decided to switch from analog to digital and begin building the ENIAC, and that perhaps it was possible to break the Sperry patents. A federal case began, one of the longest and most expensive in US history. After seven years, 32,000 exhibits, and a 20,000-page transcript of testimonies, Judge Earl Larson ruled that the patents were *invalid*. An excerpt from his ruling:

> Eckert and Mauchly did not themselves first invent the electronic digital computer, but instead derived that subject matter from one Dr. John Vincent Atanasoff.

Cited by Clark Mollenhoff in *Atanasoff: Forgotten Father of the Computer*, 1988, page 210.

The news of the verdict was buried by the timing; it came out the same day as the "Saturday Night Massacre" of the Watergate Scandal.

Eckert and Mauchly certainly deserve credit for the massive amount of work they did to commercialize Atanasoff's ideas, and Sperry remained a powerful contender among the largest computer companies, but the barrier to entry was gone. Whether you agree with Judge Larson's conclusion or not, you have to admit that the result was that *no one* owned the broad patent on electronic digital computing; anyone could enter the market without fear of infringement or having to pay hefty royalty payments. And here were Intel and Motorola and other companies making increasingly capable microprocessors with transistor counts doubling every two years.

Besides the legal barriers to entry, the technical problem of designing an entire computer had been just as daunting for a small company. But with microprocessors, that barrier was also gone. The skill set needed to hook together the processor and memory and keyboard and display was something an electrical engineering student could do as a senior project. The computing industry was emancipated as it had never been before.

4.7.2 The True Origin of IEEE Std 754

John Palmer, 1946–2017

Rivals Motorola and Intel continued to improve their microprocessors, with transistor density improving to the point where they could consider working with more than 8 bits of data at a time. Intel started work on a 16-bit processor called the i8086 that could address more than the 2^{16} ("64K") bytes of memory that limited the first generation of personal computers from Apple, Commodore, and Radio Shack (Tandy). Bruce Ravenel was the lead architect of both the i8086 and a slower version called the i8088 that had the same instruction set but worked with 8 bits at a time to lower costs.

Everyone wondered how IBM could enter the personal computing market; it certainly did not fit their traditional sales model. IBM did something unprecedented in its history: It outsourced both the processor design (Intel) and the creation of an operating system (Microsoft). Work began in the late 1970s to create what would become the IBM PC.

In 1976 an Intel engineer, John Palmer, proposed that the i8086 and i8088 could have a coprocessor, one designed specifically to perform floating-point operations, to make the IBM PC more suitable for workstation-style technical computing.

Intel executives expressed strong skepticism that there was a viable market for such a chip, but Palmer was so convinced that he said if he were paid one dollar for every such chip sold, he would forgo his salary. The project was finally given the green light in 1977, and Palmer frequently told me he wished they had taken his offer because it would have made him a very wealthy man.

The IBM PC was designed with an empty socket for an optional i8087 right next to its main processor, the i8088. Microsoft's Bill Gates expressed his opinion to Intel, "You're not going to fill many of those sockets." But the i8087 floating-point coprocessor sold many millions of copies for $150 each, continued selling for years (unlike most chips), and held its price instead of having to drop in price with Moore's law (also unlike most chips).

Berkeley professor William "Velvel" Kahan had given a guest lecture at Stanford ten years earlier, about the outrageously bad results possible in the floating-point arithmetic of some computer makers. It made an impression on Palmer, who heard the lecture as a student. Palmer hired Kahan as a consultant for the i8087 project.

William "Velvel" Kahan, ca. 1980

Kahan (rhymes with "John"; the "h" is silent) had consulted for Hewlett-Packard on the algorithms for their scientific handheld calculators that used decimal scientific notation with ten decimals of accuracy, and an exponent that ranged from 10^{-99} to 10^{99}. Decimal arithmetic is the right choice for calculators because it eliminates the problem of rounding error when converting numbers from decimal (the Human Layer) to binary (the Compute Layer), and although it runs slower, speed is not an issue with handheld calculators.

In his consulting role, Kahan immediately and adamantly set down three ground rules for the number format:

- It would use a decimal scientific notation, not binary.
- All operations and functions would be reproducible across computer systems.
- The higher-precision version of the 64-bit format would be 128 bits wide.

He lost all three arguments. Ravenel and Palmer said they were already too far down the design path to make design changes. The 64-bit format would be binary, with a sign bit, biased exponent, and a fraction with a hidden bit that is always 1 just left of the radix point. The higher-precision version of the format would be 80 bits wide, and all calculations would covertly be done with that 80-bit "extended double" precision. That way, Intel could claim that any differences between their 64-bit computations and that of other vendors was because Intel's were *better*, and this would be a selling point. So, irreproducibility and unportability were a key part of Intel's strategy.

As Intel's consultant, Intel tasked Kahan not with making design choices but instead with becoming their paid advocate for every design choice that Palmer and the chip designers had made. Kahan was stubborn and not given to compromise, and the discussions were exasperating to Palmer, but Palmer eventually prevailed.

Palmer and Kahan published the i8087 design choices as a proposed "standard" way to do floating-point in 1979, and announced the i8087 chip in April 1980. Because of Kahan's reputation as an expert in computer arithmetic, everyone incorrectly assumed he was the architect of the design.

The Intel i8087 floating-point coprocessor chip

Motorola began designing its own floating-point coprocessor chip, the Motorola 68881, to follow that proposed standard. But computer companies like IBM, Digital Equipment Corporation, and others saw the proposal as a threat and asked IEEE to form a Standards Committee so they could have input into the proposed standard and level the playing field. That was the origin of what we now call IEEE 754 (the official name is IEEE Std 754™).

4.7.3 The Two Kahans

It is an interesting exercise to go through Kahan's writings and his (unrefereed) web postings to classify his position statements into two categories that are perfect opposites. They are:

- *Anything less than perfection in computer arithmetic is not to be tolerated.*
- *Perfection is an impractical goal; speed and low cost are more important.*

To pick a few from the first category in his writings:

"Almost all True is altogether a Lie."

"Every violation of Mathematics is always punished sooner or later. (The punishment does not always befall the perpetrator.)"

"This is no place to list all the corrections *Excel* needs. It was cited here only to exemplify Errors Designed Not To Be Found."

"Routine use of far more precision than deemed necessary by clever but numerically naive programmers, provided it does not run too slowly, is the best way available, with today's mixture of popular programming languages with overtaxed underfunded education, to diminish the incidence of roundoff-induced anomalies below any level of commercial significance even if we knew about every anomaly."

And in the second category,

"A machine whose every function rounds correctly is impractical."

"Faster matrix multiplication is usually too valuable to forego for unneeded exact reproducibility."

"Rather than waste time testing every element for troubles that almost never happen, a program can run faster on average by testing one or two flags after the array has been computed and, if a flag was raised, recompute the array some better way."

"Consequently numerical anomalies go mostly unobserved or, if observed, routinely misdiagnosed. Fortunately, most of them don't matter. Most computations don't matter."

from https : // people . eecs . berkeley . edu/~wkahan/

By flipping between the two extremes, Kahan was able to defeat both those who criticized Palmer's choices for being too-perfect as well as those who thought things could be made better (such as by making the math functions correctly rounded, as Kahan originally wanted). He was a very effective "hired gun." His lengthy epistle mocking Microsoft *Excel* for using hidden extra precision to reduce rounding error in visible digits is followed, on the next page, by a touting of Intel's use of 80-bit hidden extra precision to reduce rounding error in visible digits.

4.7.4 The i8087 Design Becomes the IEEE Standard at Last

kludge [klooj] NOUN   A workaround or quick-and-dirty solution
that is clumsy, inelegant, difficult to extend and hard to maintain.

After many years of intense argument (mostly over Kahan's insistence on subnormal number support, also called "gradual underflow"), IEEE Std 754 was approved and published in 1985. It is not free; anyone wanting to use it has to pay $100 to IEEE for a copy of the Standard. Palmer once told me, chortling, that "Whatever was in the 8087, that's what's in the IEEE Standard." Intel had won its battle. While IBM and Cray resisted following the Standard for the next decade, most vendors adopted (most of) the Standard, believing it would lead to data portability (which it did), and identical results on different computer systems (which it did not).

When a format starts with a few dubious choices, such as underflow to zero, that leads to the need for other dubious choices such as "gradual underflow" that complicate the circuitry and create numerical hazards. The format ignored the 1950s lesson of *not* using sign-magnitude arithmetic, so once again there was "positive zero" and "negative zero." Are they equal? Yes, they decided. The two kinds of zero should test as equal. But their reciprocals should be *positive infinity* and *negative infinity,* so now you have a format where x can equal y but a function $f(x)$ can be as far different from $f(y)$ as you can get. When an engineer creates a hasty solution with flaws and then adds more hasty solutions to patch those flaws, it is called a *kludge.*

Palmer decided to have a special bit pattern for indeterminate results like $\frac{0}{0}$, $\log(-2)$, and he called it "NaN" for "Not a Number," even though imaginary numbers and complex numbers are, well, numbers. The bit test only applied to the exponent bits, however, leaving a vast set of bit patterns representing the same NaN. Two of the patterns were carved out to mean $-\infty$ and ∞. Here are how many bit patterns were wasted saying a result is a NaN (IEEE half was added to the Standard in 2008):

Precision	Number of NaN Patterns	
IEEE half (16 bits)	4094	(over 6 percent of the bit patterns)
IEEE single (32 bits)	16 777 214	
IEEE double (64 bits)	9 007 199 254 740 990	(slightly over 9 trillion)

How would NaN work in comparison operations if it was not a specific bit string? The first axiom of Aristotelian logic is that A is A; a thing is equal to itself. But incredibly, the decision was made that NaN is *not* equal to itself. The web got ever more tangled. The mathematically absurd concepts of "positive zero" and "negative zero" created a mess of contradictory statements about how they should be handled. The 754 Standard is almost 100 pages long.

To support the Standard in hardware, the logic circuitry has to handle this set of exception cases:

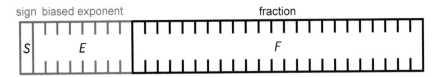

$$
\text{Value} =
\begin{cases}
\text{if } E = 11\cdots1, &
\begin{cases}
\text{if } F = 0, & (-1)^S \times \infty, \\
\text{else} &
\begin{cases}
\text{if } F \geq 0.5, & \text{quiet NaN}, \\
\text{else} & \text{signaling NaN},
\end{cases}
\end{cases} \\
\text{else} &
\begin{cases}
\text{if } E = 00\cdots0, &
\begin{cases}
\text{if } F \neq 0, & \text{subnormal}: (-1)^S \times 2^{E-\text{bias}+1} \times (0 + F), \\
\text{else} &
\begin{cases}
\text{if } S = 0, & \text{``positive zero''}, \\
\text{else} & \text{``negative zero''},
\end{cases}
\end{cases} \\
\text{else} & \text{normal}: (-1)^S \times 2^{E-\text{bias}} \times (1 + F).
\end{cases}
\end{cases}
$$

The IEEE Std 754 format, 32-bit example, and the exception cases in a nutshell

While I suspect Kahan was initially appalled at being tasked with defending signed zero, he eventually did the mental gymnastics required and pronounced it a Good Thing because it could help with a concept called "branch cuts" in complex arithmetic. To make this work, the Standard requires yet another counterintuitive exception that the square root of "negative zero" should be "negative zero." I am not making this up.

There were originally four rounding modes; more have been added in revisions of the Standard, but once you have multiple rounding modes, you remove any hope of guaranteeing the same results across different computing systems, even if all you do is plus-minus-times-divide. Scientific computing favors banker's rounding (round to nearest, with ties to even), but graphics accelerators favor truncation (round toward zero) since it is faster and usually has an imperceptible effect on image quality.

Another guarantee of Never Twice the Same Answer was that the *transcendental* math library functions (logarithms, exponentials, trigonometric and inverse trigonometric functions) do not have to be correctly rounded, according to the Standard. Correct rounding would mean different computing systems would get the same result down to the last bit, if they used the same rounding mode.

Kahan reversed his initial demand for reproducibility and wrote position papers about how difficult and utterly impractical it would be to have such a requirement. Like a lawyer, he proved able to take either side of any design decision. He advocated for Intel by citing mathematics when Intel's choice improved accuracy or correctness, and by citing practicality when the choice worsened either one.

4.7.5 A Bold Claim of Round-Trip Perfection

In his 1983 commentary "Mathematics Written in Sand," Kahan boasts that you can square an IEEE float, and if it does not overflow or underflow, you can then take the square root and get back the absolute value of the original number exactly. *Always*. What a marvelous property! Here is the relevant text from that commentary:

> ```
> Cynics, expecting nothing to survive roundoff,
> must be surprised to discover, on binary and
> quaternary machines but not on those with larger
> radix, that despite roundoff √(x²) = |x| for
> all x unless x² over/underflows. These
> surprises can be confirmed first by experiment,
> then by simple proofs.
> ```

https ://people .eecs.berkeley.edu/~wkahan/MathSand.pdf

An overflow would produce infinity and remain infinity when we take the square root, letting us know that something went wrong. Similarly, an underflow would presumably produce zero, and remain zero when we then take the square root, also informing the user that the value is wrong.

Shall we give it a try?

Let's set $x = 2.646978 \times 10^{-23}$ as a single-precision IEEE float. Now square it and round it with the default rounding mode (banker's rounding), and you get

$$y = x^2 \approx 1.401298 \times 10^{-45}.$$

So it did not underflow to zero. The Standard states that square roots must be correctly rounded, so now try taking the square root and see if we get back 2.646978×10^{-23}, the x we started with, down to the last decimal:

$$z = \sqrt{y} = 3.743392 \times 10^{-23}$$

.Not a single decimal in the answer is right. How could a mathematician like Kahan be *that* wrong? The explanation reveals something about IEEE floats that very few computer technologists seem to know: IEEE floats have *tapered accuracy*. Not the same kind of tapered accuracy as posits, but it is there.

Remember the decimal accuracy plots from Chapter 2? Here is what the decimal accuracy plot looks like for IEEE single precision:

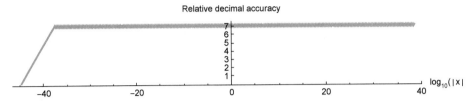

Accuracy plot for IEEE standard 32-bit precision floats, with tapered accuracy on the left

For most of the range (from about 10^{-38} to 10^{38}), decimal accuracy wobbles between about 7.2 to 7.5 decimals. But what is that ramp over on the left, from about 10^{-45} to 10^{-38}? The accuracy ramps down to almost nothing, with no decimal digits correct. Those are the *subnormal* floats, and every IEEE precision has them. So the x that I chose has its square at the low end of that ramp, which is why not a single decimal digit was correct in computing $\sqrt{x^2}$.

Realizing this could happen, Kahan came up with a "read the fine print" workaround: He created his own definition of what "underflow" means. To every other computer programmer and user I have known, "underflow" means "what happens when a result is too small to represent" and you know it happens by looking at the output, zero. But the IEEE Standard says that the underflow flag must also be set if the result of an operation lands in the subnormal region *and is not exact*. But it is *not* too small to represent; it simply had to be rounded. So now that is "underflow"? How is anyone supposed to know that a processor flag was set? By using a low-level debugging tool and stepping through the compiled code one machine instruction at a time, watching the processor flags?

For decades, Kahan raged that computer language standards should provide access to the processor flags, and he was ignored. It did not happen. Only in the last few years have the C and C++ language standards committees entertained the idea of supporting viewable processor flags, but it is probably too late. Legions of C and C++ programmers have no idea that there are versions of their language that let them know about processor flags. The Java programming language specifically *forbids* the use of processor flags, which led Kahan to write an amusing but not very mathematical diatribe titled, "How Sun's Java Hurts Everyone Everywhere."

Some have objected to the tapered accuracy of posit format, saying tapered accuracy spoils classic theorems about accuracy that apply to floats. The objectors appear to be completely ignorant of the fact that IEEE float format has tapered accuracy. It is a strange tapering because it is only on one side, and it is steep.

Researchers at Argonne National Laboratory, late in the 1990s, figured out an elegant way to have symmetrical tapered accuracy for large-magnitude numbers instead of wasting bit patterns on all the ways to represent NaN, and wrote a very cogent proposal to the IEEE 754 Committee to consider for their next revision: "gradual overflow." The committee rejected it immediately. It would spoil backward compatibility. The only revisions that tended to be accepted in the 2008 and the 2019 revisions of IEEE Std 754 were the inclusions of new precisions, like half-precision and quadruple precision. Those were completely new, so they would not spoil the behavior of the float types previously specified.

Speaking of overflow, let's try Kahan's $\sqrt{x^2} = |x|$ claim out on *large* magnitude numbers. Suppose we write a short program with $x = 10^{160}$ with all variables in IEEE 64-bit (double) precision, something resembling this:

```
double x, y, z;
x = 1.0e160;
y = x * x;
z = sqrt(y);
print z;
end
```

If you run it a few times on a machine with an Intel or AMD processor, you might very well get 1.0e160 for z, that is, 10^{160} as hoped, but you might find that it occasionally instead produces Inf to indicate that the answer overflowed to infinity.

What in the world? So you might try the quickest type of debugging, that of inserting a print statement to see what could possibly be going wrong.

```
double x, y, z;
x = 1.0e160;
y = x * x;
print y;
z = sqrt(y);
print z;
end
```

Now you run it, dozens of times, and it *always* gives the answer `Inf`. You take the "`print y`" statement out, and it goes back to producing `1.0e160` most of the time, but not always. So you get out the low-level debugger and step through the code, carefully. It always produces `Inf`.

This is one of the most dreaded and exasperating types of programming bugs: The act of debugging, that is, observing what is going on, *changes how the program behaves*. The bug has a slang term, a "heisenbug," named after physicist Werner Heisenberg who stated the quantum mechanical principle that the act of observing an event can change the event. To use an image from the early PC era when the electronics were big and clunky:

A 1980s programmer experiencing a heisenbug.

As with underflow, finding the source of this bug reveals another consequence of the choices Palmer made in designing the i8087. The squaring of 10^{160} is being done, *covertly*, in the 80-bit registers even though all the variables are declared to be 64-bit precision. The 80-bit registers can handle the square, which is 10^{320}, and the square root then returns 10^{160} and converts it back to 64-bit precision.

But if you try to observe what is happening, the processor will convert the temporary result back to 64-bit precision. IEEE 64-bit precision can only store numbers up to about 1.8×10^{308}, so 10^{320} overflows to infinity. Both `print` statements and debuggers conceal the use of covert 80-bit precision, the feature that was supposed to make Intel arithmetic "better" than the competition that actually uses 64 bits and 32 bits when that is what the program specifies.

Why would the original program sometimes produce 10^{160} and sometimes infinity? Operating systems occasionally have to swap out a running piece of code by saving the state of the processor, doing some housekeeping function, and then restoring the state. That state includes the register contents, but anything in the 80-bit temporary register is converted on-the-fly to a 64-bit float, causing it to overflow. Hence, IEEE arithmetic is not only unportable across systems, but it can also produce *different answers on the same computer with the same data*.

When Sun Microsystems was creating the Java programming language in the 1990s, an absolute design requirement was that it produce the same answer on *any* computer. A lead engineer on the project, James Gosling, figured that real number operations were easily made to behave identically by stating that they follow IEEE Std 754, done. After all, it says "Standard" right there on the label. He was horrified to discover that it was not that kind of standard at all; the only thing it really standardized were the locations and meanings of the bit fields so that float data could be interchanged between computers. When it comes to performing arithmetic, it should have been called the IEEE 754 *Guidelines*. This forced the Java designers to build their own floating-point arithmetic out of integer operations, bypassing the IEEE-float-specific hardware in everyone's processor. Kahan was livid about this, but it could have been avoided had he won his argument with Palmer in the first place, stipulating that results should be perfectly reproducible.

4.7.6 Postlude: Farewell to IEEE Floats

Not many computer technologies last more than a few years, and by the year 2005, the IEEE version of scientific notation was looking very long in the tooth, indeed. The major chip-makers decided that the exceptions were rare, so they could be handled in software instead of hardware. This simplified the hardware to processing only *normal* floats, saving chip area and execution time. Sun Microsystems made the mistake of trapping NaN exceptions to the operating system, which could take millions of times longer to handle NaN than previous versions of their processors. High-frequency stock traders found the new processors unusable, since NaN exceptions are common when the trading algorithm is missing some vital data and thus no trade should occur.

Intel, AMD, and ARM did better, trapping exceptions to be handled with microcode in their processors, which could take 200 clock cycles instead of the usual 3–8 clock cycles for normal floats. However, taking longer to do something is externally viewable, so a third party can use timers to figure out the data being processed without having permission to do so. That is called a *side-channel attack,* and researchers at UC San Diego were able to use the disparate timing to hack into web sites and steal data.

A fatal blow to IEEE floats was struck with the rise of Graphic Processing Units (GPUs), because those not only tend to turn off any exception handling, they flush subnormal numbers to zero. That meant that when you port an IEEE float to a GPU, it represents a *different number*. The data interchange part of the Standard broke.

There was nothing left to make the word "Standard" applicable.

The next stage of the farewell to IEEE floats was the rise of Machine Learning (ML), and Artificial Intelligence (AI) that actually lived up to its decades-old promise. But 32-bit IEEE floats were overkill for what was needed, and the 16-bit IEEE floats had insufficient dynamic range. Thus began a free-for-all with companies designing their own ways to represent real numbers, very much like the days before the 1985 Standard came out and every vendor had its own way of doing things. That is where things stand as I type this; full support for IEEE Std 754 in hardware is almost extinct. We need not mourn the loss, because we are once again open to a clean-slate, *mathematical* design for a real number format.

Much of the preceding history is not in previously-published form. It is from the many hours of conversations I have had with John Palmer, William Kahan, and others who were part of the process.

4.8 Lessons Learned

The mismatch between how humans like to record and process numbers and how machines can best store and manipulate them has been going on for hundreds of years, with the machines eventually becoming optimized. The most recent episode was trying to use the bits on a computer to mimic human scientific notation, instead of taking the approach that every bit counts.

Standards committees should avoid any members who have a conflict of interest, like large manufacturers of existing products. This seems like common sense, but IEEE does just the opposite and *requires* the presence of such manufacturers. The result is that we have a floating-point "standard" that is actually a proprietary Intel design from the 1970s, one that includes a raft of choices made by biased parties.

Any engineering effort should begin with a list of design goals. They often compete with each other, but at least the tradeoffs can be decided with clarity. What should be the design goals for a system of representing real numbers with a string of bits in a modern electronic computer? There should be very little objection to the following goals, starting with the most important ones:

1. The distribution of the values represented should resemble the distribution of numbers needed in calculations. That is, each bit pattern should be used approximately the same number of times (this is the principal of "maximum entropy" from information theorist Claude Shannon).
2. There should be no redundant bit patterns to mean the same thing; every bit counts.
3. The output of a program should be deterministic based on the source code and the input data, not affected by invisible "modes" of operation. The precision used must be exactly what is specified in the source code, no more and no less.
4. All math operations should be correctly rounded. This, together with **3**, ensures perfect reproducibility and portability across computer systems.
5. It should be easy to change precisions, like it is to change integer precisions.
6. If a result is not a real number, it should have a representation that propagates that condition through every operation that follows so that the final output reflects the condition.
7. It should be simple to build in hardware, and fast.

In the "Nice to Have" category:

8. It should be easy to scale the precision to any number of bits.
9. Real numbers should map monotonically to signed integers so that they can use the same hardware for comparison operations.
10. Similarly, the bit operation for negating a real representation should be exactly the same as for negating an integer, so they can use the same hardware.
11. The results of application computations should be as close to results for exact computations as possible.

Remarkably, the i8087 and thus IEEE Std 754 *failed every single one of these goals.*

The next chapter is an annotated presentation of the 12-page *Posit™ Standard (2022)* that was ratified by the Posit Working Group. That Standard meets the first ten goals above and does a better job than floats on number **11**.

5 The Posit Standard, Annotated

That is the entire Posit Standard (2022), simple and concise enough to fit on twelve pages.

Writing a standard is very different from writing an explanation, because the standard should be as concise as possible, and that means leaving out examples and the reasons that the decisions were made the way they were. Peter Lindstrom recommended that we write a document to accompany the Posit Standard that supplies those examples and reasons, which is the function of this chapter. Since the previous chapter explained the origins of IEEE Std 754, it is only fair to give the backstory on where the Posit Standard came from.

The chapters that follow do not depend on this chapter, so the reader should feel free to skim this chapter if it seems to be getting dull.

5.1 Singapore Backstory

Downtown Singapore (the Marina Bay Sands in the foreground), and Fusionopolis, home of Singapore's supercomputer centre and also the birthplace of the posit number format and its standard document.

Singapore has an amazing record of introducing big ideas to the rest of the world. The flash drive. Cell phone cameras. Ketchup. In 2015, there were two technology visionaries in Singapore, physicist Marek Michalewicz and the director of Singapore's NSCC supercomputer centre Tin Wee Tan, who saw promise in the *unum* number format I had just published. They arranged with Singapore's Agency for Science, Technology, and Research (A*STAR) and the National University of Singapore (NUS) to bring me to that country to continue development of next-generation arithmetic to replace the outdated IEEE Std 754. We knew there was a disadvantage to overcome in unum format: The unum format was *not* hardware-friendly. The unum format was a "peace offering" to IEEE floats, an attempt to patch the flaws in that design by creating a *superset* of IEEE floats with three new fields:

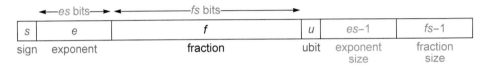

The original unum format was an IEEE-style float with metadata appended

It was a form of interval arithmetic, but extended to allow interval endpoints to be closed or open, and for the size of the exponent and fraction fields to vary as needed by the calculation. The ubit was described in Chapter 1: the presence or absence of a "⋯" to indicate there are more nonzero bits after the last one shown. This self-descriptive bit allowed the management of floating-point error to be automated (or eliminated). While they were fully supported in software and one group in China managed to build a custom chip to use unum arithmetic, working with unums proved to be much like working with character strings of varying length, with a need to allocate and free memory dynamically.

One way to use unums was to fix the exponent and fraction size and round after every operation, which allowed them to mimic, perfectly, IEEE floats. But that "backward compatibility" also prevented unums from meeting the design goals listed at the end of Chapter 4. There were redundant bit patterns, two ways of expressing the same real value. They did not map monotonically to signed integers.

The efforts in Singapore soon led to the realization that the last two properties could be achieved by discarding IEEE Std 754 altogether and beginning with the ring plots shown in Chapter 1. There were elegant ways to keep the property that flipping the ring about its horizontal axis produced the reciprocal of a number perfectly, making division as easy as multiplication. In early 2016, these were introduced as "Type II unums," rebranding the original as "Type I unums."

Type II unums always use the same number of bits, solving the objection to Type I, but introducing another difficulty: arithmetic required table look-up instead of the usual logic circuits. While Kahan had been characteristically critical of the original unum idea, in a public debate (June 2016) he stated this about Type II unums:

> "They allow open intervals. They also serve in pairs. And they typically save storage space because what you're manipulating are not the numbers, but pointers to the values. And so, it's possible to run this arithmetic very, very fast."

http ://www .johngustafson.net/pdfs/DebateTranscription.pdf, pages 17–18.

Type II unums have their uses, and Chapter 8 will present one version. But it was the idea shown in Chapter 1 (giving up the idea of perfect reciprocals), that led to "Type III unums" that map to signed integers, are fixed size, and use all the standard logic circuits that chip designers presently use to design IEEE floating-point arithmetic. If the ubit is used to indicate the open interval between exact values, that is branded as a *valid* type, which have the error-free capability of Type I unums. If they are rounded after every operation, using the default banker's rounding of IEEE floats, they are branded as *posits*. I invented posits in December 2016 with the geometrical approach in Chapter 1, with insight from Isaac Yonemoto in figuring out how regime bits are decoded, and presented them at Stanford University in February 2017. They appeared to be particularly well-suited for Machine Learning, as Yonemoto demonstrated live at the end of the Stanford lecture. Credit for the creation of posits also goes to DARPA, which provided funding for linear algebra solutions using next-generation arithmetic. Hence the footnote in the Posit Standard:

*The initial development of posit arithmetic was supported by Singapore's Agency for Science, Technology and Research (A*STAR) and by the USA's DARPA TRADES Program, Contract #HR0011-17-9-0007.

The result was an international movement. The first paper, *Beating Floating Point at its Own Game: Posit Arithmetic* accumulated hundreds of citations as people tested the idea and designed hardware and software environments to support it. Without experimental data, that first paper made this initial guess:

Precision, bits	Exponent size *eS*
8	0
16	1
32	2
64	3

Initial guess (2017) regarding how many exponent bits would be best for different posit precisions

A Posit Working Group was formed to establish a standard. It included experts from the US, Singapore, Canada, Germany, and the UK, all from universities or national research institutions; it *excluded* anyone with a commercial interest in selling computing hardware, to avoid the conflicts of interest evinced by IEEE committees:

Participants

The following people in the Posit Working Group contributed to the development of this standard:

John Gustafson, *Chair*
Gerd Bohlender
Shin Yee Chung
Vassil Dimitrov
Geoff Jones
Siew Hoon Leong (Cerlane)
Peter Lindstrom
Theodore Omtzigt
Hauke Rehr
Andrew Shewmaker
Isaac Yonemoto

It took five years to make sure the Posit Standard was as simple as possible, but no simpler. During the five years, hundreds of papers were published testing posits of various precisions on a wide variety of applications, and they all pointed to the same empirical conclusion: The best results are usually for exponent size $eS = 2$ or $eS = 3$, for all precisions. That was especially important in testing 8-bit posits for AI; an exponent size of 0 does not usually have enough dynamic range for the Machine Learning part of AI. So we changed the Draft Standard to this instead:

Precision, bits	Exponent size eS
8	2
16	2
32	2
64	2
n	2

Years of experiments showed all precisions should use the same eS.

Then we realized a huge benefit to always having $eS = 2$ for all precisions: Conversion between precisions becomes trivial. Simply append zeros to raise precision, or round the least significant bits (as a binary fraction) to lower precision. There is no need to pull apart the pieces of a posit into sign, regime, exponent, and fraction. It is like changing a 16-bit integer to a 32-bit integer by right-shifting 16 places; there is really nothing to do. That means posits can "right-size" their precision in a fine-grained way without any performance penalty.

5.2 Overview Section of the Posit Standard

Abstract

This standard specifies the storage format, operation behavior, and required mathematical functions for posit arithmetic. It describes the binary storage used by the computer and the human-readable character input and output for posit representation. A system that meets this standard is said to be *posit compliant* and will produce results that are identical to those produced by any other posit compliant system. A posit compliant system may be realized using software or hardware or any combination.

A primary goal was bitwise-perfect reproducibility, the goal that floats had failed and still fail to achieve. The last sentence is important because the Posit Standard does not specify *how* the requirements are to be met, only what the behavior must be. One designer might decide to put everything into hardware, including all the log and trig and other math library functions. Another may choose to create posit compliance entirely through software that uses integer data. Both extremes are acceptable, and so is everything in between.

1.1 Scope

This standard specifies the storage formats and mathematical behavior of posit™ numbers, including basic arithmetic operations and the set of functions a posit system must support. It describes how results are to be rounded to a real posit or determined to be a non-real exception.

That section is self-explanatory, but the reader may wonder why the ™ trademark symbol is used after the first use of the word "posit." This was done on the excellent advice of Dutch-American computer scientist Peter Braam, developer of the Lustre™ parallel file system. While posits are open-source, there is always the chance that a large company will come up with their own version of posit format that corrupts its underlying principles yet still claims to be "posit" format. The trademark provides some measure of legal protection against that happening.

1.2 Purpose

This standard provides a system for computations with real numbers represented in a computer using fixed-size binary values. Deviations from mathematical behavior (including loss of accuracy) are kept to a minimum while preserving the ability to represent a wide data range. All features are accessible by programming languages; the source program and input data suffice to specify the output exactly on any computer system.

The last sentence is a potent one, and it is crucial for portability. Suppose there are multiple rounding modes, and those modes are optionally set by a user through an operating system call. That means a program, running on a single system and using a fixed set of input data, can produce multiple answers for reasons that are *invisible* in the source code and input data. This is why posits have only one rounding mode.

Another thing that can happen is that the programmer can invoke an "optimizing compiler" that changes the operations as if the laws of algebra apply. Because of rounding, they do not. For example, an optimizer might change

$$x = (a*b) + (a*c)$$

to

$$x = a*(b + c)$$

to save a multiplication operation. Such optimizations can still be offered, but with the warning that they make the resulting compiled program non-compliant with the Posit Standard. Computer languages generally specify the order of operations (like what to do when there are no parentheses), which leads to deterministic outcomes for the arithmetic.

1.3   Inclusions and exclusions

This standard specifies

- Binary formats for posits, for computation and data interchange
- Addition, subtraction, multiplication, division, dot product, comparison, and other operations
- Fused expressions that are computed exactly, then rounded to posit format
- Mathematical elementary functions such as logarithm, exponential, and trigonometric functions
- Conversions of other number representations to and from posit format
- Conversions between posit formats with different precisions
- Function behavior when an input or output value is not a real number (NaR)

The binary number formats have already been explained in Chapters 1 and 3.

Listing addition, subtraction, multiplication, and so on means the Posit Standard includes specification of how to produce results of operations. The rule is very simple: If any input is NaR, return NaR as the output; otherwise, treat the inputs as exact real numbers, and return what the correctly-rounded result would be if performed with infinite-precision arithmetic. It is always possible to know what the correctly-rounded result would be using a finite amount of work; you simply compute to the point where you know how to round the result.

Fused expressions are interesting, and fused multiply-add was added to the 2008 version of the IEEE 754 Standard. It means returning round($a \times b + c$) instead of the usual round(round($a \times b$) + c). You can think of fused expressions as staying in the Math Layer for two (or more) operations instead of returning to the Compute Layer after every operation. A key part of the Posit Standard is that it specifies quite a few often-used functions that are fused, made of multiple operations evaluated with rounding only as the very last step. For example, it requires support for

$$\text{hypot}(x, y) = \text{round}\left(\sqrt{x^2 + y^2} \right)$$

because it occurs quite often in technical computing. If it is programmed as individual operations, you would accumulate *four* rounding errors instead of round$\left(\sqrt{x^2 + y^2} \right)$, with a single rounding only at the end. Writing code like

```
h = sqrt((x*x) + (y*y))
```

actually means asking the computer to do the following:

$$h = \text{round}(\text{sqrt}(\text{round}(\text{round}(x \cdot x) + \text{round}(y \cdot y))))).$$

That can be written more concisely using the overbar notation (Section 3.6):

$$h = \text{sqrt}(\overline{\overline{x \cdot x} + \overline{y \cdot y}}).$$

Whatever the rules for a computer language, if it supports posits, then the use of fused operations or non-fused operations must be visible in the code, never done covertly. We will see later why this prevents disasters.

The Posit Standard requires support for the functions you find on a scientific calculator: logs, exponentials, trig and inverse trig functions. We stopped short of mandating support for some of the more sophisticated functions like Bessel functions. Those can always be added to a future version of the Posit Standard.

While posits can replace floats in a program, it seems likely that both formats may coexist for a while, and there needs to be a way to convert posits to IEEE 754 floats and vice versa. So, the Posit Standard specifies how to do that for the various rounding modes and exception types of IEEE 754 floats.

The Standard also solves a subtle and long-standing issue: How should an indeterminate result (NaR for posits, NaN for floats) be converted into an integer, and back? Posits can do it properly, preserving the fact that a NaR result occurred.

It is extremely easy to convert between two posit precisions with the same exponent size *eS*, much like converting between two integer precisions. There is no need to decode the number. This is one of the reasons that Draft Posit Standard 4.0 changed to specifying that *eS* = 2 for any precision. Earlier Draft versions required that the precisions be restricted to $n = 8, 16, 32, 64, ...$ with corresponding *eS* values $eS = 0, 1, 2, 3, ...,$ in other words, $eS = \log_2(n) - 3$. The ratified version of the Posit Standard described here does not restrict n that way. Strange precisions like 5-bit posits have proved useful for the *inference* task of AI, for example. A fixed *eS* also simplifies the work of the hardware designer.

The approach to explaining when any operation or function should return NaR is to apply a single mathematical principle, uniformly. This contrasts with using different rules for different functions, as is presently done for IEEE 754.

Excluded from the standard are the specific names of the values and operations described here. The lower camel-Case naming style is used here, but naming style is excluded from this standard. Implementations may use alternative names and symbols for values and operations that match the behavior described here.[1]

---

[1] For example, the arc hyperbolic cosine is here shown as **arcCosH**, but it may be called `acosh` in the math library for C so long as it meets this standard's requirement of correct rounding for all inputs. Similarly, a language may express a sum of two posits *a* and *b* as $a + b$, though that function is here called **addition**(a, b). Rounding behavior must follow the rules in this document for any implementation to be considered posit compliant.

Because this is not a language standard, nor does it say how the functions in a math library are to be labeled, we refer to well-known math functions but only describe how they should behave, not how they should be spelled or parsed. A computer language will of course allow the writing of something like `a = b + c`, not the form $a = $ **addition**(b, c) used in the Posit Standard.

1.4 Requirements vs. recommendations, and posit-compliance

All descriptions herein are requirements of system behavior, not recommendations. The decision of how to satisfy the requirements and which precisions to support is up to the implementer of this standard, but all functionality must be provided and behave as described for a system to be posit compliant. An implementation is compliant with this standard if it supports full functionality of at least one precision. If the implementation supports more than one precision, then it must support conversions between them and every precision supported must be posit compliant.

There are no "recommendations" in the document because following recommendations or not following them leads to irreproducibility. Including recommendations makes a document a guideline, not a standard.

If someone wants to build a computing environment for 9-bit posits because they work well for AI, say, there is no need to support any other precision to be posit-compliant. The entire environment could be done in software, written using native integer operations, or (what seems more likely) the kernel operation of multiply-accumulate would be done with hardware that is as fast as possible, and most of the lesser-used operations like logarithms could be done in software.

5.3 Definitions Section of the Posit Standard

---

2 Definitions, abbreviations, and acronyms

bit field   A contiguous set of bits in a format with a defined meaning. A bit field may extend beyond explicit bits $b_{n-1} \cdots b_0$; bits beyond the format's explicit bits are considered 0 bits.

Chapter 3 introduced the concept of "ghost bits" beyond the end of a posit, which always have value **0**. Here is another figure illustrating the concept, where an "**x**" is an explicit bit and may be a **0** or a **1**:

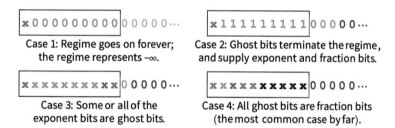

As the figure shows, while ghost bits are always **0**, they can play different roles depending on the explicit bits. Case 1 is the two exception values, either zero or NaR depending on the value of the sign bit. Case 2 is *maxPos* if the sign bit **x** is **0**, or *−minPos* if **x** is **1**. Case 3 is the "Twilight Zone" where exponent bits can be pushed off the end of the explicit bits by the length of the regime. Case 4 is the normal situation where the ghost bits are like the unexpressed bits of a decimal like "3.75", which is the same thing as "3.7500000⋯". The term "ghost bit" helps explain posit format, but for the sake of brevity, that phrase is not defined in the Posit Standard.

Chapter 4 disparaged the design of a number format by carving a word into bit fields resembling scientific notation. Posit arithmetic was invented by the geometric reasoning of Chapter 1 (the ring plots and the projective reals), which then *dictated* the bit fields, including one with a new and variable-length meaning, the regime.

> **exception** A special case in the interpretation of representations in posit format: 0 or NaR[2]
>
> ---
>
> [2]Posit representation exceptions do not imply a need for status flags or heavyweight operating system or language runtime handling.

For posits, an "exception" does not mean "throw control to an exception handler." It means the software or the circuit must do something different since the regime value, $-\infty$, is not representable with positional notation. Although the formula for zero correctly works out to $2^{-\infty}$, obviously the way to actually handle it is with a conditional branch for the case of all bits **0**.

For the case of NaR, this definition and its footnote represent one of the most profound departures from IEEE Std 754, which specifies that the arithmetic must also play the role of a debugger, constantly monitoring calculations for bad behavior and then either halting the calculation ("signaling NaN") or continuing ("quiet NaN"). But that is what debugging tools do; they can set a breakpoint when a particular value occurs as a result. Once a program is debugged, you can turn off "debug mode" and compile it for maximum speed; that is standard practice, because compiling to enable debug mode can make a program many times slower than the optimal compilation. With IEEE floats, however, you have a debugger you cannot turn off. It spends energy on every clock cycle to check for exceptions even after the programmer has guaranteed that they cannot occur.

A side bit of history, while we are on the subject of bugs:

> The slang term "bug" originated when Grace Hopper found that the computer was not working because a moth had gotten stuck in one of the relays of the Mark II that she and her team were programming in 1947.

This is a fun story, and Grace Hopper later became an Admiral and one of the most respected computer pioneers of the 20th century.

Grace Hopper (1906–1992), inventor of the concept of a programming language

Years before IBM introduced Fortran, she conceived of programs written in a high-level language that would run on any computer, and built a language environment that later evolved into COBOL.

We even have the actual moth, taped to a logbook that recorded the event:

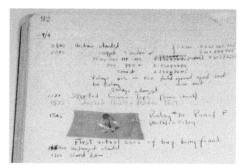

Literally the first computer bug

The etymology (and entomology) is not quite right, however. Thomas Edison, decades earlier, used to write about getting the "bugs" out of his electrical inventions, so that slang was already in popular use. The log entry shows that they found it quite funny that the bug turned out to be, literally, a bug. It almost certainly is not Grace Hopper's logbook, but her team found it amusing and she (and her team) deserve credit for spreading the term "debuggers" for the software tools used to find errors in programs.

In preventing a program from stumbling into indeterminate nonsense, there are three lines of defense:

1. The programmer is careful to think about all the possible ways things could go wrong, and makes sure the program handles exceptions in an intelligent way. Tools like debuggers and automated language analyzers assist the programmer with this task. Good compilers also issue warnings when the software does something dubious, like using a variable before it has been assigned a value.

2. Failing that, modern languages like C will check inputs to mathematical operations and produce error messages when, say, the computer is asked to take the square root of a negative number or divide by zero, and the message (via the operating system) informs the user what went wrong.

3. As a last line of defense, if somehow a processor is asked to do something that cannot produce a real number as a result, that can be handled by calling it NaR and propagating that invalidity through to the final output. It might happen that a programmer, confident the program will never misbehave, asks the compiler to disable all the exception handling of line-of-defense **2**, allowing it to run slightly faster since the math library routines do not have to first test if inputs are valid.

This is why posits make NaR the only exception that is part of the definition of the arithmetic. Some have said they really need positive and negative infinity, but that again confuses debugging and computing. It is hard to imagine anyone using a finished production code to "compute" infinity as a useful result. Line-of-defense **2** will generally provide the information when a result would be some kind of infinity, and why it happened. Nothing in the Posit Standard prevents a programmer from inserting constructs like

```
if x = 0, print "-Inf"; halt;
else y = log(x);
```

If you use the *valid* data type where posits ending in a 1 represent the real open interval between two exact numbers as described in Chapter 1, then you have a way to represent and compute with (*maxPos*, ∞) and (–∞, –*maxPos*).

---

exponent   The power-of-two scaling determined by the exponent bits, in the set $\{0, 1, 2, 3\}$.

exponent bits   A two-bit unsigned integer bit field that determines the exponent.

Why not allow the number of exponent bits to vary, independently of the precision? The main reason is that it makes hardware more complicated, more power-hungry, and after years of experimenting with various sizes of the exponent field, the value "2 bits" turned out to be optimal for a remarkably large fraction of the experiments. Software for posit arithmetic can have this flexibility, and many open-source libraries exist that support a variable number of exponent bits.

Exercise for the Reader: Can a posit with 3 bits of precision have an exponent size of 2 bits? Be careful, because this has the quality of being a trick question.

In later chapters, we will see quite a few parameters that can be used to make posit format more general, not just the precision and number of exponent bits. There is a place for both standard posits and generalized posits; standard posits always have two exponent bits; they could be explicit bits or ghost bits, but as Yoda says, always two there are.

---

format   A set of bit fields and the definition of their meaning.

fraction   The value represented by the fraction bits; $0 \leq \text{fraction} < 1$.

fraction bits   The bit field following the exponent bits.

We are careful not to use *mantissa*, which would be the appropriate term if the represented values were logarithms, but within a binade, the values are equispaced so the correct mathematical term is *fraction*.

The fovea, where relative accuracy is maximum, is when there are only two regime bits (01 or 10); that happens for fully half of all the bit patterns. In that case, taking into account the sign bit and two exponent bits, all but 5 bits are available for expressing the fraction. If n is the precision, the fraction has $n - 5$ explicit bits for the central-magnitude values, which is almost always the region where most of the computing takes place.

> **Exercise for the Reader**: For standard posits of precision at least 7, what percentage of posit bit patterns have $n - 6$ explicit bits for the fraction, just one bit less than in the fovea? What percentage have $n - 7$ explicit bits for the fraction?

> **fused** rounded only after an exact evaluation of an expression involving more than one operation.

This may be a place where the Posit Standard is a little *too* terse. Exact evaluation is not always necessary nor even possible, because it could require that an infinite number of digits be evaluated. Perhaps a better wording would be, "rounded the way an exact evaluation of an expression involving more than one operation would round" since that means the evaluation need only be accurate enough to determine how it rounds. Recall that this was the way the Math Layer was defined.

There are many more fused operations for posits than there are for IEEE floats, which allows calculations to stay in the Math Layer for more of the calculation and therefore accumulate less rounding error.

> **implicit value** A value added to the fraction based on the sign: –2 for negative posits, 1 for positive posits. Zero and NaR do not have an implicit value.

Binary float formats use a "hidden bit" to the left of the fraction bits; it is 1 for normal floats, and 0 for subnormal IEEE floats. But because posits are based on 2's complement arithmetic, the significant digits being scaled represent a value that ramps from 1 up to almost 2 for positive values, and from –2 to almost –1 for negative values. In 2's complement, 01 is 1 and 10 is –2.

For example: If the fraction bits are **011**, that means $\mathbf{0.011} = 0 \times \frac{1}{2} + 1 \times \frac{1}{4} + 1 \times \frac{1}{8} = \frac{3}{8}$, regardless of the sign. If the sign is positive, the implicit value plus the fraction is $1 + \frac{3}{8} = \frac{11}{8}$, but if the sign is negative, the implicit value plus the fraction is $-2 + \frac{3}{8} = -\frac{5}{8}$. This is a very machine-friendly way to do things, but just as with 2's complement integers, it is not very human-friendly.

LSB The least significant bit of a format or a bit field within a format.

maxPos The largest positive posit value. It is a function of *n*.

minPos The smallest positive posit value. It is a function of *n*.

MSB The most significant bit of a format or a bit field within a format.

The LSB and MSB need no further explanation. The formulas for *maxPos* and *minPos* as a function of the precision *n* are easily explained. The largest positive value *maxPos* is $0111\cdots111$, where the 1 bits in the regime go all the way to the end of the explicit bits (and are then terminated by a 0 ghost bit). The $n-1$ regime bits represent the integer one less than the run length, or $n-2$. The *uSeed* value for standard posits is $2^{2^2} = 2^4$, so $maxPos = uSeed^{n-2} = 2^{4(n-2)} = 2^{4n-8}$. We could similarly derive the smallest positive value, *minPos*, represented by $0000\cdots001$, or simply observe that $minPos = 1/maxPos = 2^{8-4n}$.

NaR Not a real. Umbrella value for anything not mathematically definable as a unique real number.

Much has already been said about NaR. "Not a Real" corrects the error in the Intel i8087 of calling it "Not a Number" when it really means not a *real* number. I was amused to see this exceptionally honest airport signage for a time of boarding:

The airport is (unintentionally, I assume) informing us that they have *absolutely no idea* when the plane is going to board. Apparently their software uses IEEE floats and converts them to integers for display.

The word "NaN" (rhymes with "ran") is a homophone with a woman's name "Nan" and a child's name for a grandmother, "nan." But "NaR" (rhymes with "far") has the advantage of being a unique word in proper English; it has no homophones.

> **n** The number of bits in a posit format. It can be any integer greater than 1.

The Posit Standard refers to the number of (explicit) bits in a posit so often, the variable n always has that meaning throughout. If you go as low as $n = 1$, all you have is zero and NaR; that does not seem very useful; there is no *minPos* nor *maxPos* value. Hence, the Posit Standard sets 2 as the smallest possible number of posit bits.

> **pIntMax** The largest consecutive integer-valued posit value. It is a function of n.

As the magnitude of a posit becomes large, the consecutive value spacings become larger than 1 and that means skipping over integers. This happens with floats as well; for example with 16-bit IEEE floats, you can count from 0 up to 2048, but the next larger value is 2050. There is no way to represent 2049. There is a non-obvious general formula for *pIntMax* as a function of n:

$$pIntMax = \left\lceil 2^{\left\lfloor 4(n-3)/5 \right\rfloor} \right\rceil$$

Let's plot the exponent, $\left\lfloor 4(n-3)/5 \right\rfloor$, as a function of n for n from 2 to 32:

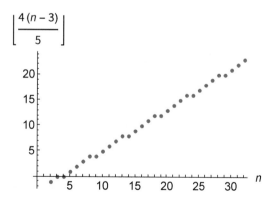

The power of 2 where standard posits start skipping over integers, as a function of posit precision

That shows why you need the ceiling function $\lceil ... \rceil$ in the formula; for the case of $n = 2$, it goes to –1 and then $2^{\lfloor 4\,(n-3)/5 \rfloor} = 2^{-1} = \frac{1}{2}$, which is not an integer. Rounding that up to 1 with the ceiling function gives the correct *pIntMax* for the extreme case of the smallest possible number of bits in a posit.

Exercise for the Reader: What is the formula for the *smallest* posit value such that it and all larger posits are guaranteed to be integers?

---

precision   The total storage size for expressing any number format, in bits. For a posit, precision is n bits.

This is consistent with terms like "single precision" and "double precision" or "8-bit precision": it is the number of bits available to express a number and is **not** the same as "accuracy." This was covered in Chapter 2, but it bears repeating because many people who should know better still confuse precision and accuracy. For example, the sawtooth pattern of relative accuracy in float formats was first called "wobbling precision," and the term stuck even though the *precision* is steady as a rock. The precision of the significant bits is also constant; no wobble there. It is the *relative accuracy* that wobbles. This was explained in Chapter 2.

---

quire value   A real number representable using a quire format described in this standard, or NaR.

The *quire* is a fixed-point format with enough precision to compute posit dot products (a sum of products) **exactly**. It is sufficiently powerful, and important, that it has an entire chapter dedicated to the ways it can be used. The idea of hardware support for an exact dot product, sometimes called a *scalar* product, comes from Karlsruhe professor Ulrich Kulisch, who introduced it in the 1970s but has been unsuccessful at getting it into any of the IEEE standards.

Notice the distinction between *quire* and *quire value*. Whereas *quire* is a format (a set of bit fields and their meanings), a *quire value* is the real number represented by a bit pattern in quire format. Similar care is taken in the Posit Standard to distinguish *posit* (a number format) from *posit value*. Our Working Group member Hauke Rehr deserves credit for noticing inconsistent wording in early drafts and making sure the wording is consistent. In this book, I frequently use "posit" to mean "posit value" where context makes it clear, like "the product of two posits."

> **quire sum limit**   The minimum number of additions of posit values that can overflow the quire format.

If the quire can store a dot product $a_1 \cdot b_1 + a_2 \cdot b_2 + \ldots + a_m \cdot b_m$ of two vectors of length m (up to some maximum length, which here is about two billion elements), then it can also compute sums exactly up to some maximum length. Just think of one of the vectors being all 1 values: $b_1 = b_2 = \ldots = b_m$; then the dot product is the sum of the a_i values. Summing numbers is much less demanding than summing *products* of numbers, so the maximum length of a sum that can guarantee no possibility of overflow or rounding errors is much larger than two billion. You would have to add *maxPos* to the sum, over and over, 2^{23+4n} times, to get the sum to overflow. Even for 8-bit posits, the quire sum limit is about 36 quadrillion summations. One of the largest sources of cumulative rounding error in floating-point programs is summations of many numbers, so the Posit Standard makes clear that the quire can eliminate that accumulation of rounding error, and also enables parallel processors to perform accumulation perfectly so the result matches that of serial execution.

> **regime**   The power-of-16 scaling determined by the regime bits. It is a signed integer.
>
> **regime bits**   A posit bit field following the MSB that uses a form of signed unary encoding (as opposed to positional notation) to represent the regime. There is always at least one regime bit R_0. For $n > 2$, there are always at least two regime bits.

Chapters 1 and 3 have already explained the regime and regime bits, but since the Posit Standard fixes the number of exponent bits at 2, there is no need to define the *uSeed*, and many of the definitions are greatly simplified. In the Posit Standard, the *uSeed* value is always $2^{2^2} = 16$.

> **rounded**   Converted from a real number to a posit value, according to the rules of this standard.

There is no need to use the phrase "correctly rounded" in the Posit Standard, since all rounding is required to be correct. Every real number rounds to exactly one posit value, and there is only one way to do it.

> **sign**   The value 1 for positive numbers, –1 for negative numbers, and 0 for 0. The NaR value has no sign.
>
> **sign bit**   The MSB of a posit or quire format.

Note the distinction between *sign*, which is –1, 0, 1, or NaR, and the *sign bit*, which can be **0** or **1**. The Posit Standard avoids oversimplifications like treating the sign as $(-1)^s$ where s is the value of the sign bit.

> **significand** The implicit value plus the fraction; $-2 \leq$ significand < -1 for negative posit values, and $1 \leq$ significand < 2 for positive posit values.

Because the implicit value is −2 for negative posit values and 1 for positive posit values, the implicit value plus a fraction in [0, 1) produces a significand in either [−2, −1) or [1, 2). Unlike floats, the significand includes the sign of the value.

5.4 Posit and Quire Formats Section

3   Posit and quire formats

3.1   Formats

This section defines posit and quire formats and their representation as a finite set of real numbers or the exception value NaR. Formats are specified by their precision, n. There is a quire format of precision $16n$ that is used to contain exact sums of products of posits of precision n. Dynamic range and accuracy are determined solely by n. This standard describes example choices for n like 8, 16, and 32. The posit format's type label is "posit" with the decimal string for n appended. The corresponding quire format's type label is "quire" with the decimal string for n appended, even though the quire has $16n$ bits.

There is a simple reason why computers are usually designed with datum sizes that are integer powers of 2, like 16, 32, and 64: It eliminates the cost of multiplying two integers when you index into an array. All you have to do is shift the index left and add the base address. This is likely to continue to be the case, so the Posit Standard gives example sizes of 8, 16, and 32 bits, even though it also applies to strange sizes like posit5, for example.

The 64-bit example is conspicuously absent; this is because there is far less need for such high precision with posit format. For the case of two exponent bits, the quire is always $16n$ bits long, and every bit counts as a significand bit because it is a fixed-point number. That means posit16, which has about 5 decimals of accuracy in the fovea, is backed by a quire with $16 \times 16 = 256$ bits of accuracy, which is about 76 decimals. That is more accuracy than an octuple-precision IEEE float. When more accuracy is needed, the quire can do more than exact dot products; it can also do extended-precision divides and square roots through a technique described later on.

3.2 Represented data

A posit value is either the exception value NaR or a real number x of the form $K \times 2^M$, where K and M are integers limited to a range symmetric about and including zero. The smallest positive posit value, *minPos*, is 2^{-4n+8} and the largest positive posit value, *maxPos*, is $1/minPos$, or 2^{4n-8}. Every posit value is an integer multiple of *minPos*. Every real number maps to a unique posit representation; there are no redundant representations. The posit values are a superset of all integers i in a range $-pIntMax \le i \le pIntMax$. Outside that range, integers exist that cannot be expressed as a posit value without rounding to a different integer; *pIntMax* is $\left\lceil 2^{\lfloor 4(n-3)/5 \rfloor} \right\rceil$.

While it is typical to describe a float or a posit as $\pm(1+f) \times 2^e$ where f is a *fraction* in the range [0, 1), it is also true that they can be expressed with two *integers* as $K \times 2^M$. Unlike IEEE floats, posits have the elegant property that the ranges for K and M are always symmetric about 0. For example, for posit16, K goes from –4095 to 4095 and M goes from –56 to 56. For precision n posits with $n > 4$,

$$-2^{n-4} < K < 2^{n-4}, \text{ and}$$
$$-4n+8 \le M \le 4n-8.$$

If $n = 2, 3,$ or 4, K goes from –1 to 1 because the fraction bits are always ghost bits, **0**. The M range shown above still holds for $n = 2, 3,$ or 4. The K range is important for hardware design because, among other things, it tells how large the integer multiplier will have to be. The multiplier is often one of the more expensive parts of the arithmetic logic in terms of chip area.

In contrast, for IEEE binary16 (half-precision floats), K goes from –2047 to 2047 and M goes from –24 to 15, and there really is no pattern for the ranges for different precisions since the number of exponent bits is not designed as a simple mathematical function of the precision.

A quire value is either NaR or an integer multiple of the square of *minPos*, represented as a 2's complement binary number with $16n$ bits. Quire format can represent the exact dot product of two posit vectors having at most 2^{31} (approximately two billion) terms without the possibility of rounding or overflow.[3]

---

[3] The product of two posit values in a format of precision n is always exactly expressible in a posit format of precision 2n, but the quire format obviates such temporary doubling of precision when computing sums and differences of products. Sums of posit values using the quire are guaranteed exact up to 2^{23+4n} terms, per the quire sum limit.

One way to describe a fixed-point format with m bits to the right of the radix point is to say it is an integer that represents a multiple of 2^{-m}. That is like the explanation in the Posit Standard. The footnote was prompted by one researcher asking the interesting question, "Does the product of two n-bit posits always fit into a posit with $2n$ bits?" It does, though the proof is tricky.

Some numerical analysts claim dot products of single-precision floats can—*almost always*—be made accurate by using double precision to hold the pairwise products and for the accumulated sum, then converting the sum back to single precision. The argument is, a double-precision IEEE float always has enough fraction bits and exponent bits to hold the product of two IEEE single-precision floats, which is true, but what about the accumulation?

Consider the dot product of these two single-precision vectors *a* and *b*:

$$a = \{3.2 \times 10^7, \ 1.0, \ -1.0, \ 8.0 \times 10^7\}$$
$$b = \{4.0 \times 10^8, \ 1.0, \ -1.0, \ -1.6 \times 10^8\}$$

Every element of each vector can be expressed with a single-precision float without rounding. In fact the values are whole numbers. For example, "8.0×10^7" is just the scientific notation for "eighty million." The pairwise products $a_i \cdot b_i$ are

$$1.28 \times 10^{16}, \ 1.0, \ 1.0, \ \text{and} \ -1.28 \times 10^{16}.$$

Try summing the numbers in single precision, and you get 0.0 exactly. Just for the sake of caution, use double precision to hold the products, and sum them in double precision as well. The straightforward way to add is left to right,

$$((1.28 \times 10^{16} + 1.0) + 1.0) - 1.28 \times 10^{16}.$$

Again, you get 0.0 exactly. Doing it left-to-right takes three addition steps. If there are two hardware adders that we can use in parallel, we might instead evaluate it pairwise to do the sum in only two steps:

$$(1.28 \times 10^{16} + 1.0) + (1.0 - 1.28 \times 10^{16})$$

and again, using IEEE double precision, we get 0.0 for the sum. Getting 0.0 for a dot product of two vectors is often very important because it means the vectors are *orthogonal*. That is, the vectors are exactly at right angles to each other.

There is only one problem, however. The correct answer is **2**.

For this simple calculation, rounding led to *infinite* error. The vectors are **not** orthogonal.

---

panacea | ˌpan-ə-ˈsē-ə |

noun

1. a single remedy that cures everything
2. a remedy that falls short of its claims

The word panacea comes from the Greek for "cure-all," so it literally means the first definition, but because things claimed to be cure-alls almost never actually work for *all* ailments, the word has come to mean an *overclaim*, something to be suspicious of. "Perpetual motion machine" is the same way; while it means a machine that runs forever, it has come to mean "a false claim of perpetual motion." In numerical computing, double precision is a panacea... the second definition. It cannot make sums of products free of error, as the example shows. So the footnote is intended to point the reader away from "double-precision posits" and instead to the quire, which can do exact sums of products up to at least two billion such products.

The properties of example and general posit format precisions are summarized in Table 1:

Property	posit8	posit16	posit32	posit*n*
fraction length	0 to 3 bits	0 to 11 bits	0 to 27 bits	0 to max$(0, n-5)$ bits
minPos	$2^{-24} \approx 6.0 \times 10^{-8}$	$2^{-56} \approx 1.4 \times 10^{-17}$	$2^{-120} \approx 7.5 \times 10^{-37}$	2^{-4n+8}
maxPos	$2^{24} \approx 1.7 \times 10^{7}$	$2^{56} \approx 7.2 \times 10^{16}$	$2^{120} \approx 1.3 \times 10^{36}$	2^{4n-8}
pIntMax	$2^4 = 16$	$2^{10} = 1024$	$2^{23} = 8388608$	$\lceil 2^{\lfloor 4(n-3)/5 \rfloor} \rceil$
quire format precision	128 bits	256 bits	512 bits	16n bits
quire sum limit	$2^{55} \approx 3.6 \times 10^{16}$	$2^{87} \approx 1.5 \times 10^{26}$	$2^{151} \approx 2.9 \times 10^{45}$	2^{23+4n}

Table 1: Properties of posit formats

This table is intended as explanatory, not essential to the definition of the format. All the entries can be derived from the other parts of the document, but it seemed too useful to be able to see the dynamic ranges and significant bits and other properties of the example sizes, and in the rightmost column, formulas for the properties as a function of a general precision *n*.

3.3 Posit format encoding

Figure 1 defines the general format for posit encoding. The regime is a variable-length field. All of its bits but the last are identical. The longer the regime, the more bits of fields to its right are not represented. These truncated bits extending beyond the LSB are treated as 0 bits. Figure 2 shows how part of the exponent field and all of the fraction field can be truncated. Figure 3 shows the extreme case where the regime extends to the LSB. The four constituting bit fields in order of decreasing significant bits are:

1. Sign bit S. S represents an integer s, its literal value, 0 or 1. The implicit value is $(1 - 3\,s)$.

2. Regime bit field R consisting of k bits identical to R_0, terminated by $\overline{R_0} = 1 - R_0$ as shown in Figures 1 and 2, or just after the LSB as shown in Figure 3. R represents $r = -k$ if R_0 is 0, or $r = k - 1$ if R_0 is 1.

3. The exponent field E has length 2 bits, but one or both bits may be beyond the LSB and thus have value 0. E represents an integer e, its bits treated as a 2-bit unsigned integer. $0 \le e < 3$.

4. The fraction bit field F has length $\max(0, n - 5)$ bits, but any number of those bits may be beyond the LSB of the posit representation and thus are taken to be 0 bits. The number of explicit bits is m. F represents the fraction f, an m-bit unsigned integer divided by 2^m. $0 \le f < 1$.

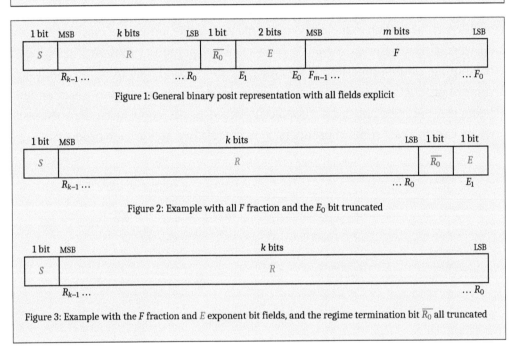

Figure 1: General binary posit representation with all fields explicit

Figure 2: Example with all F fraction and the E_0 bit truncated

Figure 3: Example with the F fraction and E exponent bit fields, and the regime termination bit $\overline{R_0}$ all truncated

This section is the first to use color-coding, similar to the color-coding used in this book to make bit fields clearer. We toyed with the idea of making the colors part of the Posit Standard, but decided against it.

The table below shows the red-green-blue (RGB) and cyan-magenta-yellow-black (CMYK) components I *recommend* to those writing papers on posits, as a guideline. These colors were determined with the help of readers who have the most common form of color blindness (red-green).

Field, Color	R G B	C M Y K
Sign, red	1.00 0.33 0.33	0.00 0.67 0.67 0.00
Regime repeating part, gold	0.80 0.60 0.40	0.00 0.25 0.50 0.20
Regime termination, brown	0.60 0.40 0.20	0.00 0.33 0.67 0.40
Exponent, blue	0.25 0.50 1.00	0.75 0.50 0.00 0.00
Fraction: black	0.00 0.00 0.00	0.00 0.00 0.00 1.00

RGB and CMYK color values for the color-coding of posit bits

The posit value p is inferred from the bit fields S, R, E, and F as follows:

1. Check if p represents an exception: if all bits except S are 0 bits (cf. figure 3),
   - if $S = 0$, then $p = 0$.
   - if $S = 1$, then p is NaR.

2. Otherwise, let $f := 2^{-m} \sum_{\ell=0}^{m-1} f_\ell \, 2^\ell$,
   - if $R_0 = 0$, then $r = -k$.
   - if $R_0 = 1$, then $r = k - 1$.

   And $p = ((1 - 3s) + f) \times 2^{(1-2s) \times (4r + e + s)}$.

In Figure 3, if R_0 is 1, then it represents *maxPos* ($S = 0$) or $-minPos$ ($S = 1$).

Chapter 1 explained the case of $eS = 0$, and Chapter 3 introduced general eS values; the "explicit bit" and "ghost bit" methods were used there. The Posit Standard uses a slightly different explanation that is specific to the case of $eS = 2$.

The observation that a string of all 0 bits, together with an infinite string of ghost bits, represents an r value of $-\infty$, did not occur to me until after the Posit Standard was ratified; otherwise this section could have been simplified. From the point of view of a designer, however, it is still best to call attention to the fact that special handling is needed when $r = -\infty$.

Spelling out the fraction bits with Σ summation notation makes unambiguous what they mean, for anyone not familiar with "decimals" that use a base other than ten.

The formula $p = ((1 - 3s) + f) \times 2^{(1-2s) \times (4r+e+s)}$ looks quite different from the formula you usually see for a normal float, which looks like $(-1)^s \times (1 + f) \times 2^{e-\text{bias}}$, so those familiar with floats often react with puzzlement. Because the implicit value is 1 for positive posits and –2 for negative posits, $(1 - 3s)$ is a simple formula that reproduces those two values as a function of the sign bit. The scaling $2^{(1-2s) \times (4r+e+s)}$ could have been written $2^{(-1)^s \times (4r+e+s)}$, but the convention of writing $(-1)^s$ is a little silly since it implies you actually have to call a power function to raise –1 to the power of s. All you need is a function that returns 1 if $s = 0$ and –1 if $s = 1$; the function $(1 - 2s)$ is a simple way to accomplish that. The s value also shows up in the $(4r + e + s)$ term, and that is because posits are based on 2's complement. To negate, flip all the bits and add 1.

3.4 Quire format encoding

Quire format is a fixed-point 2's complement format of precision $16n$, with fields as follows:

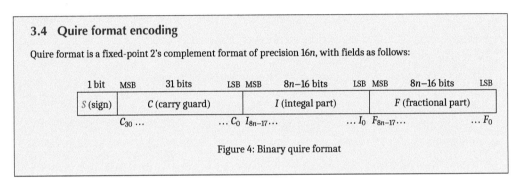

Figure 4: Binary quire format

Eleven people spent an enormous amount of time on this document, and not one of us noticed that "integal" should be "integral"; our brains tend to fill in that missing "r" so those who have studied the Posit Standard also have not noticed this typo. Or at least, no one has reported it to us. Also, if $n = 2$, the I and F fields disappear completely and we have nothing but the sign and carry guard bits. So we should have noted that, since bit annotations I_{8n-17} and F_{8n-17} make no sense when $n = 2$.

The sign S is the usual red color-coding, but the remainder of the fields are all shown in black, because they are simply positional notation like the fraction bit field. The "$8n - 16$ bits" deserves an explanation. The smallest positive posit, *minPos*, is always 000⋯001 where the run length k is $n - 2$ bits (the precision, less 1 for the sign bit and 1 for the regime termination bit). That bit string represents a regime value of $r = -(n - 2) = 2 - n$. Recall that the *uSeed* is $2^{2^2} = 16$ when $eS = 2$, so $minPos = 16^{2-n} = 2^{4 \cdot (2-n)} = 2^{8-4n}$. In computing a dot product, we might have to accumulate products that are as small as $minPos^2 = (2^{8-4n})^2 = 2^{16-8n}$. Similarly, $maxPos^2 = 2^{8n-16}$. Hence, the integral part I and fractional part F are $8n - 16$ bits long.

The quire is always 16*n* bits total, so if we subtract the sizes of the *I* and *F* fields, we are left with something that does not depend on *n*:

$$16\,n - (8\,n - 16) - (8\,n - 16) = 32 \text{ bits.}$$

After allowing for the sign bit, that leaves 31 bits for protection from carry overflow. The worst-case scenario is that you accumulate *maxPos*^{2} over and over; you could do that up to $2^{31} - 1 = 2\,147\,483\,647$ (call it two billion) times with no possibility of overflow.

> The quire value q is inferred from the bit fields S, C, I, and F as follows
>
> - If S is **1** and all other fields contain only **0** bits, then q is NaR.
> - Otherwise q is 2^{16-8n} times the 2's complement signed integer represented by all bits concatenated.

This is a first in number formats: The "exact dot product" idea also can represent NaR, and it is the same way posits do it: a **1** bit followed by all **0** bits. IEEE floats are ill-suited to something like a quire because of all their exception conditions; how would you represent all of the possible IEEE exceptions in a *fixed-point* number?

Saying the value represented by the quire q is an integer times 2^{16-8n} is the same as saying the quire is a fixed-point format with the radix point $8\,n - 16$ bits to the left of the last bit. Here are some examples of quire values for a tiny posit, $n = 3$. The quire is $3 \times 16 = 48$ bits long, and the entire posit vocabulary is just eight items: {NaR, −16, −1, −1/16, 0, 1/16, 1, 16}:

Value	Quire bits (48–bit fixed point)
1	00000000 00000000 00000000 00000000 00000001.00000000
−1	11111111 11111111 11111111 11111111 11111111.00000000
maxPos=16	00000000 00000000 00000000 00000000 00010000.00000000
maxPos^2=256	00000000 00000000 00000000 00000001 00000000.00000000
minPos=1/16	00000000 00000000 00000000 00000000 00000000.00010000
minPos^2=1/256	00000000 00000000 00000000 00000000 00000000.00000001
NaR	10000000 00000000 00000000 00000000 00000000.00000000

Examples of 48-bit quire representations of various values when the posits have only 3 bits

The quire for 3-bit posits has its radix point just 8 bits left of the LSB.

5.5 Rounding Section of the Posit Standard

4.1 Definition and method

Rounding is the substitution of a posit value for any real number. Operation results are regarded as exact prior to rounding. The method for rounding a real value x is described by the following algorithm:

Data: x, a real number
Result: Rounded x, a posit value
if x is exactly expressible in the posit format in question **then**
$\quad$| **return** x
if $|x| > maxPos$ **then**
$\quad$| **return** $sign(x) \times maxPos$
if $|x| < minPos$ **then**
$\quad$| **return** $sign(x) \times minPos$
Let u and w be n-bit posit values such that the open interval (u, w) contains x but no n-bit posit value.
Let U be the n-bit representation of u.
Let v be the $(n + 1)$-bit posit value associated with the $(n + 1)$-bit representation $U1$.
if $u < x < v$ **or** ($x = v$ **and** LSB of U is 0) **then**
$\quad$| **return** u
else
$\quad$| **return** w
end

This is a formal algorithm for posit rounding, and it expresses the process slightly differently from the description in Chapter 3. If x is exactly expressible as a posit, there is nothing to do. If x is outside the dynamic range, saturate to signed *minPos* or *maxPos*. Otherwise, x lies between two adjacent n-bit posit values u and w. A diagram may be helpful in visualizing the above algorithm. Recall the projective reals and the ring plots of Chapters 1 and 3, and zoom in on just one part of the ring:

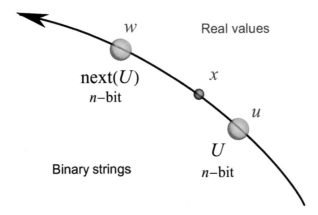

The real value in question, x, lies between posit values represented by U and one ULP higher

The real values are on the outside of the ring. Values u and w are successive posit values (no other posits lie between them). We need to round to either u or w.

When you append a **0** bit to a posit, it does not change what it represents, but appending a 1 bit represents a new in-between value, the algorithm appends a 1 bit to U and writes it as $U1$. The real value represented by $U1$ is v, as shown below:

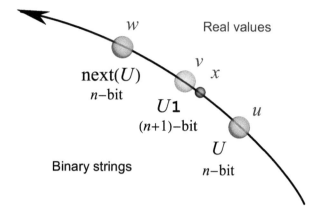

If there were $n+1$ bits instead of n, the introduced new point tells us which way to round x.

In the figure, that determines that x should round to u since x is between u and v. If x had been between v and w, it should round to w. If $x = v$, the tie case goes to whichever of u or w is represented by a bit string ending in a 0 bit ("round to even").

4.2 Fused expressions

A fused expression is an expression with two or more operations that is evaluated exactly before rounding to a posit value. Expressions that can be written in the form of a dot product of vectors of length less than 2^{31} can be evaluated exactly using quire representations and then rounded to posit format to create a fused expression, if so defined by the rules of the language. If a fused expression is computed in parallel, sufficient intermediate result information must be communicated that the result is identical to the single-processor result.[4] Fused expressions (such as fused multiply-add and fused multiply-subtract) need not be performed using quire representations to be posit compliant.

---

[4]Note that functions in Section 5 which are rounded and have two or more operations in their mathematical definition are fused expressions, such as **rSqrt**, **expMinus1**, **fMM**, and **hypot**.

Many essential operations in scientific computing can make use of the quire to stay in the Math Layer as long as possible. If we use i to represent the imaginary number $\sqrt{-1}$, then multiplying two complex numbers $a + b\,i$ and $c + d\,i$ requires evaluating two expressions for the real and imaginary parts:

$$(a + b\,i) \cdot (c + d\,i) = a \cdot c + b\,i \cdot c + a \cdot d\,i + b\,i \cdot d\,i$$
$$= (a \cdot c - b \cdot d) + (a \cdot d + b \cdot c)\,i.$$

Both $(a \cdot c - b \cdot d)$ and $(a \cdot d + b \cdot c)$ are dot products. Doing them in the quire and then converting back to posit format reduces the number of roundings from three to one.

Using the quire also solves a classic problem when using complex arithmetic based on pairs of floats: Programmers discover that if they declare a variable to be complex and let the computer take care of operations like the above multiplication, it runs significantly and inexplicably slower than writing out the product as shown above, with real variables instead of complex variables. Why? It is because routines for complex numbers are trying to avoid catastrophic underflow or overflow when computing $a \cdot c$, $b \cdot d$, and so on. The programmer may know that there is no danger of overflow, but the math library routines do not know that and assume the worst, so they go through considerable machinations to avoid overflow and underflow and produce a correctly-rounded answer.

When computing a fused operation in parallel, it might be tempting to round the intermediate results before sending them to another processor. That is not permitted, because it leads to irreproducibility. For example, a single processor might compute a sum of posits $a + b + c + d$ using a quire variable q, then round q to a posit s. That is, $s = \overline{a + b + c + d}$. If one processor computes $a + b$ and another computes $c + d$ in respective quire variables q_0 and q_1, they need to exchange q_0 and q_1, and not first round them to a local sum. That keeps fused parallel operations in the Math Layer until they are complete. Posits can restore the associative property of addition, up to some extremely large number of additions (the quire sum limit).

That said, the quire can be overkill. Doing a fused multiply-add, for example, requires fewer than $16n$ bits, so a system designer may choose to do fused multiply-add operations by some other means that is more economical. This is explicitly permitted in the Posit Standard, in the last sentence of the quoted paragraph.

The footnote points out that any operation in the list of math functions (the next section of the Posit Standard) made of composite operations must behave like a fused function. For example, computing **rSqrt**(x) means $1 \big/ \overline{\sqrt{x}}$ and not $\overline{1 / \sqrt{x}}$.

4.3 Program execution restrictions

For any language where the order of operations is well-defined, the execution order of operations in posit compliant mode cannot be changed from that expressed in the source code if it affects rounding. This includes any use of precisions or operation fusing not expressed in the source code.[5] If a language permits mixed data types in expressions, including quire and posit format types, or posits and other formats for representing real numbers, the language must specify how such expressions are evaluated in order to be posit compliant.

---

[5] Languages that offer optimization modes that covertly change rounding do so at the cost of bitwise-reproducible results and are non-compliant when used in those modes.

Many language environments offer optimizations that increase speed at the expense of portability (and compliance with a standard). Sometimes portability is much less important than going as fast as possible, and the Working Group recognized that. However, for a computing environment to be posit-compliant, there has to be at least one mode of operation that follows the Posit Standard. The Posit Standard does not specify, for example, the order of operations when expressions are not parenthesized, but it does require that there be at least one mode where the ordering is *deterministic* by the source code and the grammar rules of the language, and only those modes are posit-compliant.

5.6 Posit Standard Functions

---

5 Functions

5.1 Guiding principles for the NaR exceptional value

If an operation usually produces real-valued output, any NaR input produces NaR output, with the exception of **next** and **prior**. NaR is output when the function's value is not arbitrarily close to a unique real number for open neighborhoods of complex values sufficiently close to the input values,[6] except for discontinuous functions in Section 5.2. Functions with multiple branches such as roots and inverse trig functions apply this criterion to a single branch. A test of equality between NaR values returns True. The NaR value has no sign, so **sign**(NaR) returns NaR.

---

[6]The function may be complex-valued in the neighborhood of the inputs, but still have a real-valued limit. For example, **pow**(−1, −3) = $(-1)^{-3}$ is −1 even though the function is complex-valued in any neighborhood of the second argument. Similarly, sqrt(0) = 0.

It was difficult to write this section without bringing in concepts from calculus; our goal was to keep everything understandable to anyone with a secondary school mathematics education. The phrase "arbitrarily close to a unique real number" evokes the concept of a *limit*. One way to express the rule in simple English is this: We are comfortable saying 2 times 3 is 6 for real numbers because approximately 2 times approximately 3 is approximately 6. But what is zero divided by zero? Approximately zero divided by approximately zero is *indeterminate*. If "approximately zero" means, say, between −0.001 and 0.001, that would include $\frac{-0.001}{0.001} = -1$, and $\frac{0.0001}{0.000001} = 100$, and $\frac{-0.001}{0} = \pm\infty$. It could be any real number, as well as ±∞. That is why $\frac{0}{0}$ must evaluate to NaR. The divide operation is not *continuous* in its inputs when the denominator is zero.

The math gets a little heavy here. Instead of "approximately" we say "in an *open neighborhood* of the value in the complex plane."

The figure below shows the neighborhoods to be circles, but they can be any connected region that is open, that is, does not include its boundary.

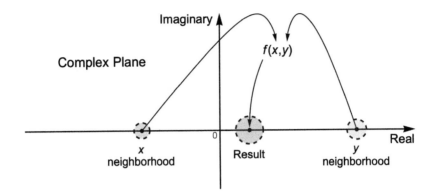

A function *f* of input arguments *x* and *y*, with their open neighborhoods in the complex plane

If we cannot make the Result region arbitrarily small about a real-valued result by making the input neighborhoods arbitrarily small, then the result is NaR. Call it "the neighborhood rule." It is not the same as saying "the function must be continuous" because it might be continuous in the complex plane yet not be real, like the arcsin function for a value greater than 1.

The power function is **pow**(x, y) = x^y. By the neighborhood rule, **pow**(0, 0) must return NaR. In any open neighborhood containing 0, y has both negative and positive values. Zero to a negative power is ±∞.

The IEEE Std 754 has a very different philosophy. It says **pow**(x, ±0) = 1, even if x is a (quiet) NaN. That is one of several ways that an indeterminate NaN result with floats can be made to disappear in a computation, and the user will never know anything went wrong. The IEEE Standard lists no fewer than *twenty* exception cases to the power function, things like, "**pow**(±0, y) is ±0 for finite $y > 0$ an odd integer." The Posit Standard handles all those exceptions with the simple rule above, which is much more careful to propagate a NaR value and not regard inputs as exact numbers, since they will in general *have come from a rounded result*.

Graphs show the reason for "except for discontinuous functions of Section 5.2":

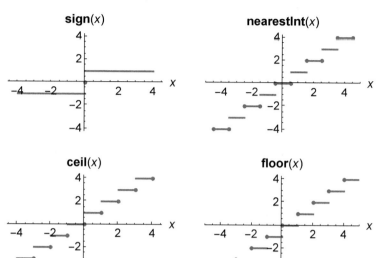

Functions designed to be discontinuous for which the neighborhood rule does not apply

The neighborhood rule does not apply to these; they should not return NaR for real input because the behavior at the discontinuity is *well-defined*.

The **next** and **prior** functions are also not covered by the neighborhood argument, because they are functions of *bit patterns*, not real numbers. As such, there is no guarantee that a NaR will propagate correctly if a programmer really wants to do something as dangerous as manipulate posit bit patterns. If you think of the posits on the ring of projective reals, **next** moves one position counterclockwise, and **prior** moves one position clockwise. So, **next**(NaR) is –maxPos and prior(NaR) is maxPos, up at the top of the ring of posits.

> The following functions shall be supported, with rounding per Section 4.1. Functions that take more than one posit value for input must have the same precision for all posit value inputs, and any posit value in the output must have the same precision as the input posits. Conversion routines may be used to make mixed-precision posit value inputs the same precision, per Section 6.1. Conversions may be explicit in source code or implicit by language rules.

This is another example of drawing the line between being an arithmetic standard or a language standard. Most languages require that inputs are promoted to the precision of the highest-precision input, to preserve as much information as possible. However, one could argue that that is like putting lipstick on a pig; if there is just one low-precision input, all inputs should be *demoted* to that low precision since the low accuracy will propagate to the output anyway. The Posit Standard does not take a side but merely says all posit inputs have to be the same precision, and whatever it is, that is the precision of the output.

5.2 Basic functions of one posit value argument

negate(*posit*)	returns −*posit*.[7]
abs(*posit*)	returns **negate** (*posit*) if *posit* < 0, else *posit*.
sign(*posit*)	returns a posit value: 1 if *posit* > 0, −1 if *posit* < 0, or 0 if *posit* = 0.
nearestInt(*posit*)	returns the integer-valued posit value nearest to posit, and
	returns the nearest even integer-valued posit value if two integers are equally near.
ceil(*posit*)	returns the smallest integer-valued posit value greater than or equal to *posit*.
floor(*posit*)	returns the largest integer-valued posit value less than or equal to *posit*.
next(*posit*)	returns the posit value of the lexicographic successor of *posit*'s representation.[8]
prior(*posit*)	returns the posit value of the lexicographic predecessor of *posit*'s representation.[9]

[7]This is the 2's complement of the posit representation. 2's complement affects neither 0 nor NaR, since they are unsigned.
[8]wrapping around, if necessary.
[9]wrapping around, if necessary.

The **negate** and **abs** functions are the only two functions in Section 5.2 that are continuous (so the neighborhood rule applies). For a hardware designer, they could hardly be simpler. To negate a posit, treat the bits as a 2's complement integer and negate it that way: flip all the bits and add one. Zero is unchanged, and NaR is unchanged, exactly the way it should be. Footnote 7 points this out.

The graphs shown previously illustrate **sign**, **nearestInt**, **ceil**, and **floor** functions. Why would you ever want to use **next** and **prior** functions? Those are useful when an algorithm does self-testing for the accuracy of its real number representation. Sometimes it is called "machine epsilon," the spacing between representable real values. If you take **next**(*posit*) − *posit* or *posit* − **prior**(*posit*), and you are not involving NaR, you will get the "epsilon" (jargon for a relatively small number) for the error in using that representation. Some programs use that to test when an iterative method has converged, for example.

5.3 Comparison functions of two posit value arguments

All comparison functions return Boolean values identical to comparisons of the posits' representations regarded as 2's complement integers, so there is no need for separate machine-level instructions. The representation of NaR coincides with the 2's complement bit string of the most negative integer, so if posit is real, **compareLess**(NaR, *posit*) returns True, etc.

compareEqual(*posit1*, *posit2*)	returns True if *posit1* = *posit2*, else False.
compareNotEqual(*posit1*, *posit2*)	returns True if *posit1* ≠ *posit2*, else False.
compareGreater(*posit1*, *posit2*)	returns True if *posit1* > *posit2*, else False.
compareGreaterEqual(*posit1*, *posit2*)	returns True if *posit1* ≥ *posit2*, else False.
compareLess(*posit1*, *posit2*)	returns True if *posit1* < *posit2*, else False.
compareLessEqual(*posit1*, *posit2*)	returns True if *posit1* ≤ *posit2*, else False.

Compared to IEEE Std 754, this is a huge simplification. If a processor supports integers at all, then supporting posit comparison does not require a single extra transistor.

In contrast, the IEEE standard goes on for two and a half pages about comparison, and includes clauses that say –0 is equal to +0, but –0 < +0 is true and +0 < –0 is false. It even introduces –NaN and +NaN. Lewis Carroll could not have done a better job of writing a standard; it goes completely down the rabbit hole.

IEEE 754 Committee, meeting to work out the rules for negative zero and positive zero

---

5.4 Arithmetic functions of two posit value arguments

addition(*posit1, posit2*)	returns *posit1* + *posit2*, rounded.
subtraction(*posit1, posit2*)	returns *posit1* − *posit2*, rounded.
multiplication(*posit1, posit2*)	returns *posit1* × posit2, rounded.
division(*posit1, posit2*)	returns *posit1* ÷ posit2, rounded.

Division by zero is always NaR, per the neighborhood rule.

5.5 Elementary functions of one posit value argument

sqrt(*posit*)	returns $\sqrt{posit}$, rounded.
rSqrt(*posit*)	returns $1/\sqrt{posit}$, rounded.
exp(*posit*)	returns e^{posit}, rounded.
expMinus1(*posit*)	returns $e^{posit} - 1$, rounded.
exp2(*posit*)	returns 2^{posit}, rounded.
exp2Minus1(*posit*)	returns $2^{posit} - 1$, rounded.
exp10(*posit*)	returns 10^{posit}, rounded.
exp10Minus1(*posit*)	returns $10^{posit} - 1$, rounded.
log(*posit*)	returns $\log_e(posit)$, rounded.
logPlus1(*posit*)	returns $\log_e(posit + 1)$, rounded.
log2(*posit*)	returns $\log_2(posit)$, rounded.
log2Plus1(*posit*)	returns $\log_2(posit + 1)$, rounded.
log10(*posit*)	returns $\log_{10}(posit)$, rounded.
log10Plus1(*posit*)	returns $\log_{10}(posit + 1)$, rounded.

This is the first occurrence of a math function that is fused: **rSqrt** must be rounded to the value it would be rounded to if it were calculated to infinite precision.

The double-rounding of first rounding the square root and then rounding the reciprocal will sometimes be off by one in the last place from a correctly-rounded $1/\sqrt{x}$. As a simple example, consider standard 8-bit posits with an input value of $x = 1/128$. The square root of x rounds to $11/128 = 0.0859375$, and the reciprocal of $11/128$ is about 11.6, which rounds up to 12. However, $1/\sqrt{x} \approx 11.3$, which rounds down to 11, as required. For 8-bit standard posits, double rounding when computing $1/\sqrt{x}$ differs from the fused result about 22 percent of the time.

The **expMinus1** function is also fused because $\overline{e^x - 1}$ can be wildly different from $\overline{e^x} - 1$. For small x, $e^x \approx 1 + x$, which can round to 1. So $\overline{e^x} - 1$ returns zero instead of x, which can be numerically disastrous. It can also be disastrous in the other direction; for example, again using 8-bit standard posits, if $x = \frac{1}{16}$, e^x is *slightly* larger than $1 + \frac{1}{16}$ so it rounds up to $\frac{9}{8}$, the next posit after 1. Then subtracting 1 returns $\frac{1}{8}$, twice as large as the correctly-rounded fused value, $\overline{e^{1/16} - 1} = 1/16$.

Why include **exp2**$(x) = 2^x$? In general, $b^x = e^{x \log b}$, which would require $\overline{e^{x \log b}}$, three rounding operations, so a fused version is certainly much more accurate. There is another reason: Computing 2^x is usually *faster* than computing e^x. Break x into its integer and fraction parts, $x = n + f$ where $n = $ **floor**(x) and f is what is left over, $0 \le f < 1$. Then $2^x = 2^{n+f} = 2^n \times 2^f$. But 2^n is encoded in the posit regime and exponent bits, and 2^f will be a number $1 + F$ where F is the fraction part of the posit result. Methods like polynomial approximations can find F accurately enough to produce correct rounding, and this will be covered in more detail in a later chapter.

The **exp10** and **log10** functions are there because they are human-friendly. And you need them to work with decibels, for example. Depending on how much math you took in school, base-10 logarithms might be the only type you have ever seen; the "natural log," log base $e = 2.71828\cdots$ usually first comes up in calculus class.

sin(*posit*)	returns sin (*posit*), rounded.
sinPi(*posit*)	returns sin ($\pi \times posit$), rounded.
cos(*posit*)	returns cos (*posit*), rounded.
cosPi(*posit*)	returns cos ($\pi \times posit$), rounded.
tan(*posit*)	returns tan (*posit*), rounded.
tanPi(*posit*)	returns tan ($\pi \times posit$), rounded.

Of course we need the traditional trigonometric functions sine, cosine, and tangent.

However, math library designers have long wished they did not have to support those functions for huge values of the input argument. Who would ever want to go around the unit circle an astronomical number of times (which requires computing with the value of π correct to many decimal places), and then want to know the cosine to ten decimals?

For 12-bit nonnegative posit values, here is a plot of what the cosine function looks like: values as a function of the posit bit pattern, treated as an integer from 0 to 2047 (that is, **00000000000** to **011111111111**)

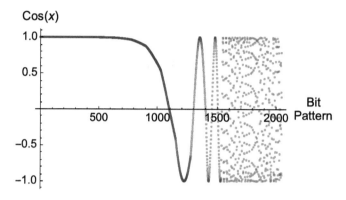

Trig functions become nonsensical for very large arguments

The function degenerates into a lacy pattern and then becomes a pseudo-random number generator beyond about 1536 (posit bit pattern **011000000000**). But now look what happens if we instead plot **CosPi**$(x) = \cos(\pi x)$:

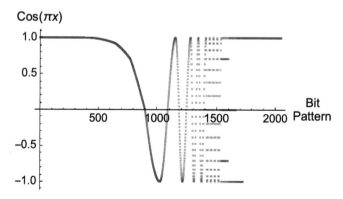

Trig functions of π times the argument are easier to compute and make more sense.

Once you get past a certain size, all the posit values are integers or half integers, and the only possible values of $\cos(\pi x)$ are –1, 0, and 1, *exactly*. Furthermore, in many important algorithms like the Fast Fourier Transform (FFT), the angles you need are π, $\pi/2$, $\pi/4$, $\pi/8$, and so on, so **cosPi** and **sinPi** are exactly what you want for the FFT.

Notice that **tan** always produces a real-valued output since the argument is a rational number that cannot land exactly on an irrational singularity like $x = \pi/2$. If the posit is close, it may produce a value that saturates to *maxPos* or *–maxPos*. However, **tanPi** *can* land exactly on such singular points, and by the neighborhood rule, it must return NaR for inputs of the form $k/2$ where k is an odd integer.

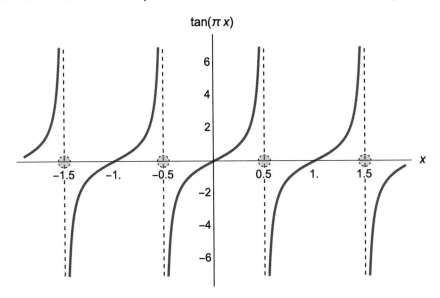

$$\tan(\pi x)$$

The tan(π x) function must return NaR for half-odd input values.

arcSin(*posit*)	returns arcsin (*posit*), rounded.	**abs**(**arcSin**) $\leq (\pi/2$, rounded).
arcSinPi(*posit*)	returns arcsin (*posit*)$/\pi$, rounded.	**abs**(**arcSinPi**) $\leq 1/2$.
arcCos(*posit*)	returns arccos (*posit*), rounded.	$0 \leq$ **arcCos** $\leq (\pi/2$, rounded).
arcCosPi(*posit*)	returns arccos (*posit*)$/\pi$, rounded.	$0 \leq$ **arcCosPi** ≤ 1.
arcTan(*posit*)	returns arctan (*posit*), rounded.	**abs**(**arcTan**) $\leq (\pi/2$, rounded).
arcTanPi(*posit*)	returns arctan (*posit*)$/\pi$, rounded.	**abs**(**arcTanPi**) $\leq 1/2$.

The inverse trig functions bring up a subtle issue regarding the endpoints of the range. Mathematically, the principal branches are chosen to be $0 \leq \arccos(x) \leq \pi$, $-\pi/2 \leq \arcsin(x) \leq \pi/2$, and $-\pi/2 \leq \arctan(x) \leq \pi/2$, but we have to use round(π) and $\pm$round($\pi/2$), which could round *up or down* depending on the posit precision. This can cause some weird things to happen to trig identities at the endpoints, so caution is needed with posits just as with floats. For example, with 16-bit posits, the closest posit to $\frac{\pi}{2}$ is $\frac{3217}{2048} = 1.5708078125$, slightly larger than $\frac{\pi}{2} = 1.570796\cdots$, which puts it on the *other side* of the singularity of tan(x) at $x = \pi/2$. That means **tan**(**arcTan**(*maxPos*)) will return *–maxPos* instead of *maxPos*. It is like trying to balance a pencil on its point; you should not be surprised at which direction it falls. Avoiding computing near singularities is the responsibility of the programmer.

sinH(*posit*)	returns sinh (*posit*), rounded.
cosH(*posit*)	returns cosh (*posit*), rounded.
tanH(*posit*)	returns tanh (*posit*), rounded.
arcSinH(*posit*)	returns arcsinh (*posit*), rounded.
arcCosH(*posit*)	returns arccosh (*posit*), rounded. $0 \le$ **arcCosH**.
arcTanH(*posit*)	returns arctanh (*posit*), rounded.

You can visualize an inverse (arc) function as flipping the function about the line $y = x$, because that reverses the roles of input and output. The reason for pointing out $0 \le$ **arcCosH** is that the inverse hyperbolic cosine is the one hyperbolic function that leads to two branches; the convention is to pick the one that is not negative:

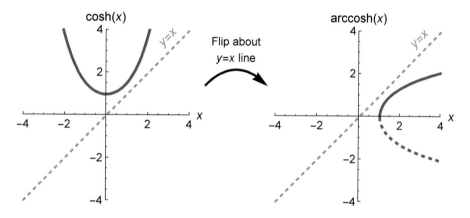

The inverse of the cosh(*x*) function is double-valued, so the non-negative value is the standard choice.

5.6 Elementary functions of two posit value arguments

hypot(*posit1, posit2*)	returns $\sqrt{posit1^2 + posit2^2}$, rounded.
pow(*posit1, posit2*)	returns $posit1^{posit2}$, rounded.[10]
arcTan2(*posit1, posit2*)	returns the argument t of $posit1 + i\,posit2$, $-\pi < t \le \pi$, rounded.[11]
arcTan2Pi(*posit1, posit2*)	returns **arcTan2**(*posit1, posit2*)$/\pi$, rounded.

---

[10]See Section 5.1 for situations that generate NaR. For example, x^y is not arbitrarily close to a single real number for any complex-valued neighborhoods of $x = y = 0$, so **pow**(0, 0) returns NaR.

[11]The apparent discontinuity in **arcTan2**(*x*, *y*) for $x \le 0$, $y = 0$ is spurious since it results from jumping between branches of a multi-valued function. It should return π, rounded, if $x < 0$, $y = 0$. arcTan2(0,0) must return NaR since the function is not arbitrarily close to a unique real value for any complex-valued open neighborhoods of the inputs.

The **hypot** (short for hypotenuse) function in IEEE Std 754 is allowed to return ∞ if the inputs are NaN and $\pm\infty$. That is another way that a NaN can be made to disappear and never warn the user that the calculation is actually indeterminate. The Posit Standard avoids this mistake; as always, any input being NaR means the output is NaR.

Computing the rounding of **hypot** is tricky; Appendix D presents an algorithm that uses the quire to make sure the function behaves as a fused operation.

The fact that **pow**(0, 0) must return NaR has already been discussed, by applying the neighborhood rule. The same rule says NaR is returned for **pow**$(x, y) = x^y$ whenever x is negative and y is not an integer. If there is a fractional part in y, the result will be imaginary, not real. That gives rise to the following pattern:

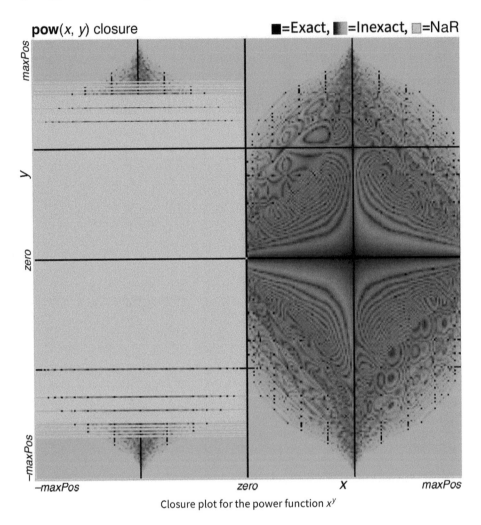

Closure plot for the power function x^y

The plot also shows the strips on the bottom and left where x or y are NaR inputs, which of course produce a NaR output.

arcTan2 has been in math libraries since the earliest days of Fortran, because together with **hypot**, it allows you to convert x-y (Cartesian) coordinates to r-θ (polar) coordinates.

The following plot shows **arcTan2** as a function of *posit1* and *posit2* (*x* and *y*) inputs:

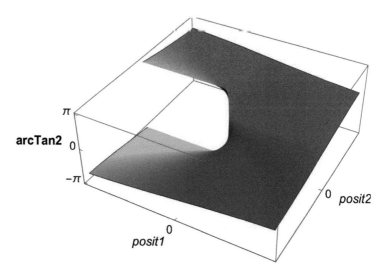

The **arcTan2** function appears discontinuous for negative *posit1* inputs, but actually it is not.

That figure makes it look like there is a discontinuity along the negative *x*-axis. But if you look at all the values of the arctangent function, it forms a spiral ramp that is continuous everywhere but *x* = *y* = 0, the one pair for which the value returned must be NaR. Everywhere else is a smooth sheet.

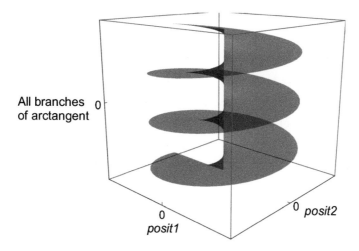

This shows several spirals of the **arcTan2** function.

Similarly to the trig functions of one variable, **arcTan2pi** divides the result by π so that many of the results can be expressed exactly, like **arcTan2pi**(3, 3) = 1/4, instead of **arcTan2**(3, 3) = round(π/4).

5.7 Functions of three posit value arguments

fMM(*posit1, posit2, posit3*)   returns *posit1* × *posit2* × *posit3*, rounded.[12]

---

[12]Because multiplication is commutative and associative, any permutation of the inputs will return the same rounded result.

"**fMM**" stands for "fused multiply-multiply." Two multiplications are done in the Math Layer, then rounded to a posit. Remember that the quire can only accumulate products of pairs of posits; you cannot multiply the quire by a posit, or at least it is not in the Posit Standard, and it would probably be best if it did not become a common extension to the Posit Standard since that is not the purpose of the quire. The quire makes it possible to guarantee that $(a + b) + c$ rounds to the same value as $a + (b + c)$. That begs the question of what to do when you need to know that $(a \times b) \times c$ rounds to the same value as $a \times (b \times c)$, and that is what **fMM** is for.

After decoding the input posits into the form $M \times 2^K$, some form of extended precision arithmetic (in software or hardware) is needed to multiply the three integers $M_{posit1} \times M_{posit2} \times M_{posit3}$ and add the three exponents $K_{posit1} + K_{posit2} + K_{posit3}$. The resulting exact value is then rounded to the nearest posit by the usual rules.

5.8 Functions of one posit value argument and one integer argument

compound(*posit, integer*)   returns $(1 + posit)^{integer}$, rounded.

rootN(*posit, integer*)   returns $posit^{1/integer}$, rounded. If *integer* is even, **rootN** ≥ 0.

These are two more examples of "Do not try this at home." That is, do not simply code the literal formula $(1 + posit)^{integer}$ or $posit^{1/integer}$ in the working precision and expect to get an accurate result most of the time. The **compound** and **rootN** functions are fused operations that need to be done with considerable extra care to avoid answers that can be wildly inaccurate. Let the numerical experts figure out how to guarantee the exact result, rounded.

5.9 Functions that do not round correctly for all arguments

Computing environments that support versions of functions in any of the above subsections that do not round correctly for all inputs must supply the source code for such functions, and use a notation for them that is distinct from the notation for the corresponding function that rounds correctly for all inputs.

There is always a place for a speed-accuracy tradeoff, and some vendors offer fast-but-less-accurate versions of standard functions.

A famous example is a square root approximation discovered by the developers of Quake, a first-person shooter game.

Quake was early in the history of three-dimensional video games.

Square roots are essential in 3D video gaming. When Quake was developed (1990s), full-accuracy square roots were so slow that they made the game unplayable. So they found a less-accurate approximation that was close enough for a video game.

The above requirement in Posit Standard Section 5.9 says that you cannot give the approximate function the same name as the one that is rounded per the Posit Standard. It has to be called something else. Like, if `sqrt(x)` is the one that is posit-compliant, then you could call the faster alternative `fauxSqrt(x)`. The source code for `fauxSqrt(x)` must then be supplied, so other systems can compile it and produce bitwise identical results.

Some programmers are very attracted to the idea of simply linking to an alternative library with the same names for the functions. That is, the "`makefile`" then contains information vital to understanding what results the program will produce, outside the source code. Doing this makes the system noncompliant with the Posit Standard, and non-portable. There is no need to covertly link a different library; just come up with a different name for the function and everyone wins. The code is portable (bitwise identical answers), fast, maintainable, and accurate enough for the task at hand.

> Do not create invisible "modes" of operation.

5.10 Functions not yet required for compliance

Special functions such as error functions, Bessel functions, gamma and digamma functions, beta and zeta functions, etc. are not presently required for a system to be posit compliant. They may be required in a future revision of this standard.

The current standard math libraries for popular programming languages do indeed have some of those less-common functions. The goal of the 2022 Posit Standard was to get to a level of support for math functions of posits so that most scientific codes could be easily ported using posits as drop-in replacements for floats. After the Posit Standard has been in use for a few years, it will become apparent which additional functions are in the highest demand, and they can be included in a revision to the Posit Standard without losing any backward-compatibility.

5.11 Functions involving quire value arguments

With the exception of qToP which returns a posit value result, these functions return a quire value result.[13] If any operation on quire values overflows the carry bits of a quire's representation, the result is NaR in quire format. Where any posit values are involved, their precisions n must agree, and the quire must be of the corresponding precision 16 n.

pToQ(*posit*)	returns *posit* converted to quire format.
qNegate(*quire*)	returns *–quire*.
qAbs(*quire*)	returns **qNegate**(*quire*) if *quire* < 0, else *quire*.
qAddP(*quire, posit*)	returns *quire* + *posit*.
qSubP(*quire, posit*)	returns *quire* – *posit*.
qAddQ(*quire1, quire1*)	returns *quire1* + *quire2*.
qSubQ(*quire1, quire2*)	returns *quire1* – *quire2*.
qMulAdd(*quire1, posit1, posit2*)	returns *quire* + (*posit1* × *posit2*).
qMulSub(*quire1, posit1, posit2*)	returns *quire* – (*posit1* × *posit2*).
qToP(*quire*)	returns *quire* converted to posit format.

Other functions of quire values may be provided, but are not required for compliance. They may be required in a future version of this standard.

It is possible to make the set of quire operations even smaller, by doing awkward things like replacing **qSubQ**(*quire1, quire2*) with **qAddQ**(*quire1*, **qNegate**(*quire2*)). If a processor is designed with native quire instructions at the machine language level, it will probably build the above list out of a more primitive set. Why is there no **qClear**(*quire*)? It seemed obvious that you do that by simply loading a 0 into the quire with **pToQ**(0).

All of these are easy to make fast in hardware (single-cycle execution), except **qToP**. The cost of **qToP** is tolerable if the quire is used as intended, because **qToP** is usually amortized over many accumulations.

The Working Group was sorely tempted to add one more required function: **qShiftRight**(*quire, integer*) would return *quire* with an arithmetic right shift by *integer* many bits, another function that can easily be accomplished in a single clock cycle. An arithmetic right shift preserves the sign of a 2's complement number by copying the MSB as things shift right. A reason you might want **qShiftRight** is if you are using the quire to sum a list of *m* positive numbers and want to know their average, accurately, as $\frac{sum}{m}$. There is no operation for dividing a quire by an arbitrary integer *m*, so you have to convert the quire to a posit and then divide that by *m*. But it is easy to see how *sum* could be a long way from the fovea and thus lose precision. Scaling by shifting *k* bits to the right could bring *sum* back to a posit in the fovea, and then it could be multiplied by $2^k/m$ to find the average. This may be added to a future version of the Posit Standard; in the meantime, it is pretty easy to write a small software routine to shift a quire right by *k* bits.

Notice that **fMA** (fused multiply-add) is not in the Posit Standard, and neither is **fAM** (fused add-multiply). That is because they are trivial to fuse using the quire. If *a*, *b*, *c*, and *y* are posit variables and *q* is a quire variable, performing $q = a \times b + c$ is exact, and $y = \textbf{qToP}(q)$ then does the correct rounding. Similarly, if we write $(a + b) \times c$ as $a \times c + b \times c$, that puts it in the form of a dot product so the quire can compute it exactly and then round it to a posit.

5.7 Conversions

We are almost done.

---

6   Conversion operations for posit format

6.1   Conversions between different precisions

Converting a posit value to a higher precision is exact, by appending 0 bits to its representation. Conversion to a lower precision is rounded, per Section 4.1.[14] In the function notation used here,

> **pmTon**(*posit*) returns the *n*-bit posit representation of an *m*-bit posit value *posit* by these conversion rules.

---

[14] Note that precision conversion does not require decoding a posit representation into its bit fields.

Because $eS = 2$ for all standard posits, they can be converted without having to figure out the regime, exponent, and fraction values. That allows changing precision to be *very* fast, easy to do in a single clock cycle.

For example, **p16To32** converts a 16-bit posit to 32-bit posit format by shifting left 16 places, so the low-order 16 bits are all **0** bits. Going the other way, **p32To16** needs to round away the rightmost 16 bits. The value of *minPos* is greater for 16-bit posits than 32-bit posits, and the value of *maxPos* is less, so those cases of "saturation" rounding can be handled by comparing the posit bit string with an integer. Otherwise, the 32-bit posit string is rounded to a 16-bit posit string by banker's rounding. Maybe an example will make this clear.

The 32-bit posit for π is **01001100100100001111110110101010**, representing the rational number $\frac{105\,414\,357}{33\,554\,432} = 3.1415926516\,056\,060\,791\,015\,625$. To round that to the nearest 16-bit posit, forget the regime and exponent and fraction bit fields and instead just look at three bits:

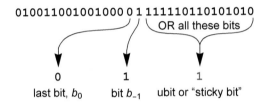

Banker's rounding needs only three bits to decide whether to round up bit b_0.

Banker's rounding says, "If bit b_{-1} is **1**, then if last bit b_0 is **1** or the ubit is **1**, round up," and we have that case here. So the 16-bit posit is the leftmost 16 bits plus 1, **0100110010010001**, and the rightmost 16 bits are discarded. If we color-coded the bits and decoded the resulting 16-bit posit, we would find it to be **0100110010010001**, representing the rational number $\frac{3217}{1024} = 3.1416\,015\,625$, a slightly less accurate approximation for π.

6.2 Conversions involving quire values

A posit compliant system only needs to support rounding from quire to posit values and conversion of posit to a quire in the matching precision, per Section 5.1.

The question came up: Do we need conversion of the quire to other formats, like integers? Or to posits of higher precision than the one with 16*n* bits that the quire serves? The answer is a definite **no**. The only conversion path for the quire is from posit to quire, and quire to posit. If someone wants to convert the quire to anything else, first convert it to a posit and then apply the other conversion routines described in the Posit Standard.

6.3 Conversions between posit format and decimal character strings

Table 2 shows examples of the minimum number of significant decimals needed to express a posit value such that the real number represented by the decimal form will round to the same posit value.

precision	posit8	posit16	posit32	posit64
Decimals	2	5	10	21

Table 2: Examples of minimum decimals in a base–ten significand to preserve any posit value

In the Human Layer, we display real numbers in decimal scientific notation. The power of ten is an integer that must be expressed exactly, but how many decimal digits must be expressed in the significand to ensure that when the decimal version is translated back into posit format, it returns bit-for-bit the same pattern? It was possible to work out what that number is for the architecture-friendly precisions of 8, 16, 32, and 64 bits, but there is no known formula for precision n. It is reminiscent of other very difficult problems, like asking how many squares on a piece of graph paper lie completely inside a circle of radius r. It is not readily amenable to a closed-form formula. The end of Chapter 2 shows how irregular the pattern can be.

6.4 Conversions between posit format and integer format

Supported posit formats must provide conversion to and from all integer formats supported in a computing environment. In converting a posit value to an integer value, if the posit value is out of integer range after rounding or is NaR, the integer value is returned the representation of which has its MSB = **1** and all other bits 0. In converting an integer value to a posit value, the integer representation with its MSB = **1** and all other bits 0 converts to NaR; otherwise, the integer value is rounded, per Section 4.1.

This rule solves a major problem with IEEE Std 754: How do you convert NaN to an integer? There is no "Error" integer form, so whatever you choose, it will destroy the information that there was an error result while computing, and the end result will look like everything went fine.

Not so for posit format. NaR converts to the integer **1000···000** (which simply means appending or trimming **0** bits of the posit to match the integer precision), and if that integer is converted back to a posit, it will turn into NaR. The integer represented by **1000···000** is really the Bad Boy of the computer integer world. It has no negative. If you try to compute its absolute value, it returns the same negative value you started with. It deserves to be the "Error" bit pattern of integer representations, but that is not the convention that is in use by computer vendors. The only thing the Posit Standard can do is provide a way for an integer to convert back into the bit pattern for NaR, and it does so with the above clause.

6.5 Conversions between posit format and IEEE Std 754™ float format

Supported posit formats must provide conversion to and from the IEEE Std 754 float formats supported in the computing environment, if any. In converting a posit value to an IEEE Std 754 float value of any type, the posit value zero converts to the "positive zero" float value, and NaR converts to quiet NaN. Otherwise, the posit value is converted to a float value per the float rounding mode in use. In converting a float value to a posit value, all forms of infinity and NaN convert to NaR. Otherwise, the float value is rounded, per Section 4.1. "Negative zero" and "positive zero" float values convert to the posit value zero.   □

The final clause recognizes the need for posits to coexist with IEEE floats. Posits convert "signed zero" to a mathematical value, zero, and if converting the other direction, "positive zero" is the float that a posit zero converts to. IEEE signaling NaN, quiet NaN, $-\infty$, and ∞ all convert to posit NaR, which converts back to quiet NaN. Other than those exceptions, a float is a rational number of the form $m \times 2^j$, where m and j are integers, so this rational number converts to the closest posit by the posit rounding rules. Other than NaR, a posit value is also of the form $m \times 2^j$. That converts to a float by whichever of half a dozen rounding modes is in use, something not generally visible in the float source code. The posit behavior is deterministic and visible to the programmer. The float behavior … not so much.

This concludes a long and perhaps tedious explanation of why decisions were made the way they were. By way of apology, the next chapter will be more fun to read than this one, I promise.

"THAT'S NOT THE WAY IT WORKS. DIFFERENT PEOPLE CRUNCHING THE NUMBERS SHOULD NOT GET DIFFERENT RESULTS."

6 Preaching About the Quire

The term *quire* is from ancient book-making: four sheets folded twice to create 16 pages.

6.1 Floats are Weapons of Math Destruction

A posit environment is not complete without a method for computing a dot product *exactly*, up to some very large number of terms m. Chapter 5 showed that the *quire*, a fixed-point accumulator, serves this purpose and is part of the Posit Standard. At first glance, that may seem like a specialized task that does not deserve hardware support. It is actually much more generally useful than you might think.

The *scalar product* of two vectors $a = \{a_1, a_2, ..., a_m\}$ and $b = \{b_1, b_2, ..., b_m\}$ is the sum of the pairwise products, $s = a_1 \cdot b_1 + a_2 \cdot b_2 + ... + a_m \cdot b_m$. Two vectors input, a scalar output, hence the term scalar product. Since it is often written $s = a \cdot b$ with a bold "·" dot, it is also often called a *dot product*. This probably started as mathematical slang, but "dot product" has become a well-understood name for the operation. Dot products arise in AI (both training and inference), signal processing, computational physics and chemistry, computer graphics ...just about any application that involves real numbers. If you know what a *weighted sum* is, that is just another name for a dot product. Yet a fourth name for it is *inner product*, a name that will be explained visually in Chapter 11.

To give you a feel for what the dot product does, you can think of it as the projection of one vector onto another, scaled by the lengths of the vectors. If **a** and **b** have unit length, then **a · b** is the cosine of the angle between them. That works in any number of dimensions, even if the ones we can visualize only go up to three dimensions.

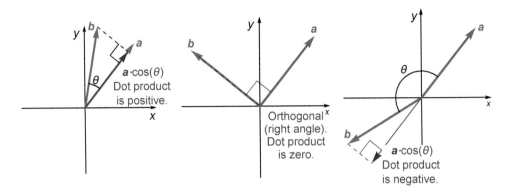

The sign of the dot product determines if two vectors make an acute, right, or obtuse angle.

Vectors are *orthogonal* (at a right angle) if their dot product is zero and both have nonzero length. If you have ever done any woodworking or origami or quilting, you know the importance of getting a "perfect 90," that is, an accurate right angle.

Orthogonality is similarly important in many scientific applications. It does not *sound* hard to get right. Try it with a pair of 3-dimensional vectors with coefficients that are small whole numbers, represented with 16-bit IEEE floats. Say,

$$\boldsymbol{x} = \{x_1, x_2, x_3\} = \{49,\ 1,\ -25\}$$
$$\boldsymbol{y} = \{y_1, y_2, y_3\} = \{75,\ -1,\ 147\}$$

To reduce rounding error, assume that we use fused multiply-adds after we get started with the first product. For example, start by finding $x_1 \times y_1$ and rounding, then add that product to $x_2 \times y_2$ with a fused multiply-add, and finally add that partial sum to $x_3 \times y_3$ with a fused multiply-add, for a total of three cumulative rounding errors. Try several different ways of doing the calculation. For float rounding, we use the underbar operator instead of the overbar used for posit rounding. The code for float environments is in Appendix B.

```
setFloatEnv[{16, 5}]
x = {49, 1, -25};
y = {75, -1, 147};
```

$$\left\{ \frac{x_{[1]} \times y_{[1]} + x_{[2]} \times y_{[2]} + x_{[3]} \times y_{[3]}}{}, \right.$$

$$\frac{x_{[1]} \times y_{[1]} + x_{[2]} \times y_{[2]} + x_{[3]} \times y_{[3]}}{},$$

$$\left. \frac{x_{[1]} \times y_{[1]} + x_{[3]} \times y_{[3]} + x_{[2]} \times y_{[2]}}{} \right\}$$

{1, -1, 0}

Three different answers, just for this tiny problem! People who major in Computer Science in college usually must take a course in Discrete Mathematics, but can graduate without ever having used a floating-point number. The correct answer is –1 for the above problem, but how would you know? No matter what precision is used, examples exist where floats cannot get the sign of a dot product right. That can lead to the error of thinking vectors are orthogonal, when they are not. Floating-point numbers are *weapons of math destruction*. Using double-precision floats cannot protect you from this hazard; see the example in Section 5.3.

The addition of real numbers is commutative because $\text{round}(a + b) = \text{round}(b + a)$. But it is not associative because it is not generally true that $\text{round}(\text{round}(a + b) + c)$ will produce the same result as $\text{round}(\text{round}(a + c) + b)$.

Exercise for the Reader:

Part 1: If we have four distinct numbers a, b, c, d, how many different ways are there of adding them if addition is commutative but not associative?

Part 2: If the four numbers are 1, 2, 2^{54}, and -2^{54} and we add them using IEEE Std 754 double-precision arithmetic (52 fraction bits) with banker's rounding, what are the possible values for the resulting sum?

Posits use the *quire* to compute dot products and sums exactly, and converting back to a posit always gets the sign right (negative, zero, or positive). Hence, this chapter is a lot of preaching about the quire.

6.2 The Kulisch Accumulator

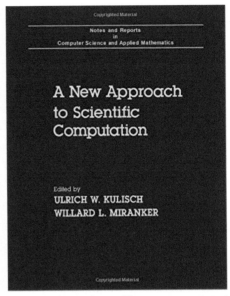

Professor Dr. Ulrich Kulisch, 1933–, and the 1983 book that *almost* changed how we compute today.

The company where I first worked in the computer industry had a name that in retrospect seems amusing: Floating Point Systems. Its founder preferred "Floating Point" with no hyphen. They built computers specialized for fast floating-point arithmetic, both single and double precision, designed to be attached to a general-purpose computer to make them more suited to high-speed technical computing. I joined in 1982, and coping with rounding error was an ever-present part of the job. When Kulisch's book (above) came out in 1983, a colleague discovered it and showed it to me. It was an epiphany. Ulrich Kulisch, at the University of Karlsruhe, was part of a group that showed how computing with floating-point numbers does not have to be fraught with cumulative errors or uncertainty about the validity of calculated results. IBM was interested and supported the effort, and even produced a software environment that supported *Karlsruhe Accurate Arithmetic* (KAA). It was based on two powerful ideas:

- Interval arithmetic; after every operation, rigorously bound the result x with a closed interval $[a, b]$ (that is, $a \leq x \leq b$) instead of rounding to a single value. That is, round *outward* to a pair of values guaranteed to contain the true answer.
- Support the exact calculation of dot products of floating-point vectors.

Unfortunately, they wedded these ideas to the use of double-precision arithmetic even though they also work for lower precision. Kulisch and his colleagues created C and PASCAL programming environments that could (usually) guarantee results accurate to within half an ULP (the best possible), as if the entire computation had taken place in the Math Layer. They made KAA available as C-XSC and PASCAL-XSC programming languages. IBM licensed the technology for a product they called ACRITH. ACRITH used KAA methods by applying interval arithmetic to bound the answer and then shrinking that bound, but the programmer of ACRITH was not exposed to any of that machinery and simply used (double-precision) floats as usual.

Kulisch also led the design of hardware support for the *EDP*, for Exact Dot Product. Floating Point Systems was intrigued enough to design and build an EDP specifically for their own 64-bit float format (which was much like Intel's i8087 64-bit format but based on 2's complement so it avoided the "negative zero" mess). Kulisch also did his best to convince the IEEE 754 Committee to include the exact dot product idea in the Standard they were still arguing about at the time. Kahan expressed concerns about the cost of perfection in counterpoint to Kulisch and succeeded in keeping the exact dot product idea out of the standard. When the IEEE Std 754 was revised in 2008, Kulisch tried again to get it into the Standard, but was again rebuffed.

If the dot products are to be exact for what later became IEEE 754 double precision, Kahan may have been right. How big would the accumulator have to be, to avoid the need to round? It has to contain the square of the smallest magnitude float, and also the square of the largest magnitude float. That is from 2^{-2148} to almost 2^{2048}, or an expanse of 4196 binades (over 1263 orders of magnitude). That takes 4197 bits plus a sign bit, but you also need bits for carry overflow protection. Would it suffice to use an array of 65 integers, 64 bits each? That would supply 38 bits to protect against overflow. But 2^{38} is about 275 billion, and it is *conceivable* that a computer could do a dot product with that many terms with every term being the maximum float squared. Even back then, it would only take a supercomputer a few minutes to do a dot product that long. So instead, they prescribed an array of 66 64-bit integers, and it became fair to claim it humanly and technologically impossible to ever overflow an accumulator that large. The total number of bits is then $66 \times 64 = 4224$ bits, which is a daunting size for a single integer now. Imagine how it looked back in the 1980s, when transistors were millions of times more expensive than they are today.

It is a psychological thing. If a programmer asks for storage for an array of 66 (64-bit) integers, that seems like a very modest request, even a tiny one. But if you think of those 66 integers as a single giant integer 4224 bits wide, hardware designers are aghast. "How can we propagate the carry bit that far in one clock cycle?" There are well-known workarounds for that problem. The 4224 size is also not an integer power of 2, like $2^{12} = 4096$, so if you have an array of EDP values in memory, it will be expensive to index into that array.

I suspect there is another reason the EDP was never accepted into the IEEE Standard: *Exceptions*. If the dot product has a value that is an infinity or a quiet NaN, what should the EDP do? It would have been quite a mess trying to integrate the EDP into all of the exception cases of IEEE Std 754.

IBM marketed its ACRITH product for a while, but it was not successful, and they eventually dropped it from their product offerings. It was one of the first tests in the marketplace of ideas of the question in Chapter 1: Is it worth it to go slower and get a perfect answer, or is it more important to go as fast as possible and get answers that are *usually* good enough? It must have surprised IBM how few customers were eager for perfect answers; apparently customers had been tolerating their "good enough" answers for a long time. When technical teams specify a benchmark to be run in a competitive procurement of a computer system, the decision is almost invariably made on the basis of quickest time to solution, not the highest accuracy in the time the customer is willing to wait.

The techniques of KAA make computing just two to three times slower in most cases. If the idea to apply the techniques of KAA to 32-bit floats had been widespread, that might have gotten traction. It would have shown the answers were (almost) always correct to 7 decimals even for situations that normally produce wildly incorrect results. That would have made it safe to use single precision, thus saving storage, bandwidth, energy, power, time, and money.

Another thing that might have made the approach more appealing is if you could apply it *selectively* to sections of the code where accuracy is most important, like the equation solver, say, but turn it off in other sections where accuracy is less important, like graphical output.

While forms of KAA were not commercially successful, the EDP is still recognized as a potentially good idea, and many refer to it as a *Kulisch Accumulator*.

6.3 Posits Revive Kulisch's Idea

Remember from Chapter 5 that standard n-bit posit values range in magnitude from $2^{-(4n-8)}$ to 2^{4n-8}, so the $a_i \cdot b_i$ products summed to form a dot product range in magnitude from $2^{-(8n-16)}$ to 2^{8n-16}. That means a fixed-point number with $8n - (-8n) = 16n$ bits can store the full range, with $16 - (-16) = 32$ bits left over for the sign bit and 31 carry protection bits. This solves two problems with any exact accumulator for floats:

- For hardware-friendly data sizes like 8, 16, 32, and 64 bits, the accumulator is also a hardware-friendly size: 128, 256, 512, and 1024 bits.
- The size of the accumulator is *much* smaller for posits than for floats: 1024 bits for 64-bit posits instead of 4224 bits for 64-bit floats.

What about exceptions? Posits only have one exception to worry about in computing dot products, and that is NaR. NaR is easily dealt with by assigning the same bit pattern **100⋯00** in the accumulator to mean NaR. That is, a **1** bit followed by all **0** bits. In an actual hardware design, the NaR exception would probably best be handled by setting a single bit as a flag so handling NaR is fast and cheap, but then when returned to posit form, the posit **100⋯00** would be returned.

These differences seem major enough that the posit accumulator deserves its own term, distinct from EDP: The *quire*. As a part of making a book, a quire is more than a page but less than a book. Originally, a quire was 16 pages, reminiscent of the $16n$ size in bits for an n-bit posit.

Another important operation is simply summing a long list of numbers. If you can do a dot product exactly, you can also sum a long list of numbers exactly. Just treat the **b** vector as {1, 1, ..., 1}, and the dot product will be the sum $s = a_1 + a_2 + \ldots + a_m$. There is a technique called *compensated summation* that is recommended by some, where for every running sum you use two additional values to track the cumulative rounding error. It triples the amount of storage and work needed to find the sum, and *it does not always work*. With the quire, you can sum extremely long lists of numbers with no possibility of rounding error. Like, with 16-bit standard posits, you would have to have well over 10^{26} values before there was any possibility of overflow, and even the current generation of supercomputers cannot do that in a tractable amount of time.

Here is a redrawing of the figure from Section 2.8 to show how the quire fits in:

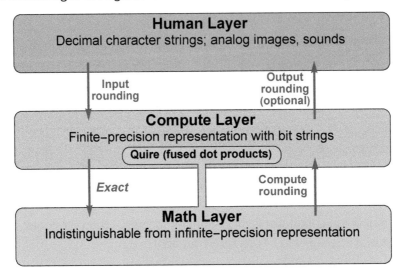

Where the quire belongs in the three layers of real number computing

While the quire lives in the Compute Layer, it is indistinguishable from infinite-precision dot products and sums, up to some very large number of terms.

Why go to so much trouble to get dot products perfect? It is time for an example.

6.4 A Demo: Two Equations in Two Unknowns

The onset of puberty is challenging enough, but to add to the trauma, that is also about the age when students are taught how to solve two equations in two unknowns. Even with a calculator, it can be tedious and error-prone. Classroom examples usually use small integers for the coefficients of the equations, but real-world problems are more likely to use decimals to several places.

Suppose we have two equations in two unknowns, and the coefficients are known to five decimals after the decimal point:

$$0.51245\,x \;+\; -0.67627\,y \;=\; -0.16382 \;\; \text{and}$$
$$-0.31054\,x \;+\; 0.41052\,y \;=\; 0.09998$$

I have chosen the coefficients so the solution is simply $x = y = 1$. We should be able to solve it by Gaussian elimination, but in doing so there will be rounding errors, especially from the divide operations. And the rounding errors will accumulate. That is why people commonly use far more precision than is needed to express the problem and its answer, as insurance.

I will instead choose to live dangerously and attempt it with 16-bit standard posits. The above equations live in the Human Layer; here is what the equations look like with all the coefficients converted to the nearest 16-bit posit, like $\overline{0.51245} = \frac{2099}{4096}$, which moves the problem into the Compute Layer:

$$\frac{2099}{4096}x + -\frac{1385}{2048}y = -\frac{671}{4096} \quad \text{and}$$
$$-\frac{159}{512}x + \frac{3363}{8192}y = \frac{819}{8192}$$

It does not always happen, but in this case we preserved the fact that the solution is $x = y = 1$. Take it in small steps so we can watch where the roundings happen. Divide the first equation by the coefficient of x, $\frac{2099}{4096}$, as if we were in the Math Layer and had access to any rational number and not just numbers of the form $m \times 2^j$:

$$x + -\frac{2770}{2099}y = -\frac{671}{2099}$$

Alas, we cannot do anything further without returning to the compute layer by rounding. The digits that differ from what they should be are shown in **red**:

$$x + -1.3198\cdots y = -0.31970\cdots$$

The relative error was not too bad: about 0.0005 and 0.0001 for the two fractions above. Now scale this equation and subtract from the second equation to eliminate x from the second equation. Assume we can do a fused multiply-subtract and thus round only once, not twice for the two terms left in the second equation:

$$x + -1.3198\cdots y = -0.31970\cdots \quad \text{and}$$
$$0x + 0.00065\cdots y = 0.00070\cdots$$

That step was a *relative error magnifier* because we subtracted two similar numbers (each of which had rounding error), and many of the most-significant digits cancelled. We know something pretty bad must have happened, because the second equation now says y must be bigger than 1 by almost 6 percent. Now backsolve by subtracting the scaled second equation from the first and rounding. In the Compute Layer, it is $x = \frac{2205}{2048}$ and $y = \frac{2167}{2048}$, but Human Layer decimals are easier to read:

$$x + 0y = 1.0766\cdots \quad \text{and}$$
$$0x + y = 1.0581\cdots$$

The computed value of *x* has even more relative error, almost 8%. Geometrically, we are trying to find the point of intersection of two lines that are almost parallel. That is called an *ill-posed problem*, which is any problem where a tiny change in the input numbers causes a large change in the result. If it seems like a contrived problem, but it is not; ill-posed problems arise all the time in computer graphics, statistical curve-fitting, and many physics problems.

A traditional numerical analyst would sneer and tell me, "Well, that is what you get for using 16-bit precision to solve a linear system. Next time use double precision. That will almost certainly get you something close to 1.0 for *x* and *y*." My reply would be, "Say hello to my little friend:"

256–Bit Quire

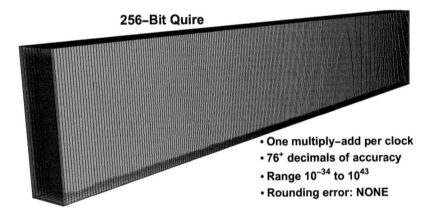

- **One multiply–add per clock**
- **76⁺ decimals of accuracy**
- **Range 10^{-34} to 10^{43}**
- **Rounding error: NONE**

But we are not going to replace all the posit values with quire values. We are going to use the quire to *measure* the error exactly. It is always a good idea when solving equations to plug in the answer and see if it really does solve the equations. We can compute the left-hand sides exactly, using the quire, because it is in the form of a dot product. The *residual* is the difference between the right-hand side values and what the equations evaluate to.

In the Compute Layer, the exact residual is as follows:

$$-\frac{671}{4096} - \left(\frac{2099}{4096} \times \frac{2205}{2048} - \frac{1385}{2048} \times \frac{2167}{2048} \right) = \frac{87}{8\,388\,608} \quad \text{and}$$

$$\frac{819}{8192} - \left(-\frac{159}{512} \times \frac{2205}{2048} + \frac{3363}{8192} \times \frac{2167}{2048} \right) = -\frac{789}{16\,777\,216}$$

Those quire values on the right-hand side (the residual values) are small and can be used to find a correction amount to the computed *x* and *y*. They convert to posit values $0.000010371\cdots$ and $-0.0000046968\cdots$.

Solve the original equations for correction amounts x_c and y_c, where we can re-use most of the arithmetic we did the first time instead of having to recompute it.

$$0.51245\,x_c \;+\; -0.67627\,y_c \;=\; 0.000010371 \quad \text{and}$$
$$-0.31054\,x_c \;+\; 0.41052\,y_c \;=\; -0.000046968$$

That works out to $x_c = -0.076548$ and $y_c = -0.058020$. Add those correction amounts to x and y and round, and you get $x = y = 1$. But we are not done yet; *again* compute the residual using this improved x and y, and you find the residual in the quire is zero. Which means we *proved* the answer, not merely guessed it. That is something floats cannot do, because they have no exact dot product capability.

Just for fun, I did try solving the equations using double-precision floats like the imaginary critic suggested. Here is what it produced:

$$x = 0.99999\,99999\,9996458\cdots$$
$$y = 0.99999\,99999\,9997324\cdots$$

This is a better guess than 16-bit posits (without using the quire), but it is still a guess, and it is wrong. Plus, there is no way to *know* it is wrong because you cannot check it with exact arithmetic. With a posit-quire environment, you can actually find out what is true and what is not, instead of being adrift in a sea of guesses.

More about the quire and linear equation solvers is ahead in Chapter 12.

6.5 Accurate Evaluation of Polynomials

Here is a simple third-degree (cubic) polynomial, $y = 9 + 3\cdot x - 5\cdot x^2 + x^3$:

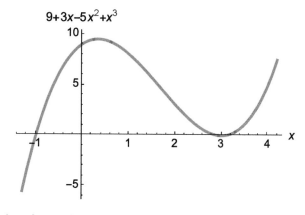

The polynomial has a simple root at $x = -1$ and double root at $x = 3$.

Computing powers is expensive, so polynomials are usually computed with *Horner's rule*, where you create a nested expression requiring only multiplies and adds:

$$9 + 3 \cdot x - 5 \cdot x^2 + x^3 \qquad = 9 + x \cdot \left(3 - 5 \cdot x + x^2\right)$$
$$= 9 + x \cdot \left(3 + x \cdot \left(-5 + x\right)\right).$$

When computing the points, we have to use a rounded add and two rounded (fused) multiply-adds, so there is some cumulative rounding error.

$$9 + 3 \cdot x - 5 \cdot x^2 + x^3 \approx \overline{9 + x \cdot \left(\overline{3 - x \cdot \left(\overline{-5 + x}\right)}\right)}$$

To exaggerate what can go wrong, try plotting using posits with only 10 bits:

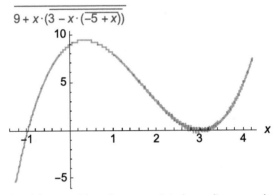

When computed with low precision, the accumulated rounding errors become apparent.

Since 10-bit posits only have 6 significant bits in the fovea, we can perhaps forgive the stair step bumpiness visible here, like for x between -1 and 1. However, something much worse is happening near $x = 3$, where the rounding is causing the result to bounce around. Zoom in on that region, where we allow x to be a fine-grained sampling of the real numbers and not restricted to the posit vocabulary:

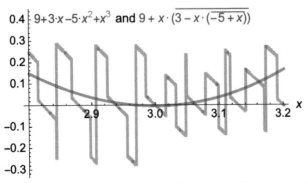

Correct value in blue, and the chaotic effects of rounding in magenta

Alarming. This happens with floats or posits *of any precision*; using higher precision just changes the scale of the vertical axis, but the chaos remains. The polynomial has a double root at $x = 3$, and Horner's rule magnifies relative error near compound roots. With *floats*, there is not much you can do about it. You know what is coming.

We can write Horner's rule as a series of assignments to temporary variables t_i:

$$t_0 = x$$
$$t_1 = -5 + t_0$$
$$t_2 = 3 + x \cdot t_1$$
$$t_3 = 9 + x \cdot t_2$$

Try it with the posit just smaller than 3, $\frac{47}{16} = 2.9375$, and round after every step:

$$t_0 = x = 2.9375$$
$$t_1 = \overline{-5 + t_0} = -2.0625 \text{ (exact)}$$
$$t_2 = \overline{3 + x \cdot t_1} = -3.0625 \text{ (exact would have been } -3.05859375)$$
$$t_3 = \overline{9 + x \cdot t_2} = 0.0390625 \text{ (exact would be } 0.01538 \cdots)$$

It is the last step that kills the relative accuracy, because it is the subtraction of two similar numbers, one of which is an input (assumed exact) and the other of which has accumulated rounding errors that could go either way in a jittery pattern.

But notice that the series of assignments is a system of linear equations in the temporary variables t_i. The t_0 and t_1 always come out right because the result lands on an exact posit near $x = 3$, but t_2 experiences rounding error. If you compute it in the quire, and then subtract that from the rounded t_2 result of -3.0625, you get ...a *residual*. Similarly with t_3. You can use those residuals to compute corrections to t_2 and t_3. After rounding the corrected t_3, that is the value to plot, at the cost of just a very few extra multiplies and adds and use of the quire. Here is what the corrected plot looks like, still jagged, but as good as you can do with 10-bit posits:

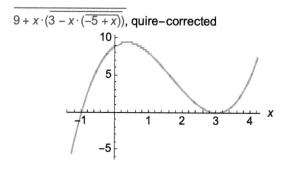

The quire allows correct evaluation to within 0.5 ULP everywhere.

Just to be clear, here is the section that was chaotic before, and this time we have to zoom in all the way to $2.999 < x < 3.001$ to be able to see the jaggedness clearly. Even though they have only 10 bits, these posits go all the way down to about 2×10^{-10}, so the curve is very smooth where it grazes the x-axis even at this high magnification.

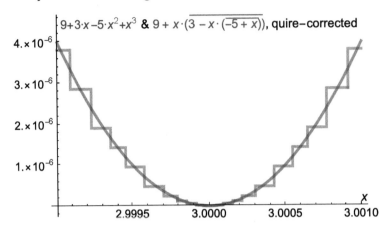

What was chaotic is now the smoothest part of the discrete plot.

6.6 The Unevaluated Sums Approach

Suppose we have fast hardware for 16-bit posits, but no higher precision, and we need a little extra precision for a few calculations involving multiplies, divides, and square roots. The quire provides a way to operate at a higher precision for those operations, not just for dot products. In the fovea, 16-bit posits have 12 significant bits (about 3.7 decimals) of relative accuracy. But suppose we want to evaluate $y = x^2 \big/ \sqrt{1 + x^3}$ to three decimals, for $x = 42$. The problem is having to evaluate $x^3 = 42 \times 42 \times 42 = 74\,088$, which is well outside the fovea. The closest posit value is 74240, represented by the posit bit pattern 0111110000010001. The 6-bit regime crowds fraction bits off the end of the posit, leaving us with only 8 bits of significance. That low relative accuracy propagates to the rest of the calculation.

The correct value for y is $6.480696\cdots$ and the closest 16-bit posit to that value is $\frac{1659}{256} = 6.4804\cdots$. But we have to round six times. Computing $y = \overline{x \cdot x} \big/ (\sqrt{1 + \overline{x \cdot x \cdot x}})$ yields $\frac{1657}{256} = 6.4726\cdots$ with only two decimals correct, off by 4 ULPs. The quire provides a way we can do better.

6.6.1 Evaluate the Cube

Computing the square $42 \times 42 = 1764$ loses no accuracy with 16-bit posits. The posit bit pattern for 1764 is **0111010101110010**, and no significant **1** bits are lost. But computing the cube of 42 will lose accuracy.

Here is a commented code that shows how the quire can help:

```
setPositEnv[{16, 2}]
pSquared = 42 * 42; (* The rounding changes nothing. *)
quire = pSquared * 42 (* The quire does this without rounding.*)
pHigh = quire ; (* Round the cube to the nearest posit. *)
quire = quire - pHigh; (* Scrape off the high-order part. *)
pLow = quire ; (* Convert the low-order part to a posit.*)
{pHigh, pLow} (* Maintain the cube as an unevaluated sum. *)
```

74 088

{74 240, −152}

This is our first example of an *unevaluated sum* result. The correct value, 74 088, is expressible exactly as the sum of two posit16 values, 74 240 + (−152), but we keep the two values separate instead of computing their sum. We have not lost any accuracy yet.

6.6.2 Add 1 to the Cube

In general, the safe way to do an addition to an unevaluated sum is as follows:

```
quire = 1 + pHigh + pLow; (* Quire sums do no rounding. *)
pHigh = quire ; (* Round the sum to the nearest posit. *)
quire = quire - pHigh; (* Scrape off the high-order part. *)
pLow = quire ; (* Convert low-order part to a posit. *)
{pHigh, pLow} (* Maintain the cube as an unevaluated sum. *)
```

{74 240, −151}

This does not assume that the low-order part, −151, can be represented exactly. As it turns out, it can because we are inside the range *−pIntMax* to *pIntMax*, which for posit16 is −1024 to 1024.

6.6.3 Take the Square Root with Extra Accuracy

The square root of 1.02 is very close to 1.01, and the square root of 1.01 is very close to 1.005. In general, if the square root of x is exactly r, then the square root of $x + h$ is very close to $r + \frac{1}{2} h/r$ if h is small relative to x. To see why, write out the expansion of $\left(r + \frac{1}{2} h/r\right)^2$ and simplify it; you get $r^2 + h$ and a lower-order term. If you apply that reasoning for the square root, you get

$$\sqrt{x_{high} + x_{low}} \approx r + \frac{1}{2}\left(x_{high} + x_{low} - r \cdot r\right)/r.$$

That is the basis for the following approach to getting a more accurate square root as an unevaluated sum. Finding $x + h - r \cdot r$ can be done in the quire with no rounding.

```
pRootHigh = √pHigh  (* Find first approximation. *)
quire = pHigh + pLow - pRootHigh * pRootHigh;  (* No rounding. *)
pRootLow = quire / (2 * pRootHigh)  (* Correction amount. *)
```

$$\frac{545}{2}$$

$$-\frac{1257}{4096}$$

The correct value of $\sqrt{74089}$ is $272.1929\cdots$ and the sum of the posit values for the high order and low order part of square root above is $272.1931\cdots$, an approximation good to five decimals.

6.6.4 Do the Final Division

We already found the numerator, **pSquared** = 1764. Dividing with unevaluated sums looks a lot like the square root method. Divide the high-order parts, then use the quire to find the correction term. Suppose r is the approximation to x_{high}/y_{high}. Then

$$\frac{x_{high}+x_{low}}{y_{high}+y_{low}} \approx r + \left(x_{high} + x_{low} - r \cdot y_{high} - r \cdot y_{lo}\right)/r.$$

In this example, the numerator does not need an x_{low} part because x_{high} was exact.

The following finishes the more accurate evaluation of the expression by performing the final division operation:

```
yHigh = pSquared / pRootHigh; (* Find quotient, high-order. *)
quire = pSquared – yHigh * pRootHigh – yHigh * pRootLow;

yLow = quire / pRootHigh; (* Correction amount. *)
{yHigh, yLow}
```

$$\left\{ \frac{1657}{256}, \frac{263}{32\,768} \right\}$$

The first posit value, $1657/256 \approx 6.4734$, was the closest posit to the exact value of $x^2 \big/ \sqrt{1 + x^3}$ for $x = 42$, so it is as if the result was a fused calculation (no rounding until the last operation). But we have something better. The unevaluated sum $\{y_{high}, y_{low}\} = \left\{ \frac{1657}{(256}, \frac{263}{32\,768} \right\}$ sums to $6.480682\cdots$ which is quite close to the correct value $6.480696\cdots$.

6.6.5 What Would Floats Do?

While it may seem unfair and repetitive to criticize IEEE floats here, it is important to point out what a half-precision IEEE float would do ("binary16" in the nomenclature of IEEE Std 754). What does $y = x^2 \big/ \sqrt{1 + x^3}$ evaluate to with $x = 42$, using standard 16-bit floats?

Zero. You will get $y = 0$. The relative error is infinite.

For example, the following code evaluates the expression for y using standard binary16 floats, which have 5 bits of exponent:

```
setFloatEnv[{16, 5}]
x = 42;

y = x * x / √(1 + x * x * x)
```

0

So "WWFD" (What Would Floats Do?) is probably *not* a good idea for a slogan cap.

The answer is all too often, "Probably fail, without telling you that they failed."

6.6.6 Duos, Trios,…Why Stop at Two?

The term *unevaluated sum* is a mouthful, and the approach works when the uneval-uated sum has more than just a high-order and low-order part. If it has only two parts, call it a *duo*. It could have a first-order, second-order, and third-order part. Call that a *trio*. Short, easy words. Here is a proposed nomenclature, inspired by music terminology:

Number of Parts	Name
1	Solo
2	Duo
3	Trio
4	Quartet
5	Quintet

The quire enables low-precision posits to "punch above their weight class." Remember that 8-bit standard posits have just over one decimal of relative accuracy, at *best*, but a quintet of them can do quite a decent job of expressing $\pi = 3.1415926\cdots$, as follows:

```
setPositEnv[{8, 2}];
pi1 = π̄;
pi2 = π - pi1;
pi3 = π - pi1 - pi2;
pi4 = π - pi1 - pi2 - pi3;
pi5 = π - pi1 - pi2 - pi3 - pi4;
{pi1, pi2, pi3, pi4, pi5}
```

$$\left\{ \frac{13}{4}, -\frac{7}{64}, \frac{1}{1024}, -\frac{1}{65\,536}, \frac{1}{262\,144} \right\}$$

If that quintet of posit8 values were summed in the quire, the quire value would be 3.14159**011** ⋯, which is π correct to six decimals, even though each posit has very low precision.

Imagine that you need to know the value of 60° in radians, which is $\pi/3$ or $\pi \cdot (1/3)$. Furthermore, you need it correct to a relative error less than 0.0005 and all you have is support for 8-bit posit arithmetic. The correct value is 1.047⋯.

First, create a trio for 1/3:

```
third1 = 1 / 3;
third2 = 1 / 3 - third1;
third3 = 1 / 3 - third1 - third2;
{third1, third2, third3}
```

$$\left\{ \frac{11}{32}, -\frac{5}{512}, -\frac{3}{4096} \right\}$$

That trio sums to 0.33325 ⋯. We can use the first three π terms to make a trio approximating π.

```
N[pi1 + pi2 + pi3]
```

3.14160 1563

Algebraically, the *evaluated* product of the trio for π and the trio for 1/3 is $pi_1 \cdot third_1 + pi_1 \cdot third_2 + pi_2 \cdot third_1 + \ldots + pi_3 \cdot third_3$, a total of nine terms. That is a dot product. Therefore, we can evaluate it exactly using the quire:

```
quire = (pi1 * third1) +
   (pi1 * third2 + pi2 * third1) +
   (pi1 * third3 + pi2 * third2 + pi3 * third1)
(*+ (pi2*third3+pi3*third2) + (pi3*third3) *)
N[quire]
```

$$\frac{67}{64}$$

1.046875

The products are grouped in parentheses by first-order, second-order, and so on, and commented out at the end because the products of small numbers are too small to contribute much to the result and can be ignored.

Find the radian answer as an unevaluated sum of posits:

```
radian1 = quire;
radian2 = quire - radian1;
radian3 = quire - radian1 - radian2;
{radian1, radian2, radian3}
```

$$\left\{ 1, \frac{3}{64}, 0 \right\}$$

Because `radian3` is zero, all we need is a duo to express the radian answer: $\left\{ 1, \frac{3}{64} \right\}$. The sum, 1.046875, has a relative error compared to $\pi/3$ of about 0.0003, satisfying the accuracy requirement.

6.7 Can the Quire Fuse Any "Basic Block"?

The term *basic block* is from the design of compilers, the software tools that turn human-readable computer languages into machine-level executable code.

> Definition: **basic block**
>
> *computer science*
> A straight-line sequence of operations with no branches. It has a single entry point and a single exit point.

Decomposing a program into basic blocks is usually the first thing a compiler does, no matter what language. If we stick to the subset of blocks where there are only plus-minus-times-divide operations and all the values involved are real numbers, then the basic block is the same as a mathematical *expression*, a formula.

Can the quire help reduce the cumulative rounding error of *any* such basic block? That sounds a little like black magic, or too good to be true. Yet, incredibly, *it can*; that is part of the foundation of Kulisch's KAA used for the C-XSC and PASCAL-XSC programming environments. To use an example from Kulisch, suppose we want to compute the following expression that uses all four basic arithmetic operations, where a, b, c, d, and e are real (posit) input values and e is not zero:

$$X = (a + b) \times c - \frac{d}{e}$$

To express that as a basic block, introduce temporary values t_i for each step:

$$
\begin{aligned}
t_1 &= a \\
t_2 &= t_1 + b \\
t_3 &= c \times t_2 \\
t_4 &= d \\
e \times t_5 &= t_4 \\
t_6 &= t_3 - t_5
\end{aligned}
$$

Notice how the divide was handled: $t_5 = t_4/e$ was multiplied by the denominator e to get $e \times t_5 = t_4$, so now there are only plus-minus-times operations in the six-step sequence. The sequence constitutes a system of six linear equations in six unknowns. To make that more obvious, rearrange each equation so the t_i values are all on the left, and the matching t_i unknowns line up in columns:

$$
\begin{aligned}
t_1 & & & & & & = a \\
-t_1 &+ t_2 & & & & & = b \\
& c \times t_2 &- t_3 & & & & = 0 \\
& & & t_4 & & & = d \\
& & & -t_4 &+ e \times t_5 & & = 0 \\
& & -t_3 & & t_5 &- t_6 & = 0
\end{aligned}
$$

Recall how we solved two equations in two unknowns, then found the residual (the defect in the temporary values) and used it to find corrections to the computed values? That same approach works here. You might have to apply it more than once, but very rarely more than twice because it usually converges very fast.

But not always. This approach is, unfortunately, not foolproof. Some computations are so ill-posed that the corrections are not small compared to the original guess, and the process does not converge. I wondered if it could be used to compute the correctly-rounded value of $a \times b \times c$, even for very low-precision posits. That is, the fused multiply-multiply **fMM** in the Posit Standard. If I use 4-bit posits with 0 bits of exponent, there are only 7 positive values in the vocabulary: $\{\frac{1}{4}, \frac{1}{2}, \frac{3}{4}, 1, \frac{3}{2}, 2, 4\}$. That means $7^3 = 343$ possible input combinations to $a \times b \times c$. If we simply compute $\overline{a \times b} \times c$, with two roundings, it differs from $\overline{a \times b \times c}$ in 59 cases of the 343, or about 17 percent of the cases.

If we use the quire, only six cases fail to round correctly: the three permutations of $\{a, b, c\} = \{\frac{3}{4}, \frac{3}{4}, \frac{3}{2}\}$ and the three permutations of $\{a, b, c\} = \{\frac{3}{4}, \frac{3}{2}, \frac{3}{2}\}$. That is a big improvement over 59 double-rounding errors, but it is not perfect. Hence, we still need the **fMM** function to perform $a \times b \times c$ entirely in the Math Layer, and then round.

6.8 The Importance of Getting the Sign Right

Unless you are one of the rare people who actually have to solve quadratic equations in your adult life after completing school, you probably have to look up the quadratic formula for roots r_1 and r_2 of a quadratic equation $a \cdot x^2 + b \cdot x + c = 0$ so that you can write the polynomial as $(x - r_1) \cdot (x - r_2)$. Here is a refresher:

$$r_1, r_2 = \frac{-b \pm \sqrt{b^2 - 4 \cdot a \cdot c}}{2 \cdot a}$$

The $b^2 - 4 \cdot a \cdot c$ part is the *discriminant* because it discriminates between three cases:

- If less than zero, there are no real roots (the roots r_1 and r_2 are complex valued and the imaginary part is nonzero).
- If equal to zero, the two roots are identical (also called a double root), $r_1 = r_2$.
- If greater than zero, there are two distinct real roots; $r_1 \neq r_2$.

To keep the number of decimals small in this example, suppose we are using 16-bit posits, with $a = 2.9590$, $b = -7.9629$, and $c = 5.3574$. The discriminant $b^2 - 4 \cdot a \cdot c$ is negative, $-0.0024\cdots$, so there are *no real roots*. But if you compute by rounding after every multiply, you find b^2 rounds to $\overline{b \times b} = 63.40625$, which is the same value you get computing the rounded $\overline{4 \times \overline{a \times c}}$. That would say you have a double root instead of no real roots at all, potentially a catastrophic numerical mistake.

Instead use the quire, where multiplying $a \cdot c$ by 4 is done with additions:

```
setPositEnv[{16, 2}];
{a, b, c} = {2.9590, -7.9629, 5.3574}; (* Round to posit. *)
quire = a * c;
quire = quire + quire; (* 2·a·c *)
quire = quire + quire; (* 4·a·c *)
quire = b * b - quire;

discriminant = quire; (* Round the discriminant to a posit *)
Print[N[discriminant, 3]] (* result to 3 significant digits *)
```

-0.00248

Besides preventing the catastrophe of getting the sign of the discriminant wrong, the quire does a pretty good job of computing its value in case we want to know what the two complex roots are.

Just out of curiosity, I tried the same problem but with 16-bit floats. Pretend we have something like a quire for floats, and do the computation in the Math Layer:

```
setFloatEnv[{16, 5}]; (* IEEE Standard 16-bit precision *)
{a, b, c} = {2.9590, -7.9629, 5.3574}; (* Round to float. *)
quire = a * c; (* Could use a 64-bit float as 'quire', maybe. *)
quire = quire + quire; (* 2·a·c *)
quire = quire + quire; (* 4·a·c *)
quire = b * b - quire;
discriminant = quire; (* Round the discriminant to a float *)
Print[N[discriminant, 3]] (* result to 3 significant digits *)
```

0.0518

Ugh. Even worse, floats flipped the sign of the discriminant to positive. Most of that error was from inaccurate conversions of the five-decimal inputs to binary16 floats.

To five decimals, 16-bit posits preserved every digit:

$$N\left[\left\{\overline{2.9590},\ \overline{-7.9629},\ \overline{5.3574}\right\},\ 5\right]$$

{2.9590, -7.9629, 5.3574}

But rounding to the nearest binary16 float was lossy for *b* and *c* input values:

$$N\left[\left\{2.9590,\ \underline{-7.9629},\ \underline{5.3574}\right\},\ 5\right]$$

{2.9590, -7.9648, 5.3555}

Another example of the importance of getting the sign right comes from computational geometry. In a plane, suppose you have a directed line that passes through point *A* to point *B*, and *C* is another point on the plane. How do you use the coordinates of the points to compute whether *C* is to the *left* of the line, *on* the line, or to the *right* of the line?

You need this, for instance, to know if a point lies inside a polygon or outside, or whether two line segments intersect.

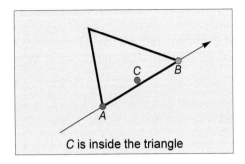

C is inside the triangle

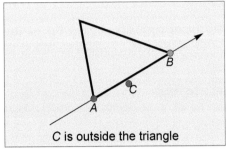

C is outside the triangle

It is crucial for computer graphics to know when a ray passes through a triangle or outside of it.

The test is to compute the *cross product* of the vector from A to B, $\overrightarrow{AB}$, with the vector from A to C, $\overrightarrow{AC}$. That produces a vector in the z direction, which is positive if C is left of the line and negative if C is right of the line. The magnitude of the cross product is the area of the parallelogram formed by the $\overrightarrow{AB}$ and $\overrightarrow{AC}$ vectors:

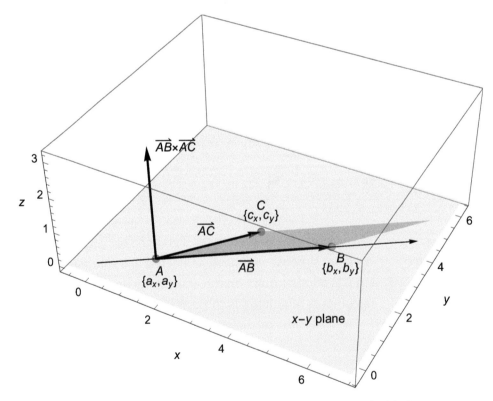

The cross product is a vector in the positive z direction; C is left of the line

The cross product G of two vectors in the x-y plane $\{x_1, y_1, 0\}$ and $\{x_2, y_2, 0\}$ is $\{0, 0, x_1 \cdot y_2 - x_2 \, y_1\}$. In this case, $\{x_1, y_1\} = \vec{AB} = \{b_x - a_x, b_y - a_y\}$. Similarly, the vector $\{x_2, y_2\} = \vec{AC} = \{c_x - a_x, c_y - a_y\}$.

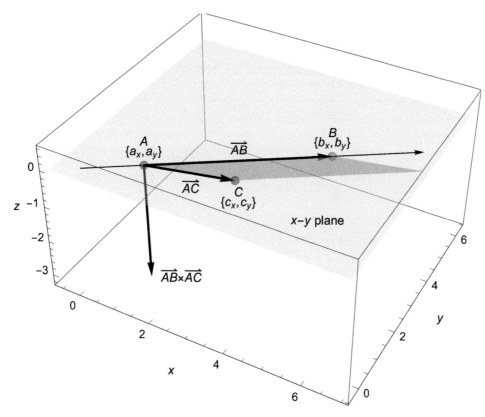

The cross product is a vector in the negative z direction; C is right of the line

$$
\begin{aligned}
G &= (b_x - a_x) \cdot (c_y - a_y) - (b_y - a_y) \cdot (c_x - a_x) \\
&= a_x \cdot b_y - a_y \cdot b_x + a_y \cdot c_x - a_x \cdot c_y + b_x \cdot c_y - b_y \cdot c_x
\end{aligned}
$$

By now, you should recognize that second line as quire fodder: nothing but multiply-accumulates, which the quire can do perfectly. It is easy to see how wrong this might be with floats, even using fused multiply-add operations, since you accumulate at least six rounding errors. Furthermore, floats underflow to zero, which gives the false result that a point close to the line is actually on it exactly. You may have seen computer graphics artifacts where light (or the wrong color) seems to peep through the crack between two polygons that butt together. You are seeing float rounding error. One approach is to crank up the precision so that the "light leaks" happen less often, but that is not an actual solution to the problem.

Set the floats to binary32 (single precision), the usual precision for computer graphics. Pick four random numbers for the coordinates of points *A* and *B*, rounded to the nearest float using the underscore operator.

```
setFloatEnv[{32, 8}]; (* 32-bit precision, 8-bit exponent *)
SeedRandom[2718281]; (* Seed the random number generator *)
{ax, ay, bx, by} = RandomReal[{-1, 1}, 4];
N[%, 8] (* binary32 is uniquely described by 8 decimals *)
```

{0.68277556, −0.47821945, −0.47052243, −0.16485038}

Now place *C* exactly 2/3 of the way from *A* to *B*, without rounding (since *Mathematica* supports exact rational arithmetic). Check that the cross product places it exactly on the line defined by *A* and *B*:

```
{cx, cy} = (2 * {ax, ay} + {bx, by}) / 3
ax * by – ay * bx + ay * cx – ax * cy + bx * cy – by * cx
```

$$\left\{ \frac{30\,032\,179}{100\,663\,296}, -\frac{37\,624\,225}{100\,663\,296} \right\}$$

0

But rounding changes the *C* coordinates ever so slightly. Zoom in on the grid of possible floating-point *x* and *y* coordinates and you can see that rounding moves *C* up and to the left, so it is now just above (to the right of) the directed line from *A* to *B* (shown as a blue arrow):

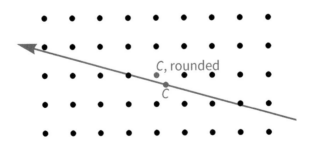

The correct point *C*, in blue, is rounded to the grid point shown in red, which is not on the ray.

Use **cxr** and **cyr** for the coordinates of the rounded *C* shown in red. Here is the value of the cross product test, to three decimals:

{cxr, cyr} = {cx, cy}

N[ax * by – ay * bx + ay * cxr – ax * cyr + bx * cyr – by * cxr, 3]

$$\left\{ \frac{5\,005\,363}{16\,777\,216}, \ -\frac{391\,919}{1\,048\,576} \right\}$$

-8.34×10^{-9}

The cross product is negative, which proves that rounded *C* is to the right of the ray from *A* to *B*. But that was an exact calculation of the cross product. What happens if we round after the first multiplication and the five fused multiply-adds?

N[ax * by – ay * bx + ay * cx – ax * cy + bx * cy – by * cx, 3]

2.39×10^{-9}

It went *positive*, indicating *C* is *left* of the line defined by *A* and *B*, which it is not. If a ray-tracing program fires a ray very close to the edge of a polygon, it can very easily incorrectly conclude that the ray hit the polygon when it did not, or vice versa, even with over seven decimals of accuracy. The same thing can happen with 64-bit floats and 128-bit floats. It *cannot* happen with posits if the cross product is computed using the quire. This suggests that a Graphics Processing Unit (GPU) using posits could produce better images with low precision than a GPU using 32-bit floats and generating light leaks and other ray-tracing mistakes.

There are some situations where you *really* do not want to be using approximations.

Man on edge of roof

Man on edge of roof, approximately

7 A Bridge to Sane Floats

"You realize, of course, that this means war."

Those invested in IEEE float format regard posit format as a threat.
This chapter shows a path to an excellent compromise.

It has been many decades since the Intel i8087 coprocessor became the definition of floating-point everywhere, and we have learned a lot from the experience. Floats have one advantage over posits: The fraction is always in the same set of bits, so hardware is slightly more parallel in that respect. It is possible to decode the exponent in parallel with decoding the fraction, which decreases the delay time for an operation.

What if we can create a format that has the fixed-field property of floats but otherwise repair the ideas of IEEE Std 754 floats that were dubious to begin with and also have not aged well?

7.1 A New Parameter: Maximum Regime Size rS

While standard posits set $eS = 2$, it is still a variable parameter we can experiment with if we want to customize a number system for a particular application.

Define *rS* as the *maximum regime size,* a number less than the posit precision:

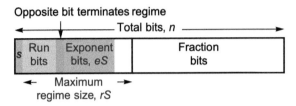

Regime is terminated the usual way, by an opposite bit.

The regime can still be terminated by an opposite bit, but it could also be terminated by hitting the maximum allowed size *rS*. Then there is no opposite bit, and the
next bit begins the exponent bit field.

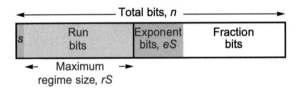

Regime is instead terminated by reaching the maximum size, *rS*.

As usual, we use low precision to make the idea easier to visualize. Create a new
version of **setPositEnv**, with three parameters: *nBitsP, rS,* and *eS*; if we make *rS* one
less than the precision, it is the same format as the posit format presented so far.

```
Clear[setPositEnv]
setPositEnv[{n_Integer /; n ≥ 2,
    r_Integer /; r ≥ 1, e_Integer /; e ≥ 0}] := Module[{t},
  {nBitsP, rS, eS} = {n, Min[r, n - 1], e};
  nPat = BitShiftLeft[1, n];
  NaR = -BitShiftRight[nPat];
  pMask = nPat - 1;
  eSizeP = BitShiftLeft[1, eS];
  uSeed = BitShiftLeft[1, eSizeP];
  t = nBitsP - rS - 1;
  {minPos, maxPos} = Which[
    t == 0, {uSeed^{1-rS}, uSeed^{rS-1}},
    t ≤ e, {uSeed^{-rS} * 2^{2^{e-t}}, uSeed^{rS} * 2^{-2^{e-t}}},
    True, {uSeed^{-rS} * (1 + 2^{e-t}), uSeed^{rS} / 2 * (2 - 2^{e-t})}];
]
```

Changes are shown in **blue**. While integer powers of two still have perfect reciprocals, whenever there are nonzero fraction bits in *minPos* and *maxPos*, they will no
longer be exact reciprocals of each other. They are *close* to being reciprocals; their
product is always in the range $1 ≤ maxPos \times minPos ≤ 1.125$.

Here is a worst-case example:

```
setPositEnv[{nBitsP = 4, rS = 2, eS = 0}]
{minPos, maxPos}
N[minPos × maxPos]
```

$$\left\{\frac{3}{8}, 3\right\}$$

```
1.125
```

The **pToX** and **xToP** functions must also be altered slightly, to switch to decoding exponent bits if there is either an opposite bit or the *rS* limit is reached. Putting a limit on *rS* has the effect of blunting the tapering of the accuracy plot. Look at 16-bit standard posits, with *rS* = 15:

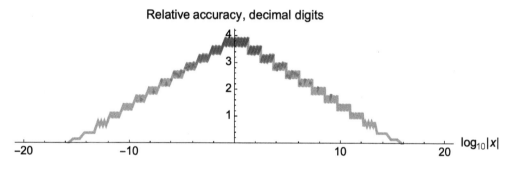

Accuracy plot for standard 16-bit posits, with no limits on the regime length

Now limit the number of regime bits to 8, which cuts the dynamic range in half. To compensate, increase the exponent size to *eS* = 3 to double the dynamic range:

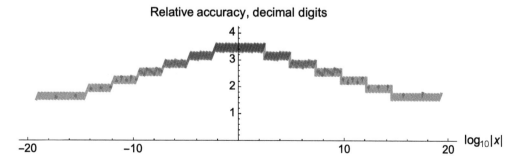

Limiting the regime to a maximum of 8 bits blunts the tapering and raises the outermost accuracy.

At the cost of 0.3 decimal accuracy in the fovea, we no longer drop all the way to zero decimal digits at the extreme magnitudes because there are always at least four fraction bits. That also has the advantage of making the "Twilight Zone" go away since the rounding bit is always a fraction bit. Notice that the width of the lowest stair steps on the left and right are twice as wide as the other stair steps outside the fovea. Here is the accuracy plot with $rS = 4$ restricting the regime length:

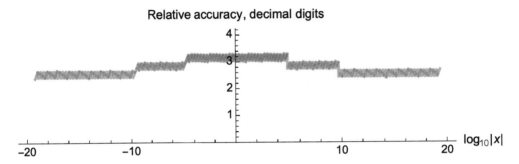

Further limiting the maximum regime to 4 bits flattens the tapered accuracy to only three levels.

If we restrict the number of regime bits to 2 (one of 00, 01, 10, or 11, representing −2, −1, 0, or 1 respectively) and increase the number of exponent bits to 5, a remarkable thing happens. We keep the dynamic range about the same, but all the tapering goes away. Except for the wobble, relative accuracy is flat:

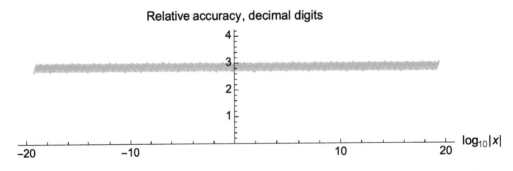

Further limiting the maximum regime to 4 bits flattens the tapered accuracy to only three levels.

Remember that an IEEE standard 16-bit float has five bits of exponent. It also has a subnormal region, so it has a ramp on the left end of its accuracy plot:

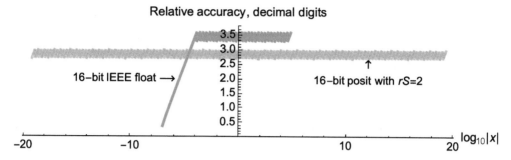

The untapered posit uses 7 bits for the exponent description, compared to 5 bits for IEEE 754 16-bit floats.

Here are the four posit graphs overlapped to make them easier to compare:

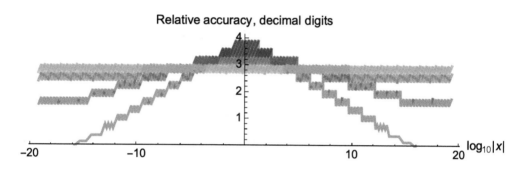

This shows the four examples of regime-limited posits; $rS = 15$ (no limit), 8, 4, and 2.

If $rS = 2$, the fraction always begins at the same place and has the same number of bits. In other words, it behaves like a float!

Definition: A *sane float* is a posit format with the regime size rS restricted to two bits, so there is no tapering of relative accuracy.

However, look how much better the 16-bit "sane float" is than an IEEE Std 754 float with the same precision:

- Accuracy is absolutely flat, without a subnormal exception region on the left.
- The only exception values are 0 and NaR. And there is only one NaR bit pattern.
- Rounding is always banker's rounding, with no hidden modes.
- Sane floats map monotonically to 2's complement integers.
- There is no "negative zero."
- Negation is performed by negating the bit string as if it were an integer.
- The dynamic range is about 10^{-19} to 10^{19}, which covers the region needed for most scientific computing and AI.
- *minReal* $\approx 1/maxReal$ and reciprocation is perfect for integer powers of 2.
- Changing precisions is as easy as with any other kind of posit: Append **0** bits to increase precision, or round off bits with banker's rounding to decrease precision.

The power-of-two scaling is still expressed with both the regime and the exponent bit fields. With $eS = 5$, regime bits express the power of $uSeed = 2^{2^5} = 2^{32}$.

Regime bits	r	Scaling represented
00	-2	$(2^{32})^{-2} = 2^{-64}$
01	-1	$(2^{32})^{-1} = 2^{-32}$
10	0	$(2^{32})^{0} = 1$
11	1	$(2^{32})^{1} = 2^{32}$

The regime-exponent combination works like an IEEE exponent field with a bias of 64, but no exceptions.

As usual, the exponent bits fill in the gaps between the jumps by 32 in the power-of-two scaling. The smallest scaling is when regime and exponent are all 0 bits and that represents –64, and the largest scaling is when regime and exponent are all 1 bits and that represents +63, so we can regard the regime and exponent fields as a single field of positional notation binary with bias of 64.

We have seen that some of the claims about float accuracy fall apart in the subnormal region, so it is an advantage not to have one; it also simplifies the hardware. Recall what originally created the need for subnormals that created such heated debate in the IEEE 754 committee meetings: Without subnormals, you can have two different float values x and y for which $x – y$ underflows to zero, making it look like $x = y$. Subnormals were a kludge created to repair the choice to underflow too-small numbers to zero. Posits (and therefore sane floats) do not underflow to zero, so a need for subnormals does not arise.

In Chapter 3, we had this formula for the value represented by a posit:

$$x = \left((1 - 3s) + f\right) \times 2^{(1-2s)\times\left(2^{eS}\times r+e+s\right)}.$$

where s is the value of the sign bit, f is the fraction, e is the value of the exponent bits, r is the value represented by the regime bits, and of course eS is the exponent size. With the new rS parameter, this formula really does not need to change other than to call out the two exception values, because a run of 0 bits to the end of the number cannot mean $r = -\infty$. So to be careful, we should change the formula:

$$x = \begin{cases} 0 & \text{if all fields are \textbf{0} bits} \\ \text{NaR} & \text{if all fields except for the sign are \textbf{0} bits} \\ \left((1-3s)+f\right) \times 2^{(1-2s)\times\left(2^{eS}\times r+e+s\right)} & \text{otherwise} \end{cases}$$

7.2 A Number Distribution for HPC Applications

Now we have quite a few ways to use n bits to represent real numbers. How do we choose? The end of Chapter 4 showed a list of design goals for a real number format, the last of which was this one:

11. The results of application computations should be as close to results for exact computations as possible.

> We need a working mathematical definition of what "close" means so that we can compute it. It is obviously dependent on the application in question; AI, signal processing, and HPC applications have very different number distributions.

Here is a straw-man proposal for a representative set of values needed for High-Performance Computing (HPC): A normal distribution (bell curve) of magnitudes, centered at zero and with a standard deviation of three decades, and equal likelihood of being positive or negative. Create a set of, say, $2^{16} = 65\,536$ values with that distribution, as high-precision reals (like 64-bit accuracy; it just has to be many bits higher than the format being tested). Call it **RRSet** for Representative Real Set.

Here is some *Mathematica* code to generate such a set, and a histogram of the values with each bin representing one order of magnitude (a decade). Some studies have found that it is very rare to use values outside the range $10^{-13} \le |x| \le 10^{13}$ in production HPC codes, and this distribution is entirely inside that range, barely.

```
len = 2^16; SeedRandom[31 416];
log10RRSet =
  Sort[Table[RandomVariate[NormalDistribution[0, 3]], {i, 1, len}]];
logRRHist = Histogram[log10RRSet, {1},
  LabelStyle → {12}, AspectRatio → .4]
RRSet = Sort[10^log10RRSet *
    Table[(2 * RandomInteger[{0, 1}] - 1), {i, 1, len}]];
```

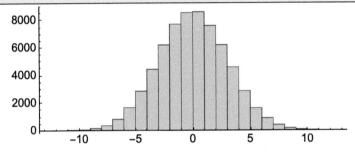

Distribution of values in RRSet, as a function of the $\log_{10}$ of the magnitude

The length was chosen as 65 536 by repeatedly doubling the number of values until further doublings did not alter the measurements to two-decimal accuracy. Why choose a normal distribution? Here is one argument for doing so. You may have seen one of these... they are popular projects for science fairs because they are easy to build. It is called a *Galton Board*:

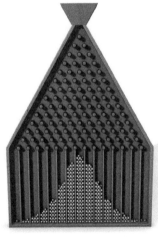

A Galton Board shows one way that bell-shaped curves arise.

It is a way of showing one way a bell curve distribution can happen. Beads fall from the top funnel and have a 50-50 chance of going left or right at each peg they hit, which converges to the normal distribution (also called a Gaussian distribution or bell curve). Now imagine we have a program like a physics simulation that has billions of input points, and as they go through all the arithmetic operations, sometimes the result is larger than the input and sometimes it is smaller. Which is more common? It is hard to guess, but you can see how it might similarly lead to a bell-shaped curve of all the values that arise in an extended calculation involving trillions of operations. It seems highly unlikely that the number magnitudes would be evenly distributed from the smallest to the largest.

With this **RRSet**, we can now start to do quantitative fidelity comparisons of different real number formats that have the same number of bits.

7.3 Fidelity: A Figure of Merit for a Format

A fundamental question to ask is how good a format is at representing the high-precision values in **RRSet**. It is not hard to test every value x in **RRSet**; convert it to the format in question by rounding to $x_{rounded}$, and find the average relative error for all of **RRSet**. Call that $r_{average}$. That lets us compute the *fidelity* in decimals, defined as follows:

> **Definition:** *Fidelity* is $-\log_{10}(r_{average})$ decimals, where $r_{average}$ is the average relative error of converting a high-precision representative set of reals to a lower precision real number format.

This often does not work for standard IEEE floats, because they can easily produce infinitely large relative error when a small x underflows to zero or a large x overflows to infinity by the rules of IEEE Std 754. Since **RRSet** magnitudes happen to be within the range 10^{-20} to 10^{20} and single precision IEEE floats range from about 10^{-45} to 10^{38} in magnitude, that should be OK, but half-precision IEEE floats would experience many infinite relative errors trying to represent **RRSet**, resulting in a fidelity of $-\infty$ decimals.

That cannot happen with posits, since the relative error caused by rounding is always finite. As specified by the Posit Standard, 16-bit standard posits have a dynamic range from about 10^{-17} to 10^{17} so some of the relative errors representing **RRSet** will be large, but that only happens for rare cases. You can start to see how this guides us in engineering a format: If it is important never to have a relative error of more than 0.1, say, then that points to adjusting *eS* and *rS* to give better representation of extreme magnitude values, at the expense of less accuracy in the fovea. Instead of having committees (or individual engineers) argue about which choice looks better based on their intuition, we can actually compute a figure of merit.

Compute the fidelity of standard 16-bit posits using the preceding definition. Here is a plot showing all the relative errors over the range of reals in **RRSet**, the average relative error, and the fidelity in decimals:

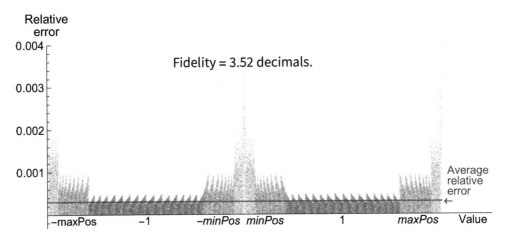

Relative errors for the sorted reals in **RRSet**, using standard 16-bit posits

The bottom of the distributions for negative and positive reals is where the magnitudes are in the fovea, which is $\frac{1}{16} \le |x| \le 16$ for the standard *eS* = 2. Adjacent to that is the region where there is one less bit of significance, so the relative error doubles. It doubles again for the regions near the extreme magnitudes, and some of the dots are off the scale in the vertical direction.

Another way to display the relative errors is to sort them from smallest to largest, and plot them along with the average relative error and the maximum relative error. Notice that the maximum relative error is well above the highest point plotted; it comes from the rare cases where the entry in **RRSet** is very near the saturation values ±*maxPos* or ±*minPos*.

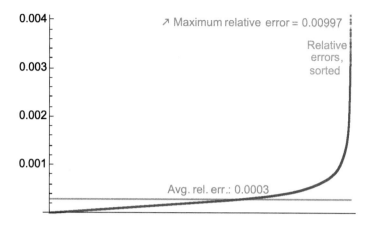

Sorted relative errors for representing **RRSet** using standard 16-bit posits

This book uses the convention that fully-tapered posits are plotted in magenta, floats of any kind are plotted in green, and now we have a way of stepping from posits to floats by adjusting rS and eS, with interpolated colors like those in the decimal accuracy plots in Section 7.1. Here is the 16-bit sane float with a dynamic range similar to the 16-bit standard posit, using $rS = 2$ and $eS = 6$:

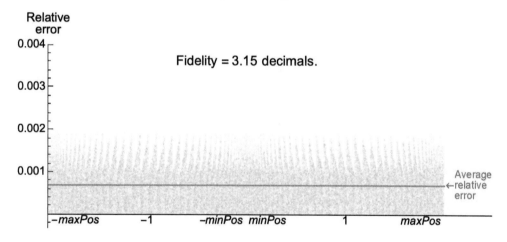

Relative errors for the sorted reals in **RRSet**, using 16-bit sane floats

Not surprisingly, relative accuracy is flat across the entire dynamic range, if we ignore the jaggedness caused by wobbling within each binade. With IEEE floats, there would be a ramped notch to zero in the center of the scatter plot because of the subnormals, and a reduced dynamic range, since the largest exponent is spent on non-real values.

The fidelity for sane floats of 3.15 decimals is significantly lower than the 3.52 decimals for posits. Put another way, posits are $10^{(3.52-3.15)} = 10^{0.37} \approx 2.3$ times more accurate than sane floats. If the methodology for testing is sound, we can say this:

> **Conclusion**: 16-bit standard posits are accurate to about one part in 3300, whereas 16-bit sane floats are accurate to about one part in 1400.

We can again sort the relative errors and plot them along with the plot of posits with no limits on rS:

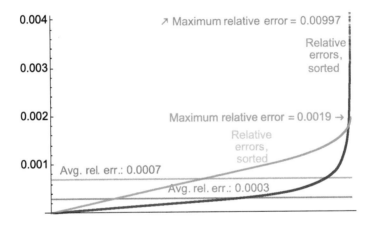

Sorted relative errors for representing **RRSet** for both standard 16-bit posits and sane floats

This makes clear why posits with no restrictions on rS have more than twice the fidelity of sane floats. *However*, notice at the extreme right that posit relative error surpasses that of floats for the rare cases of extreme magnitude values. Whether that matters or not depends on the application. Like any benchmark, using **RRSet** has the shortcoming that it cannot accurately predict fidelity for a particular application.

Now that we have a test jig, we can try various values of eS and rS to see what combination will yield the best fidelity. Increasing eS to improve the dynamic range will take bits away from the fraction but increase the width of the fovea. Decreasing rS improves the number of fraction bits available for extreme magnitudes, but it also decreases the dynamic range and therefore the incidence of very inaccurately represented reals.

Here is the table for 16-bit parameterized posits:

16 – bit Fidelity		eS						
		0	**1**	**2**	**3**	**4**	**5**	**6**
floatlike	2	−0.63	−0.50	−0.22	0.51	2.58	3.15	2.85
	3	−0.56	−0.36	0.12	1.44	3.41	3.15	2.85
	4	−0.50	−0.22	0.51	2.59	3.41	3.15	2.85
	5	−0.44	−0.06	0.95	3.46	3.41	3.15	2.85
	6	−0.36	0.12	1.44	**3.52**	3.41	3.15	2.85
	7	−0.29	0.31	1.97	**3.52**	3.41	3.15	2.85
rS	8	−0.22	0.50	2.56	**3.52**	3.41	3.15	2.85
	9	−0.14	0.72	3.04	**3.52**	3.41	3.15	2.85
	10	−0.06	0.94	3.24	**3.52**	3.41	3.15	2.85
	11	0.02	1.16	3.27	**3.52**	3.41	3.15	2.85
	12	0.11	1.38	3.27	**3.52**	3.41	3.15	2.85
	13	0.18	1.57	3.27	**3.52**	3.41	3.15	2.85
	14	0.25	1.71	3.27	**3.52**	3.41	3.15	2.85
positlike	15	0.30	1.79	3.27	**3.52**	3.41	3.15	2.85

Measured fidelity for 16-bit formats from sane float to full tapering, and a range of *eS* values

The top row is the sane floats, which do best with 5 exponent bits as shown in previous examples. The range with optimal fidelity is shown in blue, and is a surprisingly large set of possibilities. The table shows that it is catastrophic not to be able to cover the dynamic range of needed values. Columns 2 through 6 show that once the regime size *rS* gets large enough to cover all the points in **RRSet**, there is nothing gained by making *rS* larger.

Still, this narrows down the choices in a scientific way, and now we can ask: Which one is easiest to build in hardware? The likely answer is the top of the blue values where the regime size is smallest, *rS* = 6. It would also be the fastest, since the more limited the regime size is, the quicker the rest of the decoding of the exponent and fraction can start.

Generate the accuracy plot for that 16-bit format and see if it looks anything like the **RRSet** distribution. We also put the optimal sane float and the standard posit on the graph for comparison. The colors are transparent to make it clearer that the accuracy plots overlap in some regions.

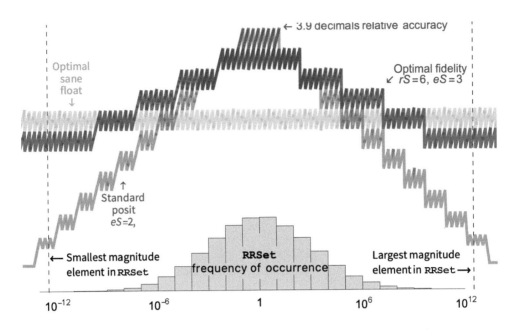

The three flavors of blunted posit on one graph, along with the distribution of RRSet.

This shows that you cannot simply look at the distribution and "eyeball" how to adjust the *eS* and *rS* values for highest fidelity. At first glance, it looks like standard posits are the best fit. But the optimal fidelity case has the same or better accuracy as standard posits once you get outside the fovea of standard posits. Having one more bit of accuracy for the top two bars of the histogram ($\frac{1}{100} \le |x| \le 100$) is outweighed by having as many as four more bits of accuracy (about 1.2 decimals more accuracy) outside that range.

We can generate a fidelity scatter plot as we did before:

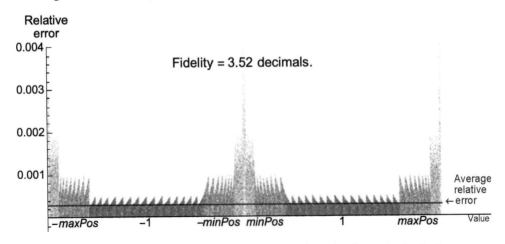

Relative errors for the sorted reals in RRSet, using the optimal case *rS* = 6, *eS* = 3.

It is a little crowded, but we can show the sorted errors together with the two cases presented earlier:

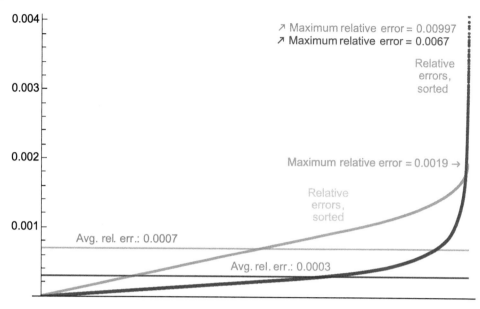

Sorted relative errors for representing **RRSet** for standard 16-bit posits, sane floats, and the optimal case

The $eS = 3$, $rS = 6$ case has the best of both worlds; the average relative error is the same as that of a pure posit, but the relative error does not go crazy for extreme magnitudes; there is still a fairly high density of represented reals in those regions. It has a dynamic range of about 3.6×10^{-15} to 2.8×10^{14}.

Most scientific computing is done with 64-bit floats presently, so we should not be surprised that it is a stretch to represent the range of needed real values with a 16-bit format.

What about 32-bit formats, though? Just as we found a sweet spot between fully-tapered and flat relative accuracy (sane float) posit types, does the optimum for 32-bit formats lie somewhere between the extremes? We can populate the average error table and find out.

32 – bit
Fidelity

		0	1	2	3	4	5	6
floatlike	2	−0.63	−0.50	−0.22	0.51	2.64	7.97	7.67
	3	−0.56	−0.36	−0.12	1.44	8.23	7.97	7.67
	4	−0.50	−0.22	0.51	2.64	8.23	7.97	7.67
	5	−0.44	−0.06	0.95	4.31	8.23	7.97	7.67
	6	−0.36	−0.12	1.44	**8.34**	8.23	7.97	7.67
	7	−0.29	−0.31	1.99	**8.34**	8.23	7.97	7.67
	8	−0.22	0.51	2.64	**8.34**	8.23	7.97	7.67
	9	−0.14	0.72	3.41	**8.34**	8.23	7.97	7.67
	10	−0.06	0.95	4.31	**8.34**	8.23	7.97	7.67
	11	0.03	1.19	8.09	**8.34**	8.23	7.97	7.67
	12	0.12	1.44	8.09	**8.34**	8.23	7.97	7.67
	13	0.21	1.71	8.09	**8.34**	8.23	7.97	7.67
	14	0.31	1.99	8.09	**8.34**	8.23	7.97	7.67
rS	15	0.41	2.30	8.09	**8.34**	8.23	7.97	7.67
	16	0.51	2.64	8.09	**8.34**	8.23	7.97	7.67
	17	0.61	3.02	8.09	**8.34**	8.23	7.97	7.67
	18	0.72	3.41	8.09	**8.34**	8.23	7.97	7.67
	19	0.84	3.80	8.09	**8.34**	8.23	7.97	7.67
	20	0.91	4.31	8.09	**8.34**	8.23	7.97	7.67
	21	1.07	6.56	8.09	**8.34**	8.23	7.97	7.67
	22	1.19	6.53	8.09	**8.34**	8.23	7.97	7.67
	23	1.32	6.53	8.09	**8.34**	8.23	7.97	7.67
	24	1.44	6.53	8.09	**8.34**	8.23	7.97	7.67
	25	1.57	6.53	8.09	**8.34**	8.23	7.97	7.67
	26	1.70	6.53	8.09	**8.34**	8.23	7.97	7.67
	27	1.83	6.53	8.09	**8.34**	8.23	7.97	7.67
	28	1.96	6.53	8.09	**8.34**	8.23	7.97	7.67
	29	2.09	6.53	8.09	**8.34**	8.23	7.97	7.67
	30	2.19	6.53	8.09	**8.34**	8.23	8.0	7.67
positlike	31	2.25	6.53	8.09	**8.34**	8.23	8.0	7.67

The top of the table is headed *eS* spanning columns 0–6.

Measured fidelity for 32-bit formats from sane float to full tapering, and a range of *eS* values

The best choice once again appears to be *rS* = 6 and *eS* = 3. All the values below that point have the same fidelity, 8.34 decimals, but restricting the regime size to no more than 6 bits means simpler hardware for counting the regime bits.

It also means a 16-bit posit optimized for **RRSet** and a 32-bit posit can be trivially converted just by appending **0** bits or rounding off bits, which will always be fraction bits. There is no Twilight Zone.

If we found higher fidelity with $eS = 3$, why does the Posit Standard (2022) specify $eS = 2$? One reason is that it halves the size of the quire, which can eliminate the problems caused by low accuracy at extreme magnitudes. Another is that **RRSet** might be too conservative, and that most applications use a smaller dynamic range of numbers or are unaffected, say, by large relative errors in tiny quantities when they are added instead of multiplied.

> **Exercise for the Reader**: If the precision is only 8 bits, find the minimal values for rS that create a posit format that can cover the dynamic range of **RRSet**, 3×10^{-13} to 3×10^{13}, for eS values 3, 4, and 5.

7.4 The *b-posit* Has Bounded Dynamic Range

Based on the experiments with **RRSet**, we should give a name to the case $eS = 3$, $rS = 6$, for which the dynamic range is bounded to 2^{-48} to 2^{48} (about 4×10^{-15} to 3×10^{14}).

> **Definition**: A *b-posit* (for bounded posit) is a posit with $eS = 3$ and $rS = 6$.

The term b-posit helps distinguish standard posits from this variant that seems particularly well-suited to HPC workloads. The precision can be any length 2 or greater, just as with any other posit format. Remember the ghost bits past the end of the number always have value **0** but can be part of the regime, exponent, or fraction fields. For example, a 6-bit b-posit would have this largest possible value:

$$011111|0000000\ldots$$

where all the bits past the vertical bar are ghost bits. The first 0 ends the regime, the next three are the exponent bits 000, and the rest are **0** fraction bits. Each regime bit represents $2^{2^{eS}} = 2^8 = 256$, and the run of five regime bits represents $k = 4$, so the value is $256^4 = 2^{32}$, which is about four billion.

Once the precision gets beyond 10 bits, there are always explicit fraction bits. A 16-bit b-posit has at least six explicit fraction bits, so the relative decimal accuracy wobbles between 1.8 and 2.1 decimals at the extreme magnitudes, instead of losing all accuracy at the extremes like standard posits do.

Hardware for b-posits should be even simpler to build than for standard posits, and faster. Decoding the power-of-2 scaling is the same task for 16-bit, 32-bit, and 64-bit b-posits. A programmer can freely switch between those three precisions to right-size the precision for the relative accuracy needed for each task, without a performance penalty for the conversion.

7.5 Customizing Posits

Henry Ford's assembly line revolutionized manufacturing. The computer equivalent is *dataflow*.

In the earliest days of automobile production, a single car would be assembled in place by one person or a small team. Each person would change tools and retrieve parts depending on what step was being done. Henry Ford applied the idea of an assembly line where the cars moved through a line of workers and dedicated machines that did just one task. This was far more efficient than having workers change tools and move around. A modern car factory can produce a car faster than once every minute.

Traditional computing is like the pre-Ford days of car production. A single processor changes instructions (tools) in running a program, retrieving data (parts) from memory depending on what step is being done, and a single processor does the entire job. The Posit Standard was designed for such processors. A dedicated arithmetic unit can perform an instruction to multiply two 32-bit real numbers, another instruction to add two 64-bit real numbers, and so on.

It would be quite complicated to have different *eS* and *rS* values supported in hardware and would also make programming a nightmare since the programmer would have to keep track not just of the precision used, but the other two parameters as well. The Posit Standard sets *eS* = 2 so that once again, a programmer only needs to keep track of what precision is used where.

Future hardware may make it simple to construct custom hardware out of many tiny integer processors that do just one thing, sort of like laying the flowchart for the algorithm onto those processors with local connections. Complicated arithmetic can then still produce one result per clock cycle. There is already at least one company, NextSilicon, that supports this kind of "dataflow" computation. Once the assembly line has been set up, it can produce one result per clock cycle, just like the pipelined hardware now does for normal float arithmetic. For such a system, it may be quite manageable to use all kinds of different parameters everywhere, customizing the arithmetic precisely for what is needed.

To give an example from NextSilicon that has nothing to do with real number arithmetic, suppose we wanted a function that reverses the order of bits in a 32-bit word. In case you do not know the C computer language or have forgotten it,

> << means shift bits left,
> >> means shift bits right,
> | means a bitwise OR, and
> & means a bitwise AND.

Also, hexadecimal (base-16) notation is indicated with a leading **0x**, and is used to describe bit patterns here, so for example **0xFF00** represents the same value as the binary **1111 1111 0000 0000**. Here is some cryptic-looking C code that somehow accomplishes the bit reversal using all integer operations (do not worry about understanding it; just take for granted that it works):

```
unsigned reverse (unsigned x) {
    x = ((x & 0x55555555) << 1) | ((x >> 1) & 0x55555555);
    x = ((x & 0x33333333) << 2) | ((x >> 2) & 0x33333333);
    x = ((x & 0x0F0F0F0F) << 4) | ((x >> 4) &0x0F0F0F0F);
    x = (x << 24) | ((x & 0xFF00) << 8) |
        ((x >> 8) & 0xFF00) | (x >> 24);
    return x;
}
```

Here is how that algorithm can be turned into an assembly line using a grid of simple integer processors:

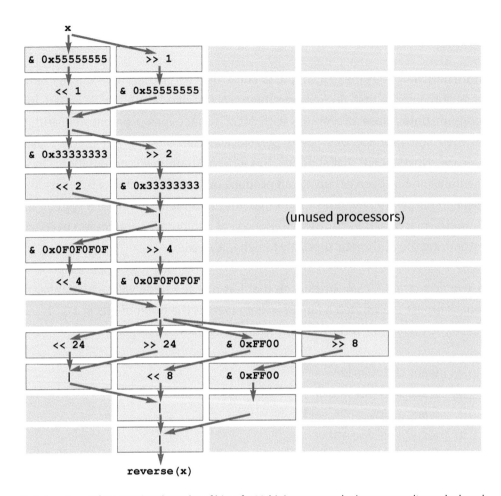

Dataflow layout for reversing the order of bits of a 32-bit integer, producing one result per clock cycle

In 2024, it is possible to print hundreds of thousands of those integer processors on a single chip, and you can see how that capability exploits things that can be done in parallel. Every processor just does one thing, over and over, so it does not have to manage an "instruction stream" at all. Each processor can throw its output to a limited number of processors in the next row, so communication is fast and inexpensive. The above assembly line takes 13 clock cycles to produce its first result, but then it produces one result every clock cycle, using parallel execution of 26 integer processors. The C compiler automatically lays out the "factory floor."

It should be possible to build software tools that collect data in high precision for a computer workload, then customize the three parameters *nBitsP*, *rS*, and *eS* for the highest fidelity. In other words, customizing the number format to fit the problem could be done automatically, by a compiler. The three parameters could be different for each phase of the execution of the program, without a performance penalty. An architecture like the one above could easily handle multiple format combinations.

7.6 Experimental Mathematics

Is there a place for *experimentation* in mathematics?

We do not normally associate math with experiments; either something is true and there is a proof, or it is false and there is a counterexample. But the design of a number format is more like engineering than math, and there is plenty of room for experiments when engineering an optimal way to do something. This chapter showed how some kinds of math can indeed be done experimentally.

The original posit definition in Chapter 1 looked simple and elegant. Just a regime and a signed fraction, and no exception cases. It is very easy to become attached to simplicity, but it became obvious it needed a bigger dynamic range (Chapter 3). If you look at the list of design goals for a real number format at the end of Chapter 4, the goal of small, fast circuits competes with the goal of matching well to the numerical vocabulary that is needed. The tent-shaped posits of Chapter 3 were a first guess at what would actually serve application needs best.

> I have seen a number of format designers who become so wedded to some elegant system that they cannot see its practical shortcomings. In more than one case, the inventor of a format refused to do the simplest experiment to find out if it really was an improvement, for fear of finding out that it was not.

The IEEE 754 float committee made the guess that a flat accuracy curve was best. I made the guess that a tent-shaped tapered accuracy curve was best. The experiment with **RRSet** showed that a compromise between the two may actually produce the best fidelity.

As an undergraduate, I had the privilege of being able to learn from one of the greatest minds of the twentieth century, and one of the things he said seems like an appropriate way to end this chapter.

"It does not make any difference how beautiful your guess is. It does not make any difference how smart you are, who made the guess, or what his name is. If it disagrees with experiment, it is *wrong*. That is all there is to it."

— *Richard Feynman* (1918–1988)

8 Logarithmic Systems, The Musical

J.S. Bach on the keyboard. I promise I will tie this in.

The idea of a *Logarithmic Number System* (LNS) for computing has cropped up many times in the previous century and now in this one. There is a resurgence of interest because an LNS seems particularly well-suited to AI, both Machine Learning (training a neural network) and Inference (applying the trained network).

8.1 A Different Kind of Binade

For positive values, what we have seen so far is numbers evenly spaced between 2^i and 2^{i+1} where i is an integer (a binade), for example, $\{16, 20, 24, 28\}$ between 2^4 and 2^5, not including the upper limit 32. The binades are of the form $2^i \times (1 + k/n)$ where k/n is the fraction and k counts up from 0 to $n-1$. Call that a *linear binade*. Rather than $(1+f)$, what if we used 2^f? That looks simpler and more elegant. Call that a *logarithmic binade*, or *log binade* for short.

To get a feel for this, consider these two ways to fill a binade from 1 to 2:

f	LinearBinade, $1+f$	Logarithmic Binade, 2^f
0.00 $= 0$	1.00	$2^0 = 1$
0.01 $= 0.25$	1.25	$2^{0.25} = 1.1892\cdots$
0.10 $= 0.5$	1.50	$2^{0.5} = 1.4142\cdots$
0.11 $= 0.75$	1.75	$2^{0.75} = 1.6817\cdots$

Two ways to turn a two-bit fraction into a range from 1 to (almost) 2.

Logarithmic binades can be the basis for an LNS. Binary numbers of the form $i + f$ are simply fixed-point format, the integer bits i and the fraction bits f. Sometimes they are written with the radix point as $i.f$ because it is just positional notation. In traditional logarithm jargon, i is the *characteristic* of a logarithm, and f is the *mantissa*. Unfortunately, many people still use the word "mantissa" for the fractional part of a *float*, and that is incorrect. The IEEE Std 754 does not make this mistake. To visualize the two approaches, here is a plot of several binades with eight values per binade (in other words, f has three bits):

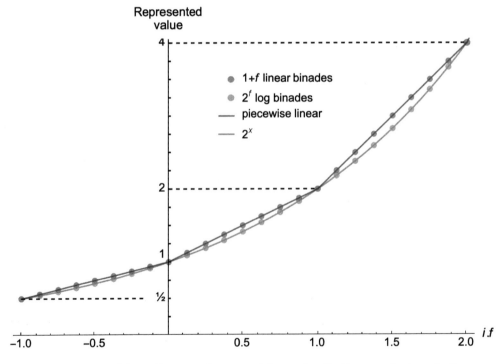

The orange curve of the LNS is a smooth exponential curve, but the blue curve is piecewise linear.

A typical way that an LNS is constructed is to start with a fixed point format, then tack on a bit to indicate the sign so both 2^{i+f} and -2^{i+f} can be represented. However, that leaves no way to represent zero, so yet another bit is tacked on that flags whether the value is zero. While it seems easy to build circuits this way, it wastes bit patterns even worse than having "negative zero" and "positive zero." Suppose we have only 5 bits for the LNS, and we use a fraction length of 1 bit, like this:

zero bit z	sign bit s	characteristic bits i	mantissa bit f

A tiny LNS with a bit tacked on to indicate if a value is zero

The color coding for the **sign, characteristic,** and **mantissa** are what we have previously used for **sign, exponent,** and **fraction** since they serve similar functions. The **zero** bit is color-coded in a new color, **purple,** since it has a meaning not seen in formats discussed so far.

What is the vocabulary of this format? The two-bit characteristic can express integers -2, -1, 0, and 1 for bit patterns **10**, **11**, **00**, and **01** as 2's complement integers. The bit string $i.f$ can be treated as a single 2's complement integer, scaled. The vocabulary is 0 and $\pm\{0, 0.25, 0.353 \cdots, 0.5, 0.707 \cdots, 1, 1.41 \cdots, 2, 2.82 \cdots\}$. Here is a plot of the values represented as a function of the bit string:

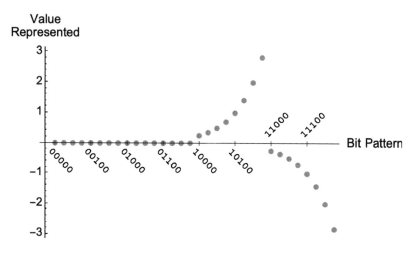

A tiny LNS system with a bit tacked on to indicate if a value is zero

You can see what happens when you tack on bits; it is very easy to get redundant bit patterns. There are only 17 distinct values represented when it should be possible to represent $2^5 = 32$ distinct values. Some LNS formats even have a bit to represent an error, such as overflow; if we tacked on a "NaN" bit to the five-bit format, the vocabulary would increase to 18, which is only about 28 percent of the $2^6 = 64$ possible bit patterns. "Bit field mentality" was mentioned in Section 4.5 as being a poor approach to format design.

Instead, use the posit approach presented in Chapter 1. Fill in the ring for the projective reals, with –1 and 1 at the west and east points of the ring, and start filling the in-between spaces by the same rules … except that when we get to the fraction bits, they are treated as part of the exponent, so the midway point between 1 and 2 (2^0 and 2^1) is $2^{0.5} = \sqrt{2} \approx 1.414$ rather than $\frac{1+2}{2} = \frac{3}{2} = 1.5$. Here is the final result:

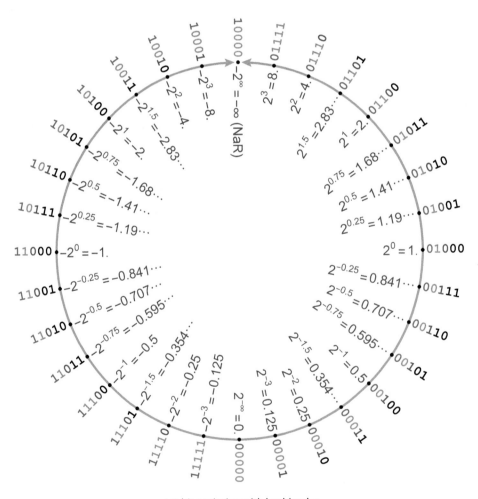

A 5-bit posit ring with log binades

This way, every bit counts. These are *log posits*, where the only distinction from the posits seen so far is at the binade level. If there is a possibility of confusion, we might call the posits seen so far *linear posits*. The cases 0 and NaR are handled elegantly and consistently, and there are 32 distinct values represented. The dynamic range is from ⅛ to 8 (almost two orders of magnitude) rather than ¼ to 2.82···, and the accuracy plot is improved quite a bit:

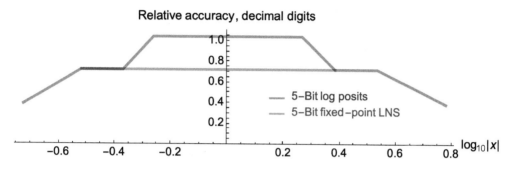

Accuracy plot for 5-bit log posits (amber) and the fixed-point LNS with a zero indicator bit

One quality of the relative accuracy plots for LNS systems is conspicuously absent: the jaggedness! When the values inside a binade represent powers of evenly-spaced logarithms, they have the same relative accuracy throughout the binade. Remember that the amount of wobble is about 0.3 decimals (since $\log_{10}(2) \approx 0.3$), and if you think of that as a wobble of ±0.15 decimal about a central value, avoiding the worst-case accuracy at the bottom of the jaggedness means the use of logarithmic spacing typically improves accuracy just a little bit, about 0.15 decimal. Logarithmic posits always have relative accuracy plots that resemble Mayan and Aztec temples:

Chichan Itza Mayan ruins in the Yucatan peninsula

With a fixed-point LNS, you get a flat line for the relative accuracy, which might be a good thing depending on the application. With a log posit, you get at least as much accuracy with the same number of bits, and even with just 5 bits we have a fovea that has more than one decimal of relative accuracy. At Facebook, Jeff Johnson was able to get satisfactory AI results (for inference) using 5-bit logarithmic posits, with a sevenfold savings in energy requirements compared to a system with linear binades.

8.2 You Already Know LNS, but by Another Name

Musicians have their own term for a binade. They call it an **octave**.

Middle C and an octave above Middle C

Suppose the pitch of Middle C is regarded as the unit pitch, 1.0; the pitch of the C one octave higher has frequency 2.0 and the pitch one octave lower has frequency 0.5. And the relative frequency of F♯ above Middle C is $2^{1/2} = \sqrt{2}$ exactly, or at least it is if the instrument is in tune. Here is a standard piano keyboard with 88 keys:

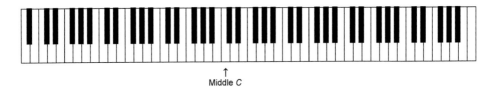

Middle C

Here is the keyboard showing the positive values in the 5-bit log posit system:

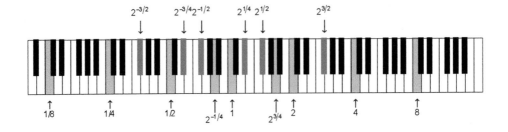

If you read piano sheet music, here is what those notes look like:

A "log posit arpeggio"

Try playing that with the damper pedal down, and you will hear the sound of a *diminished seventh chord*, which was one of J. S. Bach's favorite chords. If you have ever heard his *Toccata and Fugue in D Minor* (the first piece in Disney's movie *Fantasia*), you have heard that chord.

The analogy breaks down in that we have no "negative notes" for the negative values in an LNS, and it also breaks down because if there are more than four pitches per octave, the next step is *twelve* notes per octave (chromatic scale), not eight. Each note is higher in frequency by $2^{1/12} = 1.0596\cdots$, a semitone. Once you reach 12 values per octave, the notes are at frequency ratios that closely approximate *the ratios of whole numbers*, which is when humans perceive consonance. This observation is at least as old as Pythagoras (c. 570–495 BC).

Ratio	Interval Name	Decimal Ratio	Semitones	Log Approximation
2 : 1	octave	2	12	$2^{12/12} = 2$ (exact)
3 : 2	perfect fifth	1.5	7	$2^{7/12} = 1.498\cdots$
4 : 3	perfect fourth	$1.333\cdots$	5	$2^{5/12} = 1.334\cdots$
5 : 4	major third	1.25	4	$2^{4/12} = 1.259\cdots$
5 : 3	major sixth	$1.666\cdots$	9	$2^{9/12} = 1.681\cdots$

Small integer frequency ratios sound consonant

It is a remarkable mathematical coincidence that powers of $2^{1/12}$ line up so closely with the ratios of small integers. Had the base ratio been $2^{1/16}$, which would have been a lot more suited to base-2 computer design, you would have a very dissonant-sounding scale.

8.3 Gentlemen, Place Your Frets

Lute showing logarithmic fret placement shared across strings

As soon as you create a multi-string instrument and realize that frets will help the player hit the right note, you notice that with a very tiny compromise, all the strings can *share the same fret across all strings* with the frets spaced by a semitone (half-step). Lutes existed long before keyboard instruments, and fretted string instruments are almost as universal in human cultures as flutes and drums. Compare the fret spacings with the marks on a slide rule. They are the same thing:

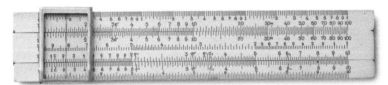

Traditional slide rule

Slide rules use logarithmic scales so that adding lengths (by positioning the slide) is an analog to adding logarithms, which means finding the product since $\log(x) + \log(y) = \log(x \cdot y)$. For very different purposes, fretted instrument designers and slide rule makers make use of the same scale.

You can "multiply" on a piano keyboard by adding *musical intervals*. For instance, if you go from Middle C to the F♯ above it (a *diminished fifth*), and then go up an octave from there, you have multiplied $\sqrt{2}$ by 2 to get $2\sqrt{2}$. Not that you would ever use a piano keyboard as a slide rule, but it is kind of cool that you can.

Because J. S. Bach wrote *The Well-Tempered Clavier* with preludes and fugues in all 24 major and minor keys, some claim that he invented the idea of an equal-tempered (logarithmic) tuning. But fretted lutes that predate Bach by hundreds of years make that theory seem a little silly.

8.4 Arithmetic with Logarithmic Systems

After that musical interlude, back to arithmetic. If there is one thing to take away from this chapter, it is this:

> Logarithmic formats makes it easy to multiply and divide, but difficult to add and subtract. In *some* situations, the tradeoff is worth it.

Remember from Chapter 1 that a "Type II unum" allows perfect reciprocation by rotating the posit ring about the horizontal axis, that is, taking the 2's complement of the bits after the sign bit. There is no need to decode the posit into its regime, exponent (characteristic), and fraction (mantissa) to find the reciprocal, and there is never any rounding error or lack of closure. Just as we compute $x - y$ using $x + (-y)$ since negating a number is usually perfect and lossless, with a Type II unum system like the logarithmic format we can compute x/y using $x \cdot (1/y)$ and there is only one rounding since the $1/y$ loses no accuracy.

Because $\log(x \cdot y) = \log(x) + \log(y)$, once you know the characteristic i and the mantissa f represented by two bit strings, finding the bit string representing $x \cdot y$ is a matter of computing $i_x + i_y + f_x + f_y$ and finding the nearest element in the vocabulary. Often, it is representable exactly. For floats, it can underflow or overflow, and for posits it can land between numbers in the vocabulary. Perhaps a closure plot will help visualize this, using the five-bit log posits (positive values only). The 15 possible inputs are $\left\{\frac{1}{8}, \frac{1}{4}, \frac{1}{2\sqrt{2}}, \frac{1}{2}, \frac{1}{2^{3/4}}, \frac{1}{\sqrt{2}}, \frac{1}{2^{1/4}}, 1, 2^{1/4}, \sqrt{2}, 2^{3/4}, 2, 2\sqrt{2}, 4, 8\right\}$:

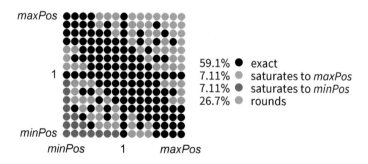

Multiplication closure plot for *logarithmic* 5-bit posits (0-bit exponent)

The closure plot for division looks similar but is reflected about the vertical axis.

If we use linear binades as usual, the multiplication closure plot has fewer exact entries. The 15 possible inputs are $\left\{\frac{1}{8}, \frac{1}{4}, \frac{3}{8}, \frac{1}{2}, \frac{5}{8}, \frac{3}{4}, \frac{7}{8}, 1, \frac{5}{4}, \frac{3}{2}, \frac{7}{4}, 2, 3, 4, 8\right\}$:

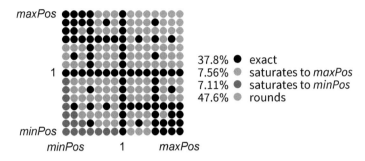

37.8% ● exact
7.56% ● saturates to *maxPos*
7.11% ● saturates to *minPos*
47.6% ● rounds

Multiplication closure plot for *linear* 5-bit posits (0-bit exponent)

Linear posit products land on an exact value only about 64% as often as do logarithmic posit products.

Exercise for the Reader: There is one more case of "saturates to *maxPos*" for linear posits than for logarithmic posits in the example above. Find it and explain why it happens.

However, the situation with addition is reversed. Here is the closure plot when we add two logarithmic posits and then have to find the logarithm nearest that sum; this time, we include all the reals from −*maxPos* to *maxPos* as inputs:

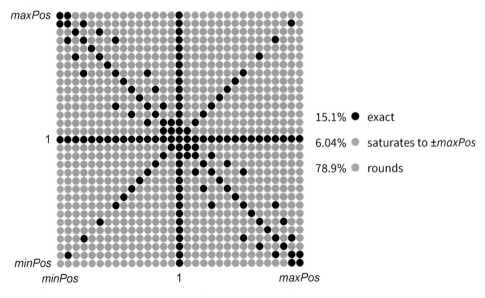

15.1% ● exact

6.04% ● saturates to ±*maxPos*

78.9% ● rounds

Closure plot for addition of logarithmic posits (5-bit precision, *eS* = 0)

For linear posit addition, the result is exact for *more than three times as many cases*. This is one reason log-type representation has not found more widespread use. Asking "What is 1 + 2?" of a computer and getting an answer of "approximately 2.82" certainly takes some getting used to.

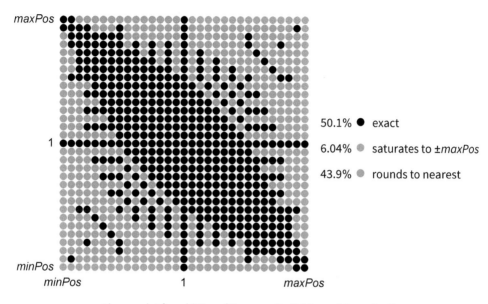

50.1% ● exact

6.04% ● saturates to ±*maxPos*

43.9% ● rounds to nearest

Closure plot for addition of linear posits (5-bit precision, $eS = 0$)

How would you even go about computing the addition of log-type numbers? A complete answer goes beyond the scope of this book, but the general approach is this: Suppose we have LNS numbers $X = 2^x$ and $Y = 2^y$ where x and y are the fixed-point representations. We seek the fixed-point z that comes closest to satisfying $2^z = 2^x + 2^y$. If $X = Y$, then $z = x + 1$. Otherwise, suppose $X < Y$; then $2^z = X + Y$ can be written as $2^z = Y \times (1 + X/Y)$. Take the log base 2 of that equation:

$$\lg(2^z) = \lg(Y \times (1 + X/Y))$$
$$z = \lg(Y) + \lg(1 + X/Y)$$
$$= y + \lg(1 + 2^{x-y})$$

Since $(1 + 2^{x-y})$ is always in the binade $[1, 2)$, you can build a lookup table for what to do for every value in the binade. This works for mantissas up to about 20 bits, at which point the table becomes too large (and slow) to be practical. Subtraction requires a different table. For many cases, one of the inputs is so much larger than the other that it is unaffected by the addition or subtraction of the smaller one. That means the table can be smaller since it only needs to hold nonzero entries.

Perhaps the world's greatest expert on LNS systems in general and especially on adding LNS values is J. Nicholas Coleman, who spearheaded The European Logarithmic Microprocessor project starting in the year 2000. Through a number of innovations, he and his colleagues were able to build a device with a 32-bit LNS that was competitive with one based on 32-bit floats, and clearly superior in some respects. If you would like more details about adding two LNS numbers (it is especially tricky if they are not the same sign and the result is small), I refer you to Coleman's excellent papers on the subject.

8.5 A Two-Dimensional Quire?

It is very difficult to build a quire for a log posit. There is an approach, but it does not scale well at all. It is presented here because AI can sometimes be done with very low precision, for which there is an approach that can work.

Consider the 5-bit log posits used as examples in the first few sections of this chapter. In the binade [1, 2) there are four values:

$$2^0 = 1. \qquad 2^{1/4} = 1.1892\cdots \qquad 2^{1/2} = 1.4142\cdots \qquad 2^{3/4} = 1.6817\cdots$$

If you add any of those to one of the other three, you get a result not in the vocabulary so you have to round it. But if you add any of them to *themselves*, you have an excellent chance of getting a result with exact representation. This suggests that we need not one quire, but *four* quires; a fixed-point exact accumulator for each of those four powers of two shown above. In other words, the quire for logarithmic posits is *two-dimensional*. You have a fixed-point number times 1, a fixed-point number times $2^{1/4}$, and so on. If you want to add 1 and $2^{-1/2}$, say, you add 1 to the quire for multiples of 1 and $\frac{1}{2}$ (in binary, **0.1**) to the quire for $2^{1/2}$. There is **no** rounding until you convert the sum of all four quires back to a log posit.

To convert the quires back to the nearest log posit, each fixed-point number other than the first must be multiplied by a fixed-point approximation to $2^{1/4}$ or $2^{1/2}$ or $2^{3/4}$ to sufficient accuracy that the correctly-rounded (that is, nearest) posit can be determined with certainty. That is expensive, but if you have long dot products to compute, you only need it once after many (very fast and low-power) multiply-accumulate operations, so it can be worth it. The guarantee of knowing if the dot product is zero can still be achieved with log posits, by testing if every one of the quires is zero, but getting the sign right is very expensive compared to linear posits.

If you need at least k bits of fraction, then you need a 2D quire with 2^k fixed-point accumulators. So this approach really does not scale very well. While we gain the marvelous property that reciprocals are perfect and quick with log binades and multiplication becomes dirt cheap, we lose the ability to do most of the things a quire can do that were described in Chapter 6.

8.6 Another Parameter for Posit Format

So far, we have three integer "knobs to turn" in posit format: the precision *nBitsP*, the regime size *rS*, and the exponent size *eS*. We can now include a button to push, *logQ*, which is **True** if the binades are logarithmic and **False** if they are linear. It is a Boolean variable and not an integer; a convention in *Mathematica* is to name Boolean variables ending in Q, so for example **IntegerQ[x]** is **True** if **x** is an integer and **False** if it is some other kind of number. The *logQ* option has a very small effect on the routines **setPositEnv**, **pToX**, and **xToP** since it merely substitutes 2^f for $1 + f$. Here is what happens to **setPositEnv**, with changes shown in blue.

```
Clear[setPositEnv]
setPositEnv[{n_Integer /; n ≥ 2,
    r_Integer /; r ≥ 1, e_Integer /; e ≥ 0, q_}] := Module[{t},
   {nBitsP, rS, eS, logQ} = {n, Min[r, n - 1], e, q};
   nPat = BitShiftLeft[1, n];
   NaR = -BitShiftRight[nPat];
   pMask = nPat - 1;
   eSizeP = BitShiftLeft[1, eS];
   uSeed = BitShiftLeft[1, eSizeP];
   t = nBitsP - rS - 1;
   {minPos, maxPos} = Which[
     t == 0, {uSeed^(1-rS), uSeed^(rS-1)},
     t ≤ e || logQ, {uSeed^(-rS) * 2^(2^(e-t)), uSeed^(rS) * 2^(-2^(e-t))},
     True, {uSeed^(-rS) * (1 + 2^(e-t)), uSeed^(rS) / 2 * (2 - 2^(e-t))}];
   ];
```

The code for **xToP** is about 40 lines long in *Mathematica*, and only about 10% is changed by allowing it to support both logarithmic and linear binades.

Remember from Chapter 7 that the combination $rS = 6$, $eS = 3$ seemed to give the highest fidelity when tested against **RRSet**. Here is a plot of the linear and log versions for 16-bit posits, showing how the calls to **setPositEnv** now have four parameters to specify:

```
setPositEnv[{nBitsP = 16, rS = 6, eS = 3, logQ = False}]
G1 = accPlot[Table[N[pToX[p]], {p, 1, -NaR - 1}], ■];
setPositEnv[{nBitsP = 16, rS = 6, eS = 3, logQ = True}]
G2 = accPlot[Table[N[pToX[p]], {p, 1, -NaR - 1}], ■];
Show[{G1, G2}, PlotRange → All]
```

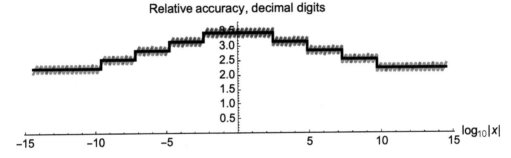

Combined plot of linear 16-bit posits (jagged, gray) and log 16-bit posits (flat steps, **black**)

You can see the jagged curve in the background for linear posits, and the darker curve in the background for log posits, with absolutely flat stair steps. We can use the experimental setup of Chapter 7 to check the fidelity. First, look at the scatter plot of the error for the values in **RRSet** as before, but with **logQ** set to **True**:

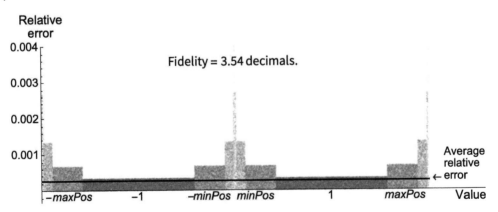

Fidelity plot for 16-bit log posits

Because the wobble is gone in the relative error, the scatter plot shows uniform blocks with heights that reflect the number of bits available for the mantissa. The fidelity has increased from 3.52 decimals for linear posits to 3.54 decimals, a small but repeatable improvement.

When we sort the errors and plot them, the average relative error changes from 0.00030 for linear posits to 0.00029 for log posits. The plot now looks piecewise linear, and the maximum relative error has dropped from 0.00670 to 0.00537.

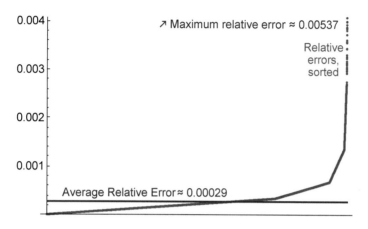

Sorted relative errors for 16-bit log posits

8.7 Logarithmic Sane Floats

This chapter started with a 5-bit log posit and no limits on the regime size rS. If we limit the regime to $rS = 2$, we get a sane float with logarithmic binades, and every value has three bits of significand (almost a decimal of accuracy). Here is the vocabulary for the positive numbers that the format can represent:

$2^{-1.75}$	=	$0.2973\cdots$
$2^{-1.5}$	=	$0.3536\cdots$
$2^{-1.25}$	=	$0.4204\cdots$
$2^{-1.}$	=	0.5
$2^{-0.75}$	=	$0.5946\cdots$
$2^{-0.5}$	=	$0.7071\cdots$
$2^{-0.25}$	=	$0.8409\cdots$
$2^{0.}$	=	$1.$
$2^{0.25}$	=	$1.189\cdots$
$2^{0.5}$	=	$1.414\cdots$
$2^{0.75}$	=	$1.682\cdots$
$2^{1.}$	=	$2.$
$2^{1.25}$	=	$2.378\cdots$
$2^{1.5}$	=	$2.828\cdots$
$2^{1.75}$	=	$3.364\cdots$

The positive 5-bit sane float values with log binades

Decimals are hard to read; that set is expressed exactly by notes on a piano keyboard if we treat Middle C as 1.0. Actually, we only need four octaves of range, so we could even fit the notes on the piano's predecessor, the harpsichord.

An Italian harpsichord

Here is a top view of that lovely instrument, with the 15 keys of the positive part of the 5-bit log sane posit shown in light blue:

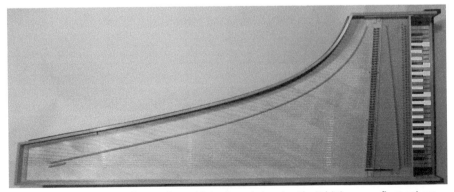

Same harpsichord, bird's-eye view with keyboard showing the 5-bit log sane float values

Sheet music for those evenly-spaced notes, as an arpeggio

There are three main ways to change the pitch you get when you pluck (or strike, or bow) a stretched string:

- Change the tension. Doubling the tension raises pitch by an octave.
- Change the weight per unit length. Doubling weight per length lowers the pitch by an octave.
- Change the length of the string. Doubling the length lowers the pitch by an octave.

It is not mechanically sound to have much higher tension on the high (treble) end of a harpsichord (or piano) than on the low (bass) end, and it will sound terrible because the low notes will sound like plucked rubber bands and the high notes will sound tinny. Changing weight per unit length, for example by adding copper windings around the string or making it thicker, is better; that is the method used for fixed-length string instruments like a guitar or violin. But there is a different *timbre* when you change the weight of the string; the quality of the pitch changes.

Harpsichord builders used some of the string-weight method to change the pitch, but the Italians in particular were quite fussy about keeping the timbre the same, so they tried to use the same tension and same gauge string for the entire instrument. That meant doubling the length of the instrument every time you go down an octave, and that explains why a lot of Italian harpsichords only had four octaves. Some historical examples survive of harpsichords that are twenty feet long. Concert grand pianos sound better than baby grands because baby grands have to rely more on the trick of making strings heavier in the bass to reduce length, and that compromises the timbre.

I plotted the function 2^x for x between −1.75 and 1.75 in light blue, and laid it with graphics tools on top of the picture of the corresponding harpsichord strings. It fits the bridge (the left end where the strings are pinned) almost perfectly.

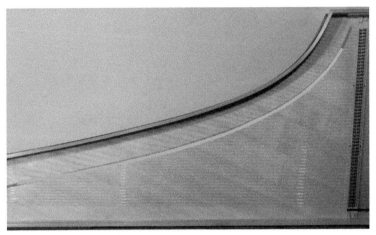

Proof that the harpsichord was designed with 2^x scaling of string length

Exponential scaling is also clearly visible in the treble strings of a harp, though like a piano, thicker strings are used in the bass or else the harp would be ridiculously tall.

Modern orchestral harp showing exponential scaling in the shorter strings

8.8 Objections, and the Round-Trip Argument

Why has LNS not caught on? It looks like it has simpler and smaller hardware since you do not have to include an integer multiplier. It has about 0.15 decimal better worst-case accuracy, and slightly better fidelity. In terms of the 11 engineering goals for a real number format, it seems to score higher than systems with linear binades. You could even take existing IEEE Std 754 floats and just replace the linear binades with logarithmic binades and gain these improvements, but no one seems to have done that. What are the objections?

One is that we know how to scale linear binades to any size and still be able to build circuits for plus, minus, times, and divide. With log binades, plus and minus get either difficult or incorrectly rounded past about 32-bit precision. The HPC community is addicted to 64-bit floats as a panacea for the problem of rounding error, so a 32-bit log binade format would be of little interest. Fair enough.

But there is another objection, one that has an anthropomorphic slant: People are uncomfortable with a system where $1+2$ is only approximately 3. There is an argument to rebut this objection, which I call the *Round-Trip Argument*. A short version of it is this: If a conversion of X introduces rounding, but the conversion back always produces X exactly, then the rounding does not create loss of information.

For example: In an LNS, suppose you want to add 5 and 7, neither of which is exactly representable in the vocabulary of the LNS. Try a 16-bit log sane float:

```
setPositEnv[{nBits = 16, rS = 2, eS = 3, logQ = True}]
{approx5, approx7} = {5̄, 7̄};
N[{approx5, approx7}, 20]
{colorCodeP[xToP[5]], colorCodeP[xToP[7]]}
```

{5.0011699270812347860, 7.0012726355354659767}

{0100100101001010, 0100101100111011}

The first line of the result expresses the approximate values of 5 and 7 to 20 decimals. The color-coded binary shows the posit representations of 5 and 7. If those are off in the last binary digit (±one ULP), the approximation for 5 turns into 4.998 or 5.005 and the off-by-one-ULP approximation for 7 is 6.997 or 7.006. So if you express the log form to four decimal places, there is no loss of information in converting back to an integer. Round trips are lossless. So 5.001 is a perfectly good proxy for the integer 5 when expressed as a log posit, and 7.001 is similarly a valid way to represent the integer 7 in the format.

If an algorithm truly requires integer values, then use integer data types. The round-trip argument is that if you can convert to a real type and back to integer and always get the original integer correctly, that is as good as you can do. There has been no loss of information. If you are using rounded real numbers on a computer, you have already left exact mathematics at the door anyway.

8.9 Postlude

The choice of whether to go between powers of two in a straight line (linear binades) or with an exponential curve (logarithmic binade) is another option in a posit-type number system. This chapter has laid out the upsides and downsides of the two choices, and there is no clear winner because each can do something the other cannot. And the log binades have a striking analogy to musical octaves and the concept of equal temperament.

I leave you with a reminder of why music teachers should probably not try to teach arithmetic.

"Correct... Now do it again, and this time, with *feeling*."

9 The Fastest Arithmetic Ever?

> A
> DESCRIPTION
> OF THE ADMIRABLE
> TABLE OE LOGA-
> RITHMES:
> WITH
> A DECLARATION OF
> The Most Plentifvl, Easy,
> and speedy vse thereof in both kindes
> of Trigonometrie, as also in all
> *Mathematicall calculations.*
> INVENTED AND PVBLI-
> shed In Latin By That
> Honorable L Iohn Nepair, Ba-
> ron of *Marchiston,*and translated into
> English by the late learned and
> famous Mathematician
> *Edward Wright.*

One of the earliest books containing tables of logarithm values. I am not sure how much trust to put in a book with thousands of numbers when there are this many dubious spellings on the title page. Spelling in the 1600s was, well, *imaginative*. I am pretty sure that is not the way John Napier ever spelled his name.

This chapter points to a completely new way to build hardware for computer arithmetic. At first it may seem like table look-up, but it is fundamentally different and can work on any set of real numbers (not necessarily posits).

9.1 The Idea of Logarithm Tables

When John Napier (1550–1617) published his table of logarithms in 1614, the idea was to rescue people from the tedium of multiplying (and dividing) numbers with many digits using the methods that Fibonacci had taught Europe, the ones you learned in elementary school.

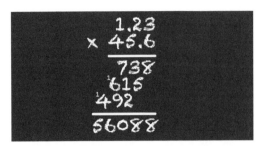

Chalkboard multiplication

It is tedious, it is error-prone, and then when you are done you have to figure out where the decimal point goes. It is the kind of exercise that makes students think they hate math, when really all they hate is manual arithmetic. When Napier realized that he could use tables of logarithms to turn multiplication into manual addition, and also provide an easy way to do divides, powers, and roots, we know he realized just how powerful the idea was. For one thing, he titled his book *Mirifici Logarithmorum Canonis Descriptio* (Latin was the language of science at the time), or *Description of the Amazing Logarithm Method*. *Mirifici* could also be translated as *Admirable* or *Wonderful*. For another thing, he spent 20 years of his life calculating 90 pages worth of logarithms by hand to create the book, clearly a labor of love. He coined the word *logarithm* from the Greek words for **log**ic and **arithm**etic.

 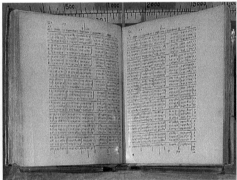

John Napier and a museum display of a 3rd edition (1620) of his book
introducing logarithms. Notice the giant slide rule in the background.

The first part of the book explains how to use logarithms, and in Section 5 he wrote

In numbers distinguished thus by a period in their midst, whatever is written after the period is a fraction, the denominator of which is unity with as many cyphers after it as there are figures after the period.

The Construction of the Wonderful Canon of Logarithms. Translated by Macdonald, William Rae. Edinburgh: Blackwood & Sons, 1889 [1620], page 8.

The word "cypher" meant "zero" back then. You see, the Hindu-Arabic positional notation that Fibonacci had introduced had only been used to represent integers. For over 400 years, if you wanted to write an approximate value for π, you had to express it something like "3 plus 14 parts per 100." The denominator 100 is "unity with as many cyphers after it as there are figures after the period," so, 3.14 for short. Napier did not invent just logarithms; he invented the decimal point. The practice soon spread all throughout Great Britain. This all happened several decades before Isaac Newton was born.

Napier must have been an interesting fellow. He thought his greatest work was not the invention of logarithms but a treatise analyzing the Book of Revelation in the New Testament, in which he claimed to have proved that the Pope was the antichrist and that the world would end in 1786. He was more successful at predicting future military weaponry, and accurately described the machine gun, the army tank ("an armored chariot capable of firing in all directions"), and the submarine, hundreds of years before they were actually created.

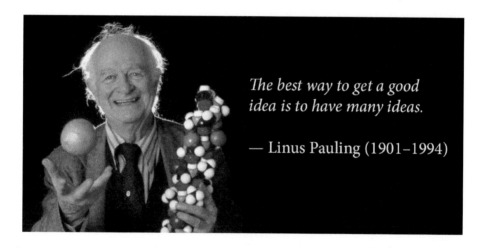

The best way to get a good idea is to have many ideas.

— Linus Pauling (1901–1994)

In the days before handheld calculators, schools used to teach the use of logarithm tables. The procedure is to first move the decimal points in the numbers you want to multiply so that the numbers are between 1 and 10, just like you do with scientific notation. For instance, 45.6 in the chalkboard example above becomes 4.56. You then look up their logarithms to several decimal places in a table that goes from 1.00 to 9.99. Except for special values, logarithms are irrational numbers like π in that you need an infinite number of decimals to represent them exactly and the digits never fall into a pattern.

x	$\log (x)$		x	$\log(x)$
⋮	⋮		⋮	⋮
1.21	0.08279		4.55	0.6580
1.22	0.08636		**4.56**	**0.6590**
1.23	**0.08991**		4.57	0.6599
1.24	0.09342		4.58	0.6609
1.25	0.09691		4.59	0.6618
⋮	⋮		⋮	⋮

Excerpt of a table used to look up the log of input values 1.23 and 4.56

That means $1.23 \approx 10^{0.08991}$ and $4.56 \approx 10^{0.6590}$. Napier's insight was that when you multiply numbers b^x and b^y for some base b, you simply add the exponents: b^{x+y}, in this case $10^{0.08991+0.6590}$. Pencil-and-paper addition is a lot easier than pencil-and-paper multiplication.

So $0.08991 + 0.6590 = 0.74891$, and now you need to find out the *antilogarithm*, that is, "0.74891 is the logarithm of what number?" Again consult the precomputed tables of logarithms:

x	$\log (x)$
⋮	⋮
5.60	0.75819
5.61	**0.74896**
5.62	0.09342
⋮	⋮

Excerpt of the same table used to look up the approximate antilogarithm of 0.74891

That log is pretty close to 0.74891, so we know that $1.23 \times 4.56 \approx 10^{0.74891} \approx 5.61$. Put back the power of 10 we took out of 45.6 to scale it to be in the table, and $1.23 \times 45.6 \approx 56.1$, which is pretty close to the chalkboard exact result of 56.088 and might be sufficiently accurate for whatever you are trying to do.

Long division is even harder to do by hand than multiplication, but with log tables all you have to do by hand is subtract, since $b^x \div b^y = b^{x-y}$. Taking a square root is also easy because you simply divide the logarithm by 2 and look up that logarithm, since $\sqrt{b^x} = b^{x/2}$. Taking numbers to a power turns into a multiplication, using the identity $x^y = 10^{y \times \log_{10}(x)}$, and you can use the log table to turn the $y \times \log_{10}(x)$ step into manual addition. Even the mechanical desktop calculators of the early 20th century could not evaluate something like $1.23^{4.56}$. For practical calculations without computers, logarithm tables are amazing. *Mirifici*, as Napier put it.

To get even more decimals means even larger tables, and over the years log tables became huge tomes full of numbers. Since this was hundreds of years before mechanical calculators, log tables with five or six significant figures were a godsend for people who needed to do a lot of number-crunching and could tolerate results that were not exact.

9.2 Mirafici Iterus

Creating the logarithm table requires rounding an infinite decimal to some number of digits. Then adding the logarithms can produce more digits than there are in the table, forcing another rounding. Finally, looking up the antilogarithm means rounding to the nearest entry in the table.

If the reader will permit me a meme:

Even four-bit (linear) posits from Chapter 1 are complicated enough to show the new approach. This shows the ring plot, and assigns x the list of the positive values:

```
setPositEnv[{4, 0}]; Show[ringPlotP, ImageSize → 285]
x = Table[pToX[i], {i, 1, 7}]
```

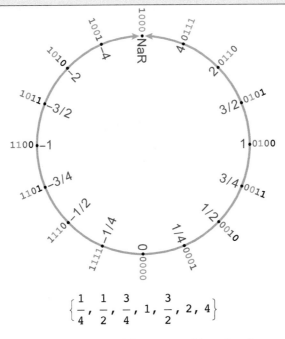

$$\left\{ \frac{1}{4}, \frac{1}{2}, \frac{3}{4}, 1, \frac{3}{2}, 2, 4 \right\}$$

Ringplot for linear posit⟨4,0⟩ and the seven positive values for an example

To figure out multiplication, we really only need to worry about the positive values, since it is trivial to deal with a 0 input or a NaN input, and the sign of the result is easy to find from the signs of the inputs. We only need a table for the seven positive values, $x_0, x_1, \ldots, x_6 = 0.25, 0.5, 0.75, 1, 1.5, 2, 4$ in decimal form. The multiplication table looks like this, using fractional form:

✕	$\frac{1}{4}$	$\frac{1}{2}$	$\frac{3}{4}$	1	$\frac{3}{2}$	2	4
$\frac{1}{4}$	$\frac{1}{16}$	$\frac{1}{8}$	$\frac{3}{16}$	$\frac{1}{4}$	$\frac{3}{8}$	$\frac{1}{2}$	1
$\frac{1}{2}$	$\frac{1}{8}$	$\frac{1}{4}$	$\frac{3}{8}$	$\frac{1}{2}$	$\frac{3}{4}$	1	2
$\frac{3}{4}$	$\frac{3}{16}$	$\frac{3}{8}$	$\frac{9}{16}$	$\frac{3}{4}$	$\frac{9}{8}$	$\frac{3}{2}$	3
1	$\frac{1}{4}$	$\frac{1}{2}$	$\frac{3}{4}$	1	$\frac{3}{2}$	2	4
$\frac{3}{2}$	$\frac{3}{8}$	$\frac{3}{4}$	$\frac{9}{8}$	$\frac{3}{2}$	$\frac{9}{4}$	3	6
2	$\frac{1}{2}$	1	$\frac{3}{2}$	2	3	4	8
4	1	2	3	4	6	8	16

Multiplication table for the seven positive values

Many products occur in multiple locations in the table. The backgrounds are color-coded by value, to make it easier to see the values that are the same. Allowing for the fact that $a \times b$ is the same as $b \times a$, there are $\frac{n(n+1)}{2} = \frac{7 \cdot 8}{2} = 28$ possible input pairs. However, there are only 18 *distinct* products since there is more than one way to produce the same product. Here are all the possible products:

$$c = \left\{ \frac{1}{16}, \frac{1}{8}, \frac{3}{16}, \frac{1}{4}, \frac{3}{8}, \frac{1}{2}, \frac{9}{16}, \frac{3}{4}, 1, \frac{9}{8}, \frac{3}{2}, 2, \frac{9}{4}, 3, 4, 6, 8, 16 \right\}$$

We can show the multiplication table in 3D by making the vertical direction the *rank* of the product in the list, from rank **0** for the smallest product, $\frac{1}{16}$, to rank **17** for the largest product, 16. The top left corner of the multiplication table becomes the corner closest to the viewer, and each vertical block is labeled by rank.

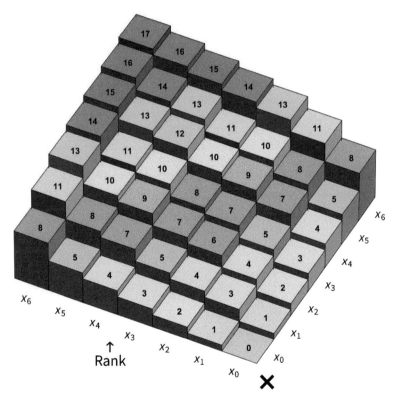

Rank plot for the multiplication table of the example set

It is like a very low-resolution contour plot. The *c* variable name for the list of possible outputs is for *contour line*.

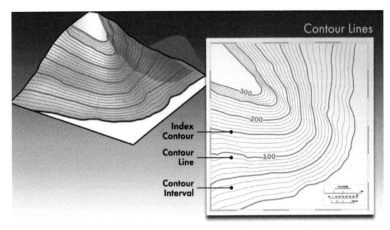

Topographic maps use contour lines to show altitude, much like a rank plot.

What if we could find a strictly increasing sequence of integers, one for each possible input, for which their *addition* table *produces the same rank plot*? It could then serve as a "logarithm table" but instead of the logarithms being decimal approximations of irrational numbers, we just use small integers that when added become pointers to the exact answer; that is, an index into a table. To cut to the chase, the answer is *yes*, we can. Call the sequence *L*, a reminder that the idea came from logarithms.

Here is one that works and uses the smallest possible integers:

$$L = \{0, 3, 5, 6, 8, 9, 12\}$$

Now build an addition table with the *L* values. Here are the two tables, multiplication of real values in *x* and addition of unsigned integers in *L*. Notice that the "contour lines" (same-colored squares) are identical.

×	$\frac{1}{4}$	$\frac{1}{2}$	$\frac{3}{4}$	1	$\frac{3}{2}$	2	4
$\frac{1}{4}$	$\frac{1}{16}$	$\frac{1}{8}$	$\frac{3}{16}$	$\frac{1}{4}$	$\frac{3}{8}$	$\frac{1}{2}$	1
$\frac{1}{2}$	$\frac{1}{8}$	$\frac{1}{4}$	$\frac{3}{8}$	$\frac{1}{2}$	$\frac{3}{4}$	1	2
$\frac{3}{4}$	$\frac{3}{16}$	$\frac{3}{8}$	$\frac{9}{16}$	$\frac{3}{4}$	$\frac{9}{8}$	$\frac{3}{2}$	3
1	$\frac{1}{4}$	$\frac{1}{2}$	$\frac{3}{4}$	1	$\frac{3}{2}$	2	4
$\frac{3}{2}$	$\frac{3}{8}$	$\frac{3}{4}$	$\frac{9}{8}$	$\frac{3}{2}$	$\frac{9}{4}$	3	6
2	$\frac{1}{2}$	1	$\frac{3}{2}$	2	3	4	8
4	1	2	3	4	6	8	16

+	0	3	5	6	8	9	12
0	0	3	5	6	8	9	12
3	3	6	8	9	11	12	15
5	5	8	10	11	13	14	17
6	6	9	11	12	14	15	18
8	8	11	13	14	16	17	20
9	9	12	14	15	17	18	21
12	12	15	17	18	20	21	24

The multiplication table for *x* has the same ranks as the addition table for *L*.

Here is the set of all possible sums of a pair of integers from *L*, and the corresponding set of all possible products of reals in *c*:

sums	0	3	5	6	8	9	10	11	12	13	14	15	16	17	18	20	21	24
products	$\frac{1}{16}$	$\frac{1}{8}$	$\frac{3}{16}$	$\frac{1}{4}$	$\frac{3}{8}$	$\frac{1}{2}$	$\frac{9}{16}$	$\frac{3}{4}$	1	$\frac{9}{8}$	$\frac{3}{2}$	2	$\frac{9}{4}$	3	4	6	8	16

It works like logarithm tables, but with all the rounding errors removed. I have shown this to enough experts in computer math and engineering to convince me that this is a result that has not been seen before. Amazing again, or *mirafici iterus*.

> It is possible to multiply real number formats without understanding anything about their exponent or fraction or having to decode their bit strings.

To spell out the process, a positive four-bit posit bit string 0001 to 0111 maps to the ith entry in the look-up table *L*, that is, L_i with indexing starting at 0. The entries are four-bit unsigned integers, since every integer in $L = \{0, 3, 5, 6, 8, 9, 12\}$ can be expressed with four bits. A second posit value maps to the jth entry, L_j. The sum $L_i + L_j$ can be between 0 and 24, so you need an adder that can hold a five-bit sum. That sum can be used to index into an array *z* that has the product. In short,

$$x_i \times x_j = z_{L_i+L_j}$$

If you did this in software, you would declare an array *z* with 25 entries and populate it like this, with the entries in whatever format you like:

z_0	z_1	z_2	z_3	z_4	z_5	z_6	z_7	z_8	z_9	z_{10}	z_{11}	z_{12}	
$\frac{1}{16}$			$\frac{1}{8}$		$\frac{3}{16}$	$\frac{1}{4}$		$\frac{3}{8}$	$\frac{1}{2}$	$\frac{9}{16}$	$\frac{3}{4}$	1	...

z_{13}	z_{14}	z_{15}	z_{16}	z_{17}	z_{18}	z_{19}	z_{20}	z_{21}	z_{22}	z_{23}	z_{24}
$\frac{9}{8}$	$\frac{3}{2}$	2	$\frac{9}{4}$	3	4		6	8			16

The software version of the idea wastes entries in storing the *z* table entries.

What you put into the result table depends on what the next use of the product is. If you want to go directly to the rounded posit representing the product, then you could put those in the table, and all the effort of rounding disappears. The destination of the result is directly back into the Compute Layer.

z_0	z_1	z_2	z_3	z_4	z_5	z_6	z_7	z_8	z_9	z_{10}	z_{11}	z_{12}	
0001			0001		0001	0001		0010	0010	0010	0011	0100	...

z_{13}	z_{14}	z_{15}	z_{16}	z_{17}	z_{18}	z_{19}	z_{20}	z_{21}	z_{22}	z_{23}	z_{24}
0100	0101	0110	0110	0110	0111		0111	0111			0111

No decoding and no rounding is needed to produce the correct posit result.

If the result is intended for the Math Layer, that is, you are about to add it to the quire, the table can store the exact products in fixed-point format:

z_0	z_1	z_2	z_3	z_4	z_5	z_6	z_7	z_8	z_9	z_{10}	z_{11}	z_{12}	
0.0001			0.001		0.0011	0.01		0.011	0.1	0.1001	0.11	1.	...

z_{13}	z_{14}	z_{15}	z_{16}	z_{17}	z_{18}	z_{19}	z_{20}	z_{21}	z_{22}	z_{23}	z_{24}
1.001	1.1	10.	10.01	11.	100.		110.	1000.			10000.

The fixed-point bit string in quire format allows exact multiply-accumulates.

If the next use for the product is the Human Layer, you could even have each table entry be a block of pixels, ready to send to a graphics display. Instant typesetting.

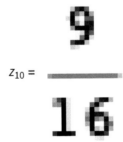

$$z_{10} = \frac{9}{16}$$

Highly magnified pixels ready to be stored in a display buffer

The empty cells, like for z_1 or z_{19}, are not populated because it is not possible to create an index that would point to that entry. In other words, the table is said to be *sparse*; it is not fully populated and that means some waste. But when the table is built into *hardware*, there is no waste, and I will show you why in a later section.

Before we leave the multiplication example, take a look at what it would have been using actual logarithms. The log table idea works no matter what the base is or how you scale the tables, so if $x = \{\frac{1}{4}, \frac{1}{2}, \frac{3}{4}, 1, \frac{3}{2}, 2, 4\}$, scale that by four to get $4x = \{1, 2, 3, 4, 6, 8, 16\}$. Then take the log base $x_1 = 2$ so that the log sequence will start $\{0, 1 \ldots\}$: $\log(4x) = \{0, 1, \log_2(3), 2, \log_2(6), 3, 4\}$. Scale that by $L_1 = 3$ and compare that to our L integer sequence:

	$x:$	$\frac{1}{4}$	$\frac{1}{2}$	$\frac{3}{4}$	1	$\frac{3}{2}$	2	4
$3 \times \log_2(4x):$		0	3	4.754 $\cdots$	6	7.754 $\cdots$	9	12
$L:$		0	3	5	6	8	9	12

That makes it look like all we have to do is find the right scale factors and we can get the L integer sequence simply by rounding the logarithms. But it is not that simple. First, the scaling by L_1 is not obvious since we do not know what L_1 is. Secondly, the roundings that produce the exact same ranking are not always to the nearest integer, and the trial-and-error guessing starting from rounded logarithms as a first guess quickly becomes intractable as you increase the precision of a format.

9.3 It Works for More than Multiplication

Compared to adding integers, adding or subtracting real numbers in a computer format is quite complicated. You have to scale the inputs to line up their radix points, and the sum has to be re-scaled and re-encoded back into the format with rounding. The method just described is powerful enough to dispense with all that complexity as well.

This time, assume the inputs are any real number in the 4-bit posit set, from −4 to 4. The addition table would have 15-by-15 entries. Here are all the exact sums possible when adding two such posits, again assigned to c:

$$c = \Big\{-8, -6, -\frac{11}{2}, -5, -\frac{19}{4}, -\frac{9}{2}, -\frac{17}{4}, -4, -\frac{15}{4}, -\frac{7}{2}, -\frac{13}{4},$$
$$-3, -\frac{11}{4}, -\frac{5}{2}, -\frac{9}{4}, -2, -\frac{7}{4}, -\frac{3}{2}, -\frac{5}{4}, -1, -\frac{3}{4}, -\frac{1}{2}, -\frac{1}{4}, 0, \frac{1}{4}, \frac{1}{2}, \frac{3}{4},$$
$$1, \frac{5}{4}, \frac{3}{2}, \frac{7}{4}, 2, \frac{9}{4}, \frac{5}{2}, \frac{11}{4}, 3, \frac{13}{4}, \frac{7}{2}, \frac{15}{4}, 4, \frac{17}{4}, \frac{9}{2}, \frac{19}{4}, 5, \frac{11}{2}, 6, 8\Big\}$$

Perhaps it is easier to read that set in decimal form:

$$c = \{-8, -6, -5.5, -5, -4.75, -4.5, -4.25, -4, -3.75,$$
$$-3.5, -3.25, -3, -2.75, -2.5, -2.25, -2, -1.75, -1.5, -1.25,$$
$$-1, -0.75, -0.5, -0.25, 0, 0.25, 0.5, 0.75, 1, 1.25, 1.5, 1.75, 2,$$
$$2.25, 2.5, 2.75, 3, 3.25, 3.5, 3.75, 4, 4.25, 4.5, 4.75, 5, 5.5, 6, 8\}$$

There could be as many as $\frac{n(n+1)}{2} = \frac{15 \cdot 16}{2} = 120$ different sums in the table, but because there are multiple ways to produce some of the sums, there are only 47 distinct real-valued sums.

Here is the contour plot of the rankings.

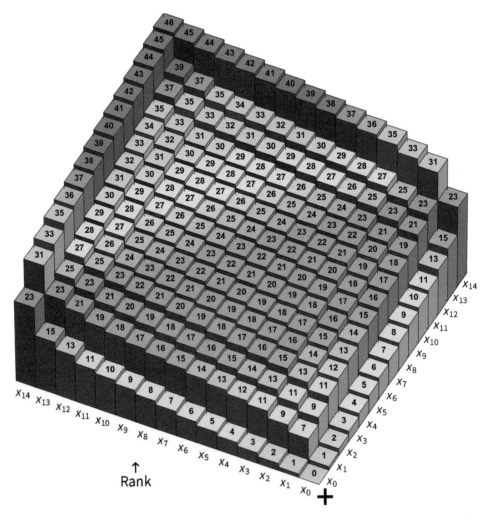

The rank plot for posit addition is not a simple tilted plane like the addition table for counting numbers.

Notice the stripe of orange rank **23** blocks; that is where the two inputs are negatives of each other and the sum is zero. If we were adding integers, the plot would look like a simple tilted plane, but the fact that we have real numbers unevenly spaced over a dynamic range makes the contour lines quite complicated. For example, the six blocks of rank **16** lie along an arc bent towards the viewer. See if you can find them in the figure. They are the places where $x_i + x_j = -\frac{7}{4}$:

$$x_1 + x_8 = \quad -2 + \frac{1}{4} \quad = -\frac{7}{4} \quad x_2 + x_6 = -\frac{3}{2} + \left(-\frac{1}{4}\right) = -\frac{7}{4} \quad x_3 + x_4 = -1 + \left(-\frac{3}{4}\right) = -\frac{7}{4}$$

$$x_4 + x_3 = -\frac{3}{4} + (-1) = -\frac{7}{4} \quad x_6 + x_2 = -\frac{1}{4} + \left(-\frac{3}{2}\right) = -\frac{7}{4} \quad x_8 + x_1 = \quad \frac{1}{4} + (-2) \quad = -\frac{7}{4}$$

As with multiplication, we need to find a set of integers L whose sums duplicate the ranking of posit addition. You are probably wondering how to find these sets of integers; I will explain that later, but for now, here is the set of smallest integers that can do that task:

$$L = \{0, 8, 10, 12, 13, 14, 15, 16, 17, 18, 19, 20, 22, 24, 32\}$$

Exercise for the Reader: The spacings between integers in L are symmetrical about L_{16}. Why is that? What property of the x list makes that the case in general?

The L table saves us from having to sum real numbers (in this case fractions); we only have to add a pair of unsigned integers. The blocks correspond perfectly to the ones in the 3D image of contours for real addition; the number they add to happens to be 25, but that is not important. It is only important that the integer sum has the same ranking as the real sum, rank **16**.

$$L_1 + L_8 = \quad 8 + 17 = 25 \quad L_2 + L_6 = 10 + 15 = 25 \quad L_3 + L_4 = 12 + 13 = 25$$

$$L_4 + L_3 = 13 + 12 = 25 \quad L_6 + L_2 = 15 + 10 = 25 \quad L_8 + L_1 = 17 + \quad 8 = 25$$

9.4 The Power to Compute Powers, x^y

Here is one more example to show the generality of the approach. One of the most difficult things to do in computer arithmetic is compute one number raised to the power of another. Unlike addition and multiplications for which $x + y = y + x$ and $x \times y = y \times x$, there is no such "commutative property" for x^y, so any table for x^y will not be symmetric about the diagonal.

The classic algorithm used to compute x^y is to express it as $2^{y \cdot \log_2(x)}$ using extra-high precision approximations for the log and the exponentiation base 2, round, and hope for the best. The approach is slow and sometimes rounds incorrectly.

As we explain later, the exact table technique is simplest when the contour plot is all uphill or all downhill in an input region. For x^y, we have that if $x > 1$ and $y > 0$. From the four-bit posit set, that means the base x is one of three values $\{1.5, 2, 4\}$ and the exponent y is one of seven values $\{0.25, 0.5, 0.75, 1, 1.5, 2, 4\}$. For that set, x^y ranges from $1.1\cdots$ to 256. There are $3 \times 7 = 21$ possible pairs x^y, but as with addition and multiplication, there can be more than one way to arrive at the same value for x^y, like $4^{0.25} = 2^{0.5} = \sqrt{2}$. Here are the 16 possible values for x^y, to three significant digits:

$$\{1.11, 1.19, 1.22, 1.36, 1.41, 1.50, 1.68,$$
$$1.84, 2.00, 2.25, 2.83, 4.00, 5.06, 8.00, 16.0, 256.\}$$

The rank plot for x^y looks like this:

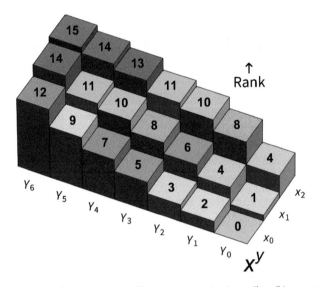

The rank plot for computing x^y is not symmetric since $x^y \neq y^x$ in general.

The asymmetry simply means that we need separate tables, L_x for x and L_y for y.

$$L_x = \{0, 3, 7\}$$
$$L_y = \{0, 4, 6, 8, 10, 12, 16\}$$

Populate a table z for indices from 0 to 23 with the precomputed values, and you can compute a power x^y in a single computer clock cycle.

With my apologies for having subscripts nested so deeply, here is the procedure in a single formula (with a large font to make it more readable):

$$X_i{}^{y_j} = Z_{L_{x_i} + L_{y_j}}$$

Perhaps it is easier to see in tabular form like the multiplication example:

x^y	$\frac{3}{2}$	2	4
$\frac{1}{4}$	1.11	1.19	1.41
$\frac{1}{2}$	1.22	1.41	2.00
$\frac{3}{4}$	1.36	1.68	2.83
1	1.50	2.00	4.00
$\frac{3}{2}$	1.84	2.83	8.00
2	2.25	4.00	16.0
4	5.06	16.0	256.

$+$	0	3	7
0	0	3	7
4	4	7	11
6	6	9	13
8	8	11	15
10	10	13	17
12	12	15	19
16	16	19	23

The table for x^y has the same ranks as the addition table for $L_x + L_y$.

The map-add-map method is easiest to apply, and the most efficient, when

- The operation is strictly monotone in the inputs. It can be strictly monotone increasing or strictly monotone decreasing, but the contour map has to look like a slope with no hills or dips or level spots in the *x* or *y* directions.
- There are many pairs of inputs that land on identical values; that keeps the integers in the *L* lists small.

9.5 Why It Is Really *Not* a Table Method

The IBM 1620 computer, introduced in 1959, used true table look-up.

Binary-Coded Decimal (BCD) refers to the use of binary numbers **0000** to **1001** to represent decimals 0 to 9. There are variations, but that is the most common way to do it. Computer calculations then are done with decimal digits just like grade school arithmetic. The IBM 1620 Model 1 computer used BCD. It used transistors, not vacuum tubes, but it still weighed more than a concert grand piano. Its cost was over a million present-day dollars after adjusting for inflation.

When the machine was properly initialized, it was loaded with the same ten-by-ten addition table and multiplication table you learned in grade school, with the sums for number 0–9 and the products for number 0–9. It was very much a "look-up table." The internal codename for the development of the IBM had been "CADET," perhaps a reference to "space cadet," but engineers on the project who realized it used table look-up instead of Boolean logic decided that it stood for "Can't Add, Doesn't Even Try." That joke spread throughout the fledgling computer industry of the time. The CADET, or IBM 1620, embodied "table look-up" as it is normally understood: Look up the block of values for the first input using memory addressing, then use the low-order bits to look up the addition table or multiplication table entry within that row; fetch the digits for use in completing the arithmetic task.

That is not even remotely like the method this chapter is talking about. If you are not familiar with logic gates, you need to grasp just three circuit symbols to understand this section. Those of you who know the basics of digital circuits can skip ahead.

9.5.1 AND Gates

Consider this very simple lightbulb circuit:

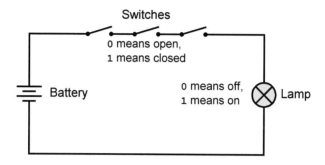

A three-input AND gate illustrated with switches in a lightbulb circuit

The lamp is only on (a **1** binary output) if the first AND second AND third switches are all closed (in their **1** state). It has to be unanimous for however many switches you have in a row. That is a three-input AND gate. In practice, microelectronics puts a limit on the number of inputs (sometimes as few as two), but it depends on the technology. There are techniques that can easily take the number up to ten inputs with only a slight increase in the delay.

If you have trouble thinking about electricity, imagine some plumbing with open or shut valves:

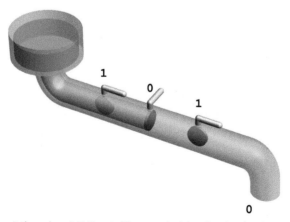

A three-input AND gate illustrated with valves in plumbing

The valves are in series, so all it takes is one shut valve to stop the flow of water.

With valve 1 AND valve 2 AND valve 3 open (their **1** state), water can flow (a **1** state).

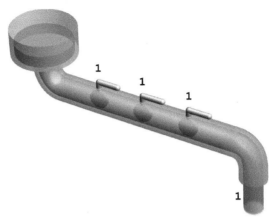

The AND gate plumbing analogy with all valves open, allowing water to flow

All of the valves have to be on (open) for the output to be flowing (on). The analogy is imperfect in that with computer logic, the output **0** or **1** values are in the same form as the inputs.

The standard circuit diagram symbol for an AND gate looks like this, though the **blue** coloring is not standard:

A standard two-input AND gate symbol

Making it blue helps clarify which lines mean logic and which lines mean wires.

9.5.2 OR Gates

Again, a simple lightbulb circuit can show how an OR gate behaves. The switches are in parallel, so any one of them being closed (the 1 state) allows the light to turn on.

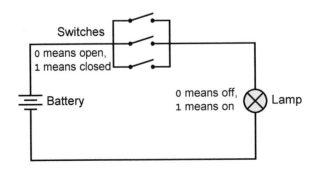

A three-input OR gate illustrated with switches in a lightbulb circuit

Similarly, here is a bit of plumbing that delivers no output unless the first OR second OR third valve is open. Unanimous **o** states deliver zero water flow.

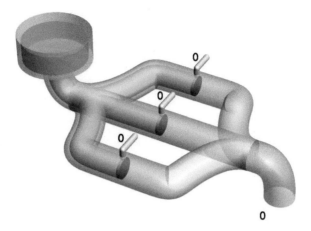

A three-input OR gate illustrated with valves in plumbing

The valves are in parallel, so all it takes is one open valve (**1** state) for the output to go to its **1** state (water flowing).

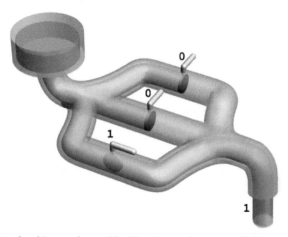

The OR gate plumbing analogy with at least one valve open, allowing water to flow

The standard symbol for an OR gate is shield-shaped like the AND gate symbol, but with a curved line on the input side and a pointed tip for the output:

A standard two-input AND gate symbol

In diagrams, AND and OR gates can have any number of input bits, but the practical number, called the *fan-in*, depends on the technology. With present-day chip logic, increasing the fan-in reduces the speed of the logic gate, so at some point it becomes advantageous to build a tree.

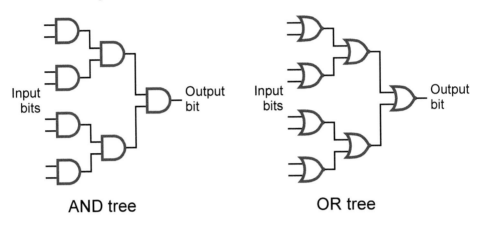

The AND tree will produce a **1** bit only if all input bits are **1**. The OR tree produces a **1** bit if any of the input bits are **1**. By the way, remember that posits have the two exception values when all but the first bit is a **0** bit. Hardware for posits can use an OR tree to test for that condition, while concurrently starting to decode the regime, exponent, and fraction.

9.5.3 The NOT Gate, or Inverter

We need one more logic element to explain how integer mapping works in hardware. An *inverter* turns a **1** into a **0**, and a **0** into a **1**. Its circuit diagram symbol looks like this:

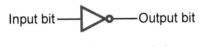

A standard inverter symbol

It is so simple that it does not really need a switching circuit or plumbing analogy. Imagine an electromagnet holding a spring-loaded switch open (no current flows, its **0** state) when the electromagnet is on (**1** state). Turn off the electromagnet (**0** state), and the switch snaps shut and allows current to flow (**1** state).

9.5.4 An Integer Mapping for the Example

The discovery of exact log tables has consequences for the way computer hardware for low-precision arithmetic is designed. Here is the table of input values and output values we need to store in a chip, for the 4-bit posit example of Section 9.2:

ARTIST STATEMENT

Inspired by John's vision and deep knowledge of all things mathematical, this artwork came fully formed in my mind's eye. Imbued with subliminal messages, it is pregnant with ideas and insights drawn from the writings in the book as we invite, ignite, and engage the curiosity of everyone who comes across the book.

John always wanted an artwork for his book that is engaging, beautiful, and done in a painterly manner. That way, the cover is open to interpretation by its viewers, its audience. Hence, this

piece of artwork is a collaborative visual piece that reflects both John's and my own vision and ideas. For relevance, coherence, and continuity, many of the motifs and forms in the artwork are drawn from those used in his charts, diagrams, and illustrations.

As a friend and advocate, it has been a pleasure as well as a privilege to be part of his book and his life.

—Sumei Chew, 14 August 2024

	x index				*L* values			
	MSB		LSB		MSB			LSB
1:	0	0	1	3:	0	0	1	1
2:	0	1	0	5:	0	1	0	1
3:	0	1	1	6:	0	1	1	0
4:	1	0	0	8:	1	0	0	0
5:	1	0	1	9:	1	0	0	1
6:	1	1	0	12:	1	1	0	0

Bit values needed to design a circuit that maps *x* to *L*

The value $x = 0$ is left out because its *L* value is also always 0. The following circuit shows the three bits of *x* introduced in the upper left, and the four bits of *L* generated on the bottom right. Here is a circuit that maps the input integers 1–6 to the *L* set {3, 5, 6, 8, 9, 12}:

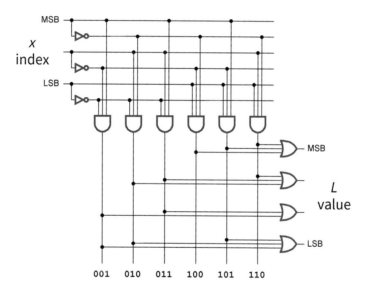

13 gates suffice to map the seven input posits to their 4-bit *L* values.

On an actual chip, the layout would not look like that because it is possible to place the gates much closer together and make the wires much shorter. This shows why the idea is not "table look-up." Sometimes, an integer mapping like this is called "wired ROM." It is much faster than a ROM (Read-Only Memory) where a grid of bits is stored with some kind of physical state, like charge or magnetization. The small black dots indicate what wires are connected together, and the location of those connections determines what the circuit does. The unused values in the "table" cost nothing because they do not exist.

At the time of writing, gates like AND have a delay time of a few picoseconds. That is followed by the integer add of the *L* values, and then a similar integer-to-integer map produces the product. So the map-add-map method can happen in a fraction of a clock cycle, and consumes very little electricity.

9.6 How Do We Find the *L* Sequences?

There are several approaches:

- Use logic to reason out the integers starting from the smallest ones and working up to the largest ones.
- Use exhaustive search (brute force) to gradually increase the integers until they match the behavior of the table.
- Treat the problem as a *linear programming* problem that satisfies a set of inequalities, and apply the powerful solvers that have been developed for such problems.

9.6.1 The Logic Approach

We know the *L* addition table has to produce ranks the same as the *x* multiplication table, so it must look like this:

```
L = {0, 3, 5, 6, 8, 9, 12};
sumRankPlot[L, L]
```

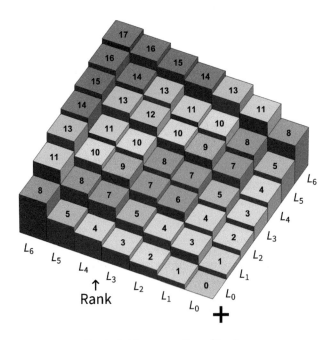

Rank plot for summation of *L* values

The ranks on the diagonal seem to be the richest source of information. Assume that $L_0 = 0$ to start. We do not yet know l_1 but we can try to express ranks L_2 and higher in terms of L_1. The notation $\in$ means "is an element of."

- The rank **3** block on the diagonal and the rank **3** block at the edge edge show that $L_0 + L_3 = L_1 + L_1$. Since $L_0 = 0$, this proves that $L_3 = 2L_1$.
- Strict ordering of the L means $L_1 < L_2 < L_3$, so L_2 is in the open interval $(L_1, 2L_1)$. Factor out L_1 and write that open interval as $L_2 \in L_1 \times (1, 2)$.
- The two ways to create a rank 5 block are $L_0 + L_5$ and $L_1 + L_3$. Apply the facts that $L_0 = 0$ and $L_3 = 2L_1$ to determine that $L_5 = 3L_1$.
- Strict ordering means $L_3 < L_4 < L_5$. That means L_4 is in the open interval $(L_3, L_5) = (2L_1, 3L_1)$, which we write as $L_4 \in L_1 \times (2, 3)$.
- Rank **8** blocks on the two corners and in the center tell us that $L_0 + L_6 = L_3 + L_3$, which means $L_6 = 2L_3 = 4L_1$ exactly.
- Rank **6** is only on the diagonal, $L_2 + L_2 = 2L_2 \in L_1 \times (2, 4)$. But we can narrow that interval because rank **5** < rank **6** and rank 5 is $3L_1$ exactly. So $2L_2 \in L_1 \times (3, 4)$, and this narrows the bound on L_2 to $L_2 \in L_1 \times \left(\frac{3}{2}, 2\right)$.
- That in turn lets us tighten the bound on L_4, because rank **4** says $L_4 \in L_1 + L_2 = L_1 \times \left(\frac{5}{2}, 3\right)$.

That says the L sequence can be expressed as $L_1 \times \left\{0, 1, \left(\frac{3}{2}, 2\right), 2, \left(\frac{5}{2}, 3\right), 3, 4\right\}$. Now the question becomes: What is the smallest integer L_1 such that L can be expressed as an integer sequence, and that satisfies $L_6 = L_2 + L_4$? Try $L_2 = 2$ and you get $\{0, 2, (3, 4), 4, (5, 6), 6, 8\}$. Remember, those are *open* intervals, so for $(3, 4)$ you cannot elect to use the endpoint 3; it is not in the set. So try $L_1 = 3$, and you get open sets big enough to contain an integer: $\left\{0, 3, \left(\frac{9}{2}, 6\right), 6, \left(\frac{15}{2}, 9\right), 9, 12\right\}$. The integer 5 is inside $\left(\frac{9}{2}, 6\right)$ and the integer 8 is inside $\left(\frac{15}{2}, 9\right)$. This proves that the smallest L set is $L = \{0, 3, 5, 6, 8, 9, 12\}$.

Exercise for the Reader: Suppose x is the set $\left\{1, \frac{3}{2}, 2\right\}$. Find the set of all possible products $x_i \times x_j$, and sketch the rank plot. Then find a sequence of increasing integers $L = \{0, L_1, L_2\}$ such that the set of all possible sums $L_i + L_j$ has the same rank and L_1 is as small as possible.

9.6.2 The Brute Force Approach

We can eliminate much (but not all) of the thinking by writing a program that searches all possible sets of increasing integer sequences until it finds one that meets all the equalities and inequalities of a rank plot. As usual, assume $L_0 = 0$, and write a software engineer's nightmare, deeply nested loops over every L_i that you cannot derive exactly from smaller L_j just by looking at the rank plot.

A linear binade from 2^2 to 2^3 with just two bits of fraction is $x = \{4, 5, 6, 7, 8\}$, if we include both endpoints. Here is what the rank plot looks like for multiplication:

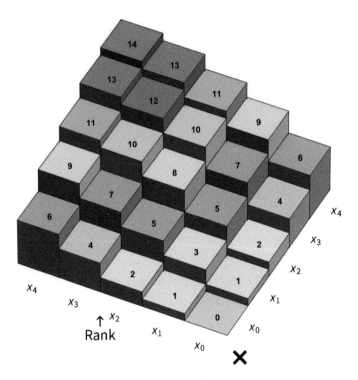

Rank plot for multiplication of $\{4, 5, 6, 7, 8\}$ inputs

Here is a set of nested loops that aborts when it finds a solution, which is the solution with smallest possible integers in L.

```
For[L1 = 1, L1 < ∞, L1++,  (* We don't know how big L1 will be.*)
  For[L2 = L1 + 1, L2 < 2 L1, L2++,  (* rank 1 < rank 2 < rank 3 *)
    For[L3 = 2 * L1 + 1, L3 < L1 + L2, L3++ ,
      (* rank 2 < rank 3 < rank 4 *)
    If[L1 + L3 < 2 * L2,  (* rank 7 < rank 8 *)
      For[L4 = L1 + L2 + 1, L4 < L1 + L3, L4++,
        (* rank 5 < rank 6 < rank 7 *)
      If[L2 + L4 < 2 * L3,  (* rank 11 < rank 12 *)
      L = {0, L1, L2, L3, L4};
      Abort[]]]]]]]]
L
```

$Aborted

{0, 6, 11, 15, 18}

This only partially automates the process, because you only discover things like the two **If** statements after you run the program and notice that some of the block heights are equal and should not be. The instant giveaway that something is wrong is if the highest rank is not the same as that of the desired rank plot.

Check that the *L* sequence {0, 6, 11, 15, 18} works:

```
sumRankPlot[L, L]
```

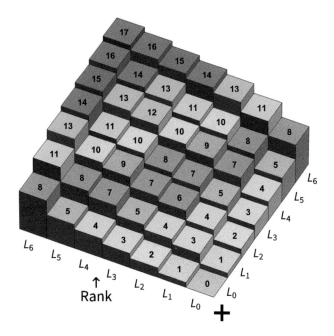

The rank plot for addition of *L* inputs matches the rank plot for multiplication of *x* inputs.

Incidentally, what if we instead wanted to find the *L* sequence for the multiplication table of the next larger binade, $x = \{8, 10, 12, 14, 16\}$? *It is the same sequence.* Multiplying *x* by a positive number (in this case, 2) does not change the multiplication rank plot. You can also take a subset; suppose the binade only had one bit of fraction so it is $x = \{4, 6, 8\}$. Simply take the corresponding entries in *L* (the first, third, and fifth entries), so $L = \{0, 11, 18\}$. That will not be the *L* with the smallest integer values, but it will work.

A table of *L* for multiplication of 8-bit standard posits is in Appendix D.

The identity $\log(\frac{1}{x}) = -\log(x)$ works for these *L* tables as well.

```
x = Reverse[1 / {4, 5, 6, 7, 8}]
mulRankPlot[x, x]
```

$$\left\{\frac{1}{8}, \frac{1}{7}, \frac{1}{6}, \frac{1}{5}, \frac{1}{4}\right\}$$

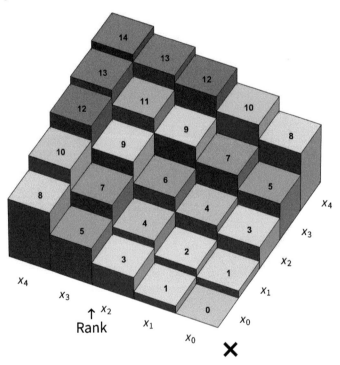

The rank plot for multiplication of the reciprocals, $X = \left\{\frac{1}{8}, \frac{1}{7}, \frac{1}{6}, \frac{1}{5}, \frac{1}{4}\right\}$

```
L = Reverse[18 - {0, 6, 11, 15, 18}]
sumRankPlot[L, L]
```

{0, 3, 7, 12, 18}

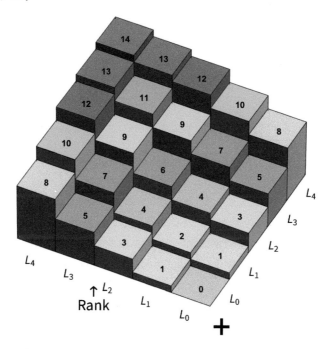

Subtract *L* from its largest value to create the *L* table for reciprocals, {0, 3, 7, 12, 18}

9.6.3 The Integer Linear Programming Approach

After I showed the idea to Marco Cococcioni and his team at the University of Pisa in 2021, they found a way to express the problem of finding *L* sequences as a *linear programming* problem, which is when a set of weighted sums must satisfy *inequalities*, not equalities. This means that a vast set of optimized tools exists to find *L* sequences, since linear programming has been an important job for computers for many decades.

They also helped me crack this problem: How do you prove that *L* always exists? Are there unsolvable situations?

Imagine that we have two sequences $L_x = \{0, 1, 2, 3, 4, 5, 6, 7, 8, 9\}$ and $L_y = \{0, 10, 20, 30, 40, 50, 60, 70, 80, 90\}$. What would the addition table look like? You might be able to do it in your head, but here it is, spelled out:

+	0	1	2	3	4	5	6	7	8	9
0	0	1	2	3	4	5	6	7	8	9
10	10	11	12	13	14	15	16	17	18	19
20	20	21	22	23	24	25	26	27	28	29
30	30	31	32	33	34	35	36	37	38	39
40	40	41	42	43	44	45	46	47	48	49
50	50	51	52	53	54	55	56	57	58	59
60	60	61	62	63	64	65	66	67	68	69
70	70	71	72	73	74	75	76	77	78	79
80	80	81	82	83	84	85	86	87	88	89
90	90	91	92	93	94	95	96	97	98	99

An addition table that produces every value exactly once

It is simply every integer 0–99. Each can be considered a pointer to the value of whatever two-input function you like, so yes, a solution always *exists*. But that would only be optimal if no two entries in the table were the same. If there are "contour lines" of values or the function is commutative (x **op** y is the same as y **op** x for operation **op**, as it is for + and × operations), then there will be different ways of producing the same result, and the above table will not use the smallest possible integers.

When I showed this table to Professor Cococcioni, he pointed out that there is a theorem in linear programming that says *an optimal solution always exists, if there is any solution at all.* Since we can always create a table like the one above as proof of the existence of at least one solution, that shows that yes, there is always an optimal L that uses the smallest possible integer values.

9.7 How Well Does the Approach Scale?

The approach does not scale well to high precision formats, and that is the catch. If you want to use it for plus-minus-times-divide, the integers in the L sequences get quite large even for an 8-bit representation. Getting up to 16-bit precision appears impractical because the L lists would take up so much circuitry that you are better off building conventional hardware for integer addition, multiplication, shifting, and rounding as building blocks.

This is a powerful approach for the kind of low precision used in AI, however. As pointed out in Section 9.2, the output of a multiplicaton using L sequences could be a fixed-point number ready to add to the quire. Most of an AI workload, whether doing training or inference, consists of multiplying two low-precision numbers and adding the result to a higher-precision accumulator. Presently, a typical neural network might be trained using some form of 16-bit floating-point for the numbers being multiplied, and a 32-bit floating-point format used for the accumulator.

At the time of writing, a 16-bit integer add takes only about 60 picoseconds. Hence, the exact map-add-map approach described in this chapter might well be the fastest (and most energy-efficient) arithmetic for AI.

"EVERY ONCE IN A WHILE I JUST LIKE TO UNWIND WITH A LITTLE ADDITION AND SUBTRACTION."

10 The Minefield Method

Canals on Mars and taste zones on the tongue were persistent myths in the scientific community. In this chapter, I will expose some similarly persistent myths about computer arithmetic.

10.1 The "Table-Maker's Dilemma"

When Napier was making his tables of logarithms over 400 years ago, he noticed that for some entries it was very difficult to know which way to round the last decimal, up or down. If the table shows the log to four decimal places, say, and the method used to compute it gives you 0.84235000 ± 0.0000001, should you say that it is 0.8423 or 0.8424? If it is 0.84234999, it rounds down to 0.8423, but if it is 0.84235001, it rounds up to 0.8424. For some inputs, you can compute dozens of zeros in a row (or dozens of nines) and still not know whether to round up or round down. The same thing happens with tables of trigonometric functions like sine and cosine and exponential functions like exp(x). William Kahan coined the term "Table Maker's Dilemma" for this problem, sometimes abbreviated TMD.

Why worry about such a tiny thing? Clearly, both values are very close to the true value, and they are off by very close to the same amount, so it might seem pedantic to make an issue of it. However, on computers it can be a Big Deal, because it can lead to math functions that give *different answers on different computers*, or even on the same computer using two different math libraries for the special functions. The IEEE Std 754 committee faced the ugly choice between allowing different answers on different computers (wait, was it supposed to be a *standard*?) or requiring some extremely expensive ultra-high precision for a few rare cases.

As mentioned in Chapter 4, Kahan initially favored the latter choice so that there would be absolute bitwise reproducibility, but that went against the design of the Intel i8087 coprocessor. So, irreproducibility was written into IEEE Std 754, and it persists even to this day (the last revision was in 2019). Many computer users are completely unaware that following IEEE Std 754 to the letter still will not guarantee that you get the same answer with different computing environments.

I visited one of the large financial institutions in London a few years ago where they manage a portfolio worth billions, and they ran a sophisticated program to assess the value of the portfolio and which assets they should buy more of, or sell, or hold. They had a supercomputer cluster based on Intel microprocessors and another cluster based on microprocessors from Advanced Micro Devices (AMD), running exactly the same software. But in assessing the portfolio value, the two clusters disagreed by about $45,000. What was worse, some stocks computed as **SELL** on one cluster and **BUY** on the other one. That was not something they wanted to have to explain to their clients.

The cause was differences in the math libraries each system used. They had been hit by the TMD, and the consequences were well beyond academic.

10.2 Intel's Pentium Divide Bug

Fun fact: After Intel had released the 286, 386, and 486 versions of their x86 processor, people bought trademark rights to web sites and branding using "586" in anticipation of being able to gouge Intel for the usage rights. Intel simply changed their branding to "Pentium." Problem solved.

From the viewpoint of a numerical analyst, one of the most hilarious debacles ever to hit the computer industry was when someone discovered a tiny and rarely-occurring error in the floating-point divide instruction (FDIV) on what was then Intel's latest version of its x86 family of chips, the Pentium. The bug was discovered not by Intel but by Thomas R. Nicely, a math professor at Lynchburg College. Analysis showed that if you were dividing one random number by another, the answer would be off in the ninth or tenth decimal in about one in 9 billion cases. It also showed that if you went out of your way to construct a particularly awful case, it could be wrong in the fifth digit.

Word spread like a political sex scandal. An Intel processor can make mistakes! IBM paused the sales of its PCs. Intel's stock crashed. This was very amusing to anyone who knows anything about floating-point arithmetic; users were clutching their proverbial pearls at the horrifying idea that a floating-point operation could produce an error. Even though floating-point computations produce an ocean of rounding errors, billions per second, Intel was dragged over the coals for having added a few drops of water to that ocean. Customers demanded replacement chips with the bug fixed. Intel had to set up a call center and train employees how to handle the angry customers, and it took many months to resolve. The cost to Intel was $475 million in 1994, which is over $1 billion as I type this, thirty years later.

Intel initially tried to explain how minor and inconsequential the bug was, which backfired as a public relations tactic. It made the angry customers angrier. There was an easy software patch that solved the problem, but no, customers wanted it to be treated like the recall of a car with brakes that could sometimes fail suddenly and without warning. The joke spread that the ![Intel inside logo] sticker on a PC was a warning label.

The entire episode illustrated the amazingly widespread ignorance of how bad floating-point arithmetic is. While some people realized that floating-point arithmetic produces rounding errors, they wanted those rounding errors to be *identical* from one system to another. Very few people realized that getting different answers on different computers had been codified into the IEEE standard since 1985.

10.3 The TMD Hits the Posit Standard (2022)

> **5 Functions**
>
> ...
>
> The following functions shall be supported, with rounding per Section 4.1...

Those "following functions" included the hard-to-round functions like $\log(x)$, $\sin(x)$, and $\exp(x)$, so that seemingly innocuous requirement actually caused me to sweat bullets. But I knew it had to be there for bitwise reproducibility, and I knew computing environments existed (including *Mathematica*) that could certainly produce correct roundings. But that perfection did not come cheap at the time.

10.3.1 Two Ways to Be Irrational

In common English, the primary meaning of "irrational" is unreasonable or illogical. That word was co-opted from mathematics, however, and describes a number that is not expressible as the ratio of two integers, $\frac{p}{q}$.

It may seem like only a mathematician would care, but please bear with me. There are two categories of irrational numbers: *algebraic* and *transcendental*.

> **Definition**: An *algebraic* number is one that is the root of a nonzero polynomial of finite degree with integer coefficients.

For example, the solution of $x^2 - 2 = 0$ is $x = \sqrt{2}$, so $\sqrt{2}$ is algebraic. So is the Golden Ratio, $\phi = \frac{1+\sqrt{5}}{2} \approx 1.618$ since it is the solution of the polynomial equation $x^2 - x - 1 = 0$. Algebraic numbers can be expressed in finite closed form using radical signs like $\sqrt[n]{}$ where n is an integer. Algebraic numbers contain the rational numbers, which contain the integers, so it is a hierarchy. They can be rational or irrational.

> **Definition**: A *transcendental* number is one that is not algebraic.

These are the numbers like $\pi = 3.14\cdots$ and Euler's constant $e = 2.718\cdots$. There is no way to express them with a finite expression involving integers and radicals.

Why should anyone care? Here is why: algebraic numbers are much easier to calculate and round, and you can tell in advance how much storage is needed and how much work it will take to compute them in a math library.

With transcendental numbers, there is no way to know in advance how much work it will take to compute the correctly-rounded result. And it so happens that trig, exponential, and log functions produce transcendental numbers in general.

10.3.2 The Method of Ziv

An IBM mathematician, Abraham Ziv, proposed (and in 1999 built) a simple strategy for dealing with the TMD. First try a method for computing the transcendental math function that optimistically uses a few extra bits of precision than needed. Most of the time, that will be enough to know how to round. For the less common cases, use much higher precision (like double what was used initially) and re-calculate to many more bits of precision. If you hit a crazy case where the computation seems to land exactly on the midpoint for over a hundred bits in a row, then invoke a very expensive method to keep going out even farther (Ziv used 768 bits for that third tier of the approach). There is no proof that three tiers always suffice, so I presume Ziv's software produces an error message for that extremely unlikely case.

When working with 64-bit reals, this is still one of the most viable approaches. Its main drawback is that it can produce surprises in performance that depend on the input data. Asking for the function of a particular number could cause that brief part of the code to run orders of magnitude slower, inexplicably.

10.3.3 The Most Common Approach

Perhaps the two most common math libraries in use currently are the ones for GNU (`libc`) and for Intel (`icc`). They opt for correct rounding in *most* cases, and admit that the answers will be off by one ULP for tough-to-round cases. Unfortunately, there is no similarity in the cases of incorrect rounding between the two, so changing compilers will in general change the answers.

Studying all possible inputs is intractable with 64-bit floats, but not that difficult with 32-bit floats (about 4 billion possible inputs) at modern computer speeds. Paul Zimmermann has done a careful study of the two libraries which he periodically updates: "Accuracy of Mathematical Functions in Single Precision." Here is an excerpt from an update he generated on May 26, 2020:

	GNU			Intel		
Function	≥ 1 ULP	≥ 2 ULPs	Max ULPs	≥ 1 ULP	≥ 2 ULPs	Max ULPs
atan	21 089 464	0	1	505 178	0	1
cos	28 209 642	0	1	15 732 888	0	1
cosh	17 868 534	3558	2	139 708	0	1
exp	170 648	0	1	250 299	0	1
log	416 908	0	1	1045	0	1
sqrt	0	0	0	0	0	0

A sampling of the many incorrectly-rounded results from the GNU and Intel math libraries

Notice that the square root function **sqrt** does not have any problem producing correct rounding. As explained above, the square root is an algebraic function and thus much easier to get right. All the others are transcendental functions. Both GNU and Intel had some functions that were off by as much as 2 ULPS, but they were less commonly-used functions.

There is a more relaxed accuracy goal called "faithful rounding."

> **Definition:** *Faithful rounding* of a real value *x* is rounding it to one of the two representable values that flank it, not necessarily the closest one.

With faithful rounding, you can never be off by more than 1 ULP. Some of the functions in both libraries can claim to follow faithful rounding, but not all. The two main drawbacks of settling for faithful rounding are, first, you can get different answers from different math libraries, and second, there is no guarantee that a monotone function will be approximated with a monotone set of values. Behavior like this can happen:

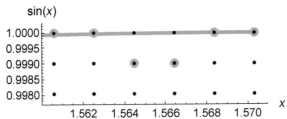

Non-monotone behavior can ruin some algorithms by preventing convergence.

The black dots are at (*x*, *y*) locations where both *x* and *y* are representable in the number format. The green line is the correct value of the sin(*x*) function just a little bit to the left of $x = \frac{\pi}{2}$. The light purple dots are all faithful roundings of the function, since sin(*x*) is a tiny bit less than 1.0 in that region. To my ear, "faithful" sounds like a euphemism for "incorrect." Incorrect but well-intentioned.

Zimmerman notes that even using the same compiler and same operating system, he got different answers on two different Intel processors. An Intel Xeon E7-4850 result was 1 ULP different from that of an Intel Xeon E7540. He attributes this to one using a fused multiply-add. It is because of such cases that the Posit Standard forbids covert "help" in reducing rounding error; whether fusing is done must be visible in the source code.

With correct rounding, a result is never more than 0.5 ULP from the exact value. It is identical across systems, and monotone where the exact function is monotone. The Posit Standard requirement of perfect rounding of transcendental functions was set in the hope of finding an inexpensive solution to the TMD. Fortunately, I found one, and I call it the Minefield Method.

10.4 Mapping the Minefield

A common way of expressing the TMD graphically is to zoom in on the grid of possible inputs x and possible outputs y so that the ULP spacings are visible, and then show the exact function that must be rounded to the nearest value. For example, here is the cos(x) function in green, with a short piece of it magnified:

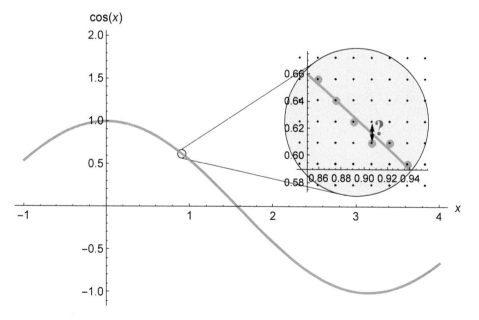

Cosine function with a zoom-in to show the discrete grid and a hard-to-decide rounding with a "?"

This is with a (linear) binade where expressible values are spaced 1/64 apart in the range $1/2 \leq x < 1$, like a 10-bit standard posit. Low precision helps make visible what the problem is. In the zoomed-in circle, the black points are the expressible values, and the light purple dots are the closest values. But notice how difficult the decision is for the point to the left of the "?" symbol. The green line of the cosine function passes almost exactly halfway between the two possible roundings, forcing the use of a very high-accuracy approximation to cos(x) to break the tie.

The diagram is a bit hard to understand. Suppose we instead plot the grid of expressible values minus the cosine, so that the x-axis becomes the definition of an exact evaluation. It is like driving on a cosine-shaped road and seeing expressible points from the driver's frame of reference.

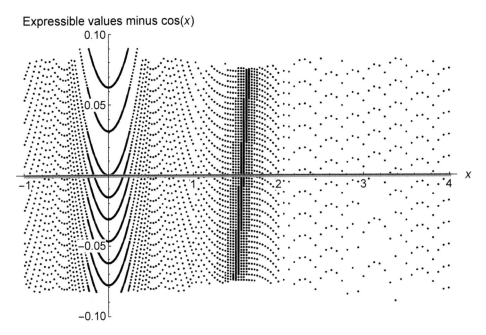

Grid of representable discrete points relative to the function in question, in this case cos(x)

That is the big picture. We are interested in which points are very close to the green line, so zoom in for the vertical axis:

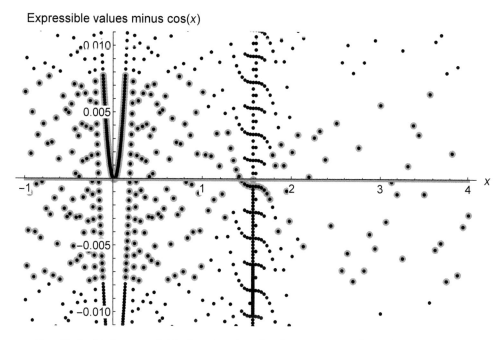

Expressible values minus cos(*x*)

Magnified grid of representable discrete points showing correct roundings of cos(*x*) in light purple

The points highlighted in light purple are again the correctly rounded choices for cos(*x*). When there is just one point that lies close to the green line, it is easy to pick the right one. The problem is when there are two choices almost the same distance from the green line, one above and one below. For every expressible *x* value, there is a column of points spaced one ULP apart above and below it. Suppose *y* is the correctly-rounded value. If your approximation goes above $y + 0.5$ ULP, you are in trouble because then the approximation will round to $y + 1$ ULP and be just as wrong as if you computed $2 + 2$ and got 5.

The point $y + 0.5$ ULP is a *mine*. So is the point $y - 0.5$ ULP, because if the approximation goes below that, then it will round to $y - 1$ ULP and therefore be incorrectly rounded. Boom.

> **Definition:** The *minefield* is the set of all mines, that is, the set of all (x, y) points where *x* is an expressible input to a function and *y* is a value either 0.5 ULP too high or 0.5 ULP too low to round to the correct output value. The *y* values are always halfway between adjacent expressible numbers in the number format.

See what the minefield looks like for the example function. The following code finds the `tooHigh` mines and the `tooLow` mines:

```
tooHigh[x_] := Module[{p},
   p = xToP[Cos[x]];
   (pToX[p] + pToX[p + 1]) / 2 - Cos[x]]
tooLow[x_] := Module[{p},
   p = xToP[Cos[x]];
   (pToX[p - 1] + pToX[p]) / 2 - Cos[x]]
minesHigh =
   Table[{pToX[p], N[tooHigh[pToX[p]]]}, {p, xToP[-1], xToP[4]}];
minesLow = Table[{pToX[p], N[tooLow[pToX[p]]]},
   {p, xToP[-1], xToP[4]}];
```

Here is a plot of the `tooHigh` mines in **blue** and the `tooLow` mines in **red**. Notice how very close they come to the *x*-axis (the green line):

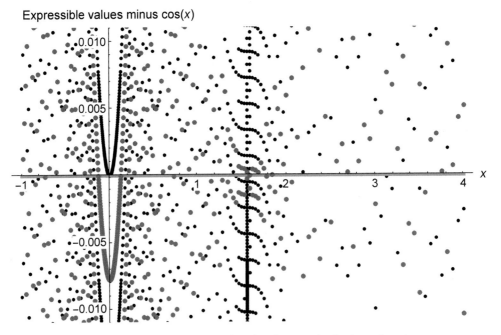

Discrete grid points with `tooHigh` (**blue**) and `tooLow` (**red**) mine points shown

Even though the number system is very coarse, and the vertical scale here goes from –0.01 to 0.01, the mines come as close to the line as 2.4×10^{6}. To make that visible, leave out the black dots for the expressible values, and blow up the vertical axis by a large amount:

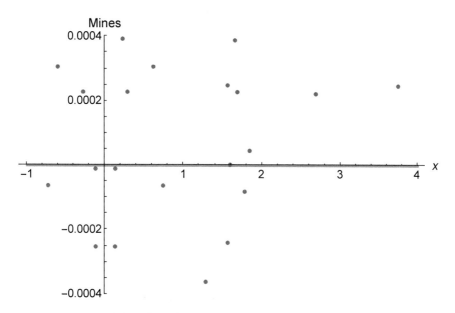

Minefield without the discrete grid points (expressible values)

The approximation is the green line. But why does it have to be a line, not a *curve*?

10.5 The Biggest Lie in Numerical Analysis

In the dozens of books and hundreds of papers about how to approximate functions on a computer, they all say the same thing about perfect rounding for every input value, and please notice that this statement is highlighted in yellow:

> If you want an approximation to a function $f(x)$ that is correctly rounded for *every* input, it will have to do calculations accurate to hundreds of bits and methods that use a massive number of calculations. You should therefore settle for a few points that are off, and for results that differ from one math library to another.

That is the position of the GNU and Intel (and many other) math libraries, as we saw earlier. And it is a universally-repeated fallacy. Many person-years have been spent finding ways to make the right tradeoff between being fast and being sufficiently accurate that customers will not complain too much, like they did about the Pentium Divide Bug.

I suspect the reason no one ever saw the right way to approach the problem is because they never visualized the problem as shown in the previous section, by making the function the frame of reference instead of the grid of expressible values. I show it again here, and connect the dots that show the mines:

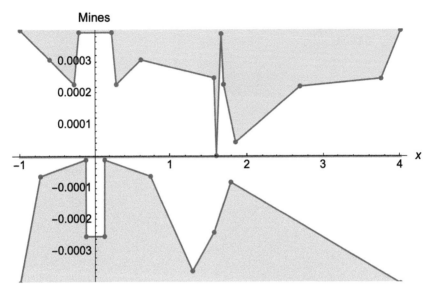

An exact green line comes *very* close to nicking two red mines and a blue mine.

Look at all that white space between the **tooHigh** and the **tooLow** mines. Any function that stays inside that white space will round the same way the cos(x) function does, even though it weaves and wanders like a drunken driver, far from the high-accuracy approximation. Like this one:

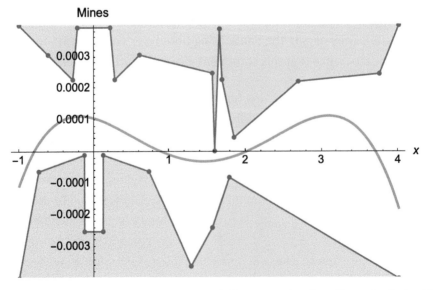

There is ample white space between mines in which an approximation will round as it should.

That is a very simple fourth-degree polynomial, computable at low precision. The biggest lie in numerical analysis is that the green line has to lie almost perfectly on the *x*-axis to avoid incorrect rounding.

> For an approximation to achieve correct rounding of $f(x)$ for every input x, it is not necessary that the approximation be extremely close to $f(x)$. It needs only to round the same way $f(x)$ rounds, and that is a *much* looser restriction.

You do not have to drive straight along the *x*-axis. In driving a military vehicle through a minefield, you can steer left and right as necessary, just so long as you do not hit a mine. The only catch is that you have to know in advance how $f(x)$ rounds. You need an "oracle" that can tell you where the hard-to-round points are so you can map the minefield. One way to compute $\cos(x)$ to arbitrarily high precision is to use this infinite series:

$$\cos(x) = 1 - \frac{x^2}{2} + \frac{x^4}{24} - \frac{x^6}{720} + \frac{x^8}{40\,320} - \frac{x^{10}}{3\,628\,800} + \frac{x^{12}}{479\,001\,600} + \ldots$$

The denominators are the *factorials* of the even numbers, where the factorial of a positive integer n is $n \times (n-1) \times \ldots \times 2 \times 1$, usually written "$n!$". If you have n different objects, $n!$ is the number of ways you can arrange them in a row. for example, $4! = 4 \times 3 \times 2 \times 1 = 24$. With enough terms and enough precision, you can always resolve the TMD with that formula by finding the handful of input cases that are going to be hard to round, and then steering the approximation to avoid them.

10.6 One Rounding to Rule Them All

In 2021, I shared the Minefield Method with a numerical interest group on the web. At Rutgers University, Professor Santosh Nagarakatte and his team of exceptional graduate students (Jay Lim in particular) were able to state my method more formally as a problem in linear programming. That allowed us to apply established linear programming solvers to automatically construct approximations that round perfectly. They work for any defined real number format. The software is open-source and available on Github, and quite easy to use.

This was a relief. The method of Ziv is not necessary. The functions required in the Posit Standard can be computed quickly and with fairly uniform speed, at least up to 32-bit precision for which you can find all the mines in a reasonable time.

For 64-bit precision, INRIA researcher Vincent LeFèvre has done pioneering work in finding and publishing the hard-to-round points for the transcendental functions. A little sophistication in the algorithm goes a long way toward knowing where to look. A 64-bit posit math library that meets the Standard is certainly possible.

Remember that standard posits can be any precision 2 bits or greater. Does this mean that we need a different math library for every precision? Thanks to a fiendishly clever trick that was published by Silvie Boldo and Guillaume Melquiond in 2004 (https://www.lri.fr/~melquion/doc/05-imacs17_1-article.pdf,) we do not. First, you need to know about a pitfall called *double rounding*.

> If you round to a lower precision, and then to a still lower precision, you can get an incorrect rounding compared to rounding in a single step.

For example, suppose you round 2.6 to the nearest integer, which would be 3. Try doing it in two stages: First round to the nearest multiple of 0.25. The options are 2.5 and 2.75, so use the closer value of 2.5. Now round that to the nearest integer, and you get 2 (since ties go to even) instead of 3. Oops. That is a double-rounding error.

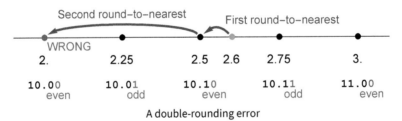

A double-rounding error

It is possible to do two roundings and always get a proper banker's rounding, if you do a not-very-obvious trick: Make the first rounding *round to odd*. That is not one of the standard roundings. The binary for 2.5 is **10.10** (even) and the binary for 2.75 is **10.11** (odd), so round-to-odd would round 2.6 up to 2.75, not the closest value. If a value is already exact in the format, round-to-odd does nothing to change the value. This cleverly stores the information needed for banker's rounding in the last bit of the number. If you then round again to a lower precision, it will be the same answer as if you used banker's rounding in the first place.

Using "round to odd" for the first rounding prevents double-rounding errors

The problem happens if the first rounding changes a value to land on a tie point. By instead bumping such cases to the nearest *non*-tie point, that records which side of the tie point it used to be.

Now imagine that we create a math library for all 33-bit posits, where the last bit was determined by round-to-odd. That is, the minefield finds functions that round-to-odd the same way the exact function to very high precision would round-to-odd. Now it is safe to do second roundings down to any precision from 2 bits to 32 bits.

Of course, in the interest of speed, it would be better to craft simpler approximations for the lower precisions. The minefields for lower precisions are a subset of the 33-bit minefield, so lower-order approximations and fewer subregion cases are needed. But the round-to-odd trick gives you something that is compliant with the standard, easily.

Intel's proprietary `icc` library has usually been the fastest of the math libraries with faithful rounding. The Rutgers team used the Minefield Method to create a 32-bit library for the usual math functions, and timed it. It was significantly *faster* than `icc`, since it was able to use much simpler approximations. But unlike `icc` and GNU, *every* result was round-to-nearest. Correctness with higher speed.

10.7 What about the Power Function, x^y?

This statement by William Kahan appears in many places on the web:

```
Nobody knows how much it would cost to compute y^w correctly
rounded for every two floating-point arguments at which it
does not over/underflow... No general way exists to predict
how many extra digits will have to be carried to compute a
transcendental expression and round it correctly to some preas-
signed number of digits. Even the fact (if true) that a finite
number of extra digits will ultimately suffice may be a deep
theorem.
```

posted at http://www.cs.berkeley.edu/~wkahan/LOG10HAF.TXT

In other words, x^y (or as Kahan writes, y^w) falls victim to the TMD. And that looks like very bad news for the Minefield Method because it is a two-input function, so even for 32-bit precision a (naive) exhaustive search for all the mines would mean computing 2^{64} cases, which is intractable. Fortunately, Kahan's statement is false.

Warning: Algebra ahead. For floats (and posits with linear binades), we can always express the value they represent as $m \times 2^j$ for integers m and j. Say $x = m \times 2^j$ and $y = n \times 2^k$, and also assume $m > 0$ to avoid getting non-real (complex) answers. If $n = 0$ then $x^y = 1$, so the following assumes $n \neq 0$. Computing $z = x^y$ means that

$$z = \left(m \times 2^j\right)^{n \times 2^k}.$$

The idea is to apply algebra steps to that expression until it expresses a polynomial in z with integer coefficients. That would mean z is algebraic, and the TMD does not apply as Kahan asserts. The steps depend on the sign of j, the sign of k, and the sign of n, so there are eight cases to consider. It is always possible to express it as such a polynomial, for which $z = x^y$ is the root. I will work through just two of the eight cases here, those where $j \geq 0$ and $n > 0$.

We know $m \times 2^j$ is a positive integer. If $k \geq 0$, then $n \times 2^k$ is an integer, so z is simply an integer to an integer power, which is an integer and not transcendental. If $k < 0$, then 2^{-k} is an integer. If your algebra is rusty, remember that $\left(a^b\right)^c = a^{b \times c}$. Raise both sides of the equation to the integer power 2^{-k}, and you get this:

$$z^{2^{-k}} = \left(m \times 2^j\right)^n.$$

That means z to the power of an integer equals a positive integer, and the above equation can be written

$$z^{\text{integer1}} - \text{integer2} = 0.$$

Both cases are expressible as finding the roots of polynomials in z with integer coefficients. Remember the two kinds of irrational numbers?

The power function for floats and posits is *algebraic*, not *transcendental*.

Exercise for the Reader: Complete the proof that z is algebraic for the other cases of the signs of j, k, and n (assuming $m > 0$). Find the polynomial in z with integer coefficients that corresponds to computing $1.25^{0.875}$.

10.8 A Summary with the Help of Arthur C. Clarke

The classic comedy *Grumpy Old Men* features Jack Lemmon and Burgess Meredith as a son and father with the same name as mine. The latter character has been used to make some excellent memes on the web, like this one:

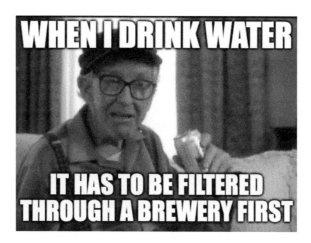

Imagine my surprise when I did a web image search for "john gustafson" "meme" and the top hit was this one:

I did not realize my technical disagreements with Professor Kahan had reached that level of public interest.

Arthur C. Clarke may be best known for his science fiction like *2001: A Space Odyssey*, but he was also a very capable technologist and futurist. For example, he conceived of the communication satellite, and his friend John R. Pierce who was Director of Bell Labs then made Clarke's invention a reality.

Clarke made an observation that seems appropriate as a way to end this chapter.

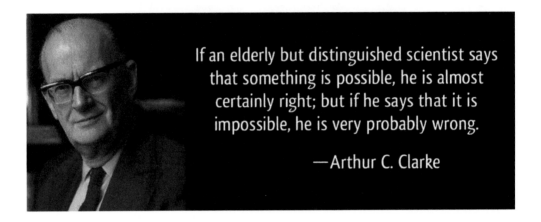

If an elderly but distinguished scientist says that something is possible, he is almost certainly right; but if he says that it is impossible, he is very probably wrong.

—Arthur C. Clarke

11 Matrix Multiply, Visualized

The transition from 2D to 3D video games became possible once the game consoles could perform thousands of 4-by-4 matrix multiplies per frame at video frame rates.

11.1 Matrix Multiplication as Origami

Matrix-matrix multiplication is at the heart of so many computer applications that it may be *the* most important basic operation of technical computing. It is the main task of AI, both for training and for inference. It is essential for the solution of systems of linear equations, the subject of the next chapter. And as the above image shows, any time you play a video game with true 3D graphics, you are asking a chip to perform a huge number of 4-by-4 matrix multiplies per second.

The matrices do not have to be square; they can be rectangular. The way matrix multiplication is usually taught, unfortunately, is with diagrams that look something like this for $A \cdot B = C$ where A, B, and C are rectangular matrices and we use a boldface "·" instead of a "×" or "·" symbol for the multiplication operation:

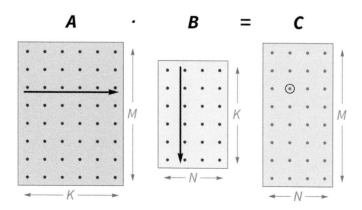

The traditional way matrix-matrix multiplication is explained

That figure shows the dot product of the third row of *A* with the second column of *B* to produce the *C* entry at the third row, second column. By doing that for every combination of row of *A* and column of *B*, you get every entry in *C*. In my opinion, that explanation is confusing and makes you wonder *why* that is the way it is done. For example, why traverse *A* horizontally but *B* vertically?

Instead, arrange the matrices as if they were on an L-shaped piece of paper, with the matching dimensions edge-to-edge:

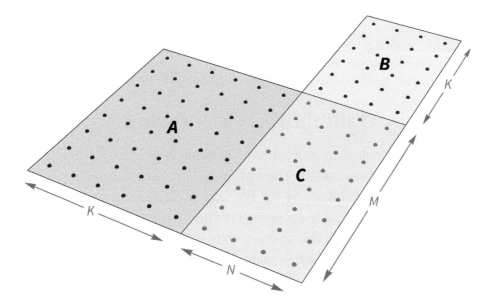

Layout for the origami explanation of matrix-matrix multiplication

Fold up the the sides to form a box corner:

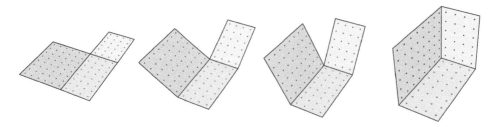

Two folds create a 3D visualization.

Now the dot product becomes a vertical column of multiply-add operations that collapse into the **C** entry. That seems to me like a more straightforward way to describe matrix-matrix multiplication.

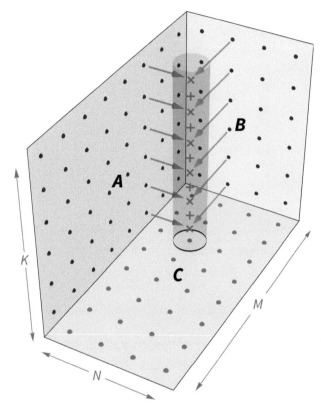

The additions inside the tube may be done in any order to form the **C** entry.

I have avoided the convention of using subscripts like $a_{3,1}, a_{3,2}, ..., a_{3,K}$ and $b_{1,2}, b_{2,2}, ..., b_{K,2}$ (the third row of **A** and second column of **B** shown in the figures) because that makes it look like the matrix entries are ordered. They simply have to match up, so interchanging two rows of **A** means interchanging the corresponding columns of **B**. Interchanging two columns of **A** means interchanging the corresponding rows of **B**. After doing the origami, rows for both input matrices are horizontal, and columns for both matrices are vertical in the z direction. That makes a lot more sense.

The box view also makes it obvious how many operations it takes to perform a matrix-matrix product: it is simply the volume of the box, $K \cdot M \cdot N$ multiplies and almost as many adds, $(K - 1) \cdot M \cdot N$.

11.2 Matrix Ordering Is an Illusion

Consider the following diagram of the product of 2-by-2 matrices **A** and **B** to produce a 2-by-2 matrix **C**:

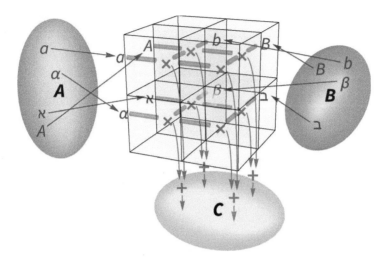

The inputs to matrix-matrix multiply are sets, not ordered lists.

The figure uses four kinds of letter A: Roman (capital, lower case), Greek (alpha α), and Hebrew (aleph $\aleph$), and four kinds of corresponding letter B so that there is no numbered ordering. We could write it four ways:

$$\begin{pmatrix} A & \aleph \\ a & \alpha \end{pmatrix} \cdot \begin{pmatrix} b & B \\ \beta & \beth \end{pmatrix} \quad \begin{pmatrix} a & \alpha \\ A & \aleph \end{pmatrix} \cdot \begin{pmatrix} B & b \\ \beth & \beta \end{pmatrix} \quad \begin{pmatrix} \aleph & A \\ \alpha & a \end{pmatrix} \cdot \begin{pmatrix} \beta & \beth \\ b & B \end{pmatrix} \quad \begin{pmatrix} \alpha & a \\ \aleph & A \end{pmatrix} \cdot \begin{pmatrix} \beth & \beta \\ B & b \end{pmatrix}$$

The result is always the same four dot products, in some order:

$$A \cdot b + \aleph \cdot \beta \qquad A \cdot B + \aleph \cdot \beth \qquad a \cdot b + \alpha \cdot \beta \qquad a \cdot B + \alpha \cdot \beth$$

But there really is no order to the inputs. The scattered input values in the figure are to emphasize there they are a *set* and not an ordered list. Once you pick an order for the inputs, though, that does determine the order of the results in **C**, so the diagram shows them arranged in a square.

Because addition is commutative and associative (Math Layer addition, not the kind done using floats), *there is no ordering* to the products. Numbering is a convenience done in the Human Layer and the Compute Layer.

In terms of computer hardware, the inputs can be fed directly into eight multipliers that operate in parallel. The blue (b, B, β, ב) and red (a, A, α, א) tubes inside the cube structure indicate that an input is used more than once. That can be done either by broadcasting the value or by re-using it locally.

> Humans write *sets* as ordered sequences in the Human Layer, and then forget that sets have no ordering. The Compute Layer disguises their layout as a linear address space. Both lead us to make unnecessary restrictions on what is possible.

We express matrices as two-dimensional ordered arrays for easier human under-standing, not for any mathematical reason. In computer programs, they are usually declared as a two-dimensional ordered array. This has major consequences for computer performance, especially when the matrices are large; in computer programs, matrices are usually expressed as an array with two indices and stored in linearly-addressed memory. But computer memory is *not* really linear. Older textbooks will tell you that an *n*-by-*n* matrix

$$\begin{pmatrix} a_{1,1} & a_{1,2} & \cdots & a_{1,n} \\ a_{2,1} & a_{2,2} & \cdots & a_{2,n} \\ \vdots & \vdots & \ddots & \vdots \\ a_{n,1} & a_{n,2} & \cdots & a_{n,n} \end{pmatrix}$$

is stored in computer memory either as

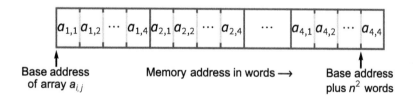

Base address
of array $a_{i,j}$

Memory address in words $\longrightarrow$

Base address
plus n^2 words

or as

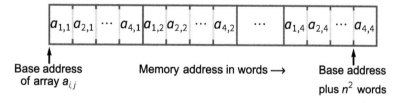

Base address
of array $a_{i,j}$

Memory address in words $\longrightarrow$

Base address
plus n^2 words

depending on whether the storage uses row-ordering or column-ordering.

However, memory is not actually built that way. Computer designers go to a great deal of trouble to make memory *look* like a linear set of addresses, but the actual structure is much more complicated, something like this:

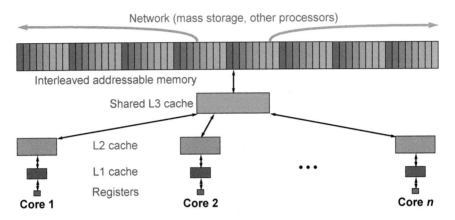

Physical computer memory can be hierarchical, shared, interleaved, and networked.

Even that is an oversimplification. Computer data storage is a non-linear hierarchy of memory (registers, cache, main memory, mass storage). The main addressable memory can be in multiple banks that are interleaved to improve time to access consecutive items. If you pick a matrix dimension that strides through memory so as to hit the same bank every time, performance can be alarmingly reduced.

Square n-by-n matrix multiplication takes order n^3 multiply-adds, but only order n^2 storage. You can break a matrix-matrix multiply into submatrix multiplies that fit into the lower level parts of the memory hierarchy that are smaller but faster, and get reuse of every value moved. With planning, the multiplier-adder hardware can be kept performing at close to its peak theoretical speed, even though main memory cannot possibly bring in data directly at the speed demanded by the arithmetic unit. In every generation, it seems that someone rediscovers this fact and writes a paper on it as if it were a new idea, but it has been known since as far back as the 1950s.

11.3 Inner Products and Outer Products

Chapter 6 defined the dot product and showed how the quire can compute very long dot products with no rounding errors. The dot product similarly does not necessarily have ordered inputs; the pairs to be multiplied must stay matched, but their summation is not ordered. Chapter 6 also mentioned that there are many synonyms for "dot product." Some call it a scalar product. Some call it a weighted sum. And some people, mostly mathematicians, call it an *inner product*. Which makes you wonder what an outer product would be.

The figure at the beginning of Section 11.2 was for a 2-by-2 matrix-matrix multiply. Suppose the input matrices were instead 4-by-1 and 1-by-4. Here is the origami version and the set-type visualization:

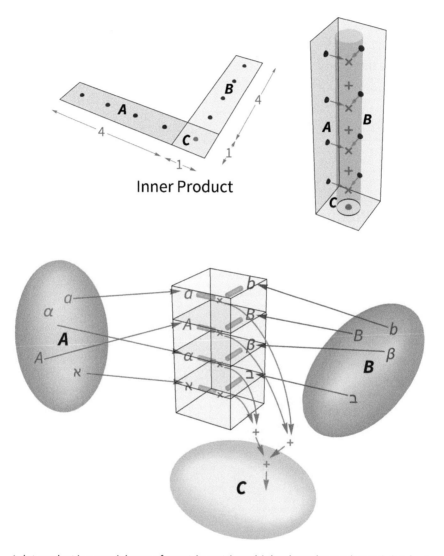

A dot product is a special case of a matrix-matrix multiply where the result matrix is 1-by-1.

The sums can be done in any order, but the above diagram does it with as much parallel computing as possible. The sums are done in pairs, so if we have 2^k values to start with, we can collapse the sums to a single total in k steps. If you want the right answer, that has to be done with an exact accumulator like the quire. The answer to the Exercise in Section 6.1 shows how terribly wrong the sum can be if you round after each add operation. It seems strange to call the result a "matrix" since it is just a single real number, but yes, it is the special case of a 1-by-1 matrix.

There is no re-use of operands. There are just order N inputs and order N multiply-adds. When the ratio of arithmetic to data motion is low, that is low *computational intensity*. Square matrix multiplies have high computational intensity, with a high ratio of arithmetic to data motion.

The *outer product* is what you get if you do the opposite of the inner product, in the sense that the dimensions are transposed.

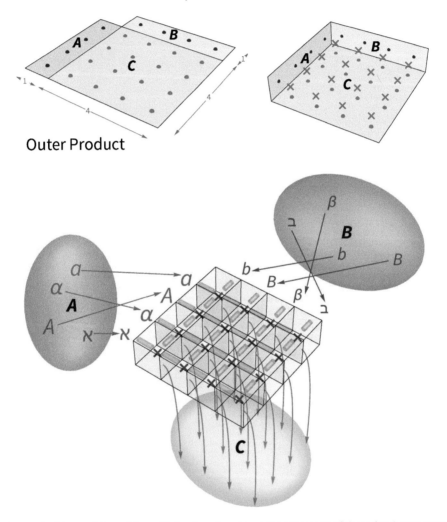

An outer product is a matrix-matrix multiply where inputs are the transpose of those for the inner product.

This is even more compute-intensive than the products of 2-by-2 square matrices; moving order N data results in order N^2 operations.

Depending on what **C** is used for next, there is no need to use the quire to control rounding error, because there are no additions.

You have seen this before, in elementary school, though it was not called an outer product back then. It is simply a *multiplication table*, like the ones in Chapter 9.

×	$\frac{1}{4}$	$\frac{1}{2}$	$\frac{3}{4}$	1	$\frac{3}{2}$	2	4
$\frac{1}{4}$	$\frac{1}{16}$	$\frac{1}{8}$	$\frac{3}{16}$	$\frac{1}{4}$	$\frac{3}{8}$	$\frac{1}{2}$	1
$\frac{1}{2}$	$\frac{1}{8}$	$\frac{1}{4}$	$\frac{3}{8}$	$\frac{1}{2}$	$\frac{3}{4}$	1	2
$\frac{3}{4}$	$\frac{3}{16}$	$\frac{3}{8}$	$\frac{9}{16}$	$\frac{3}{4}$	$\frac{9}{8}$	$\frac{3}{2}$	3
1	$\frac{1}{4}$	$\frac{1}{2}$	$\frac{3}{4}$	1	$\frac{3}{2}$	2	4
$\frac{3}{2}$	$\frac{3}{8}$	$\frac{3}{4}$	$\frac{9}{8}$	$\frac{3}{2}$	$\frac{9}{4}$	3	6
2	$\frac{1}{2}$	1	$\frac{3}{2}$	2	3	4	8
4	1	2	3	4	6	8	16

A multiplication table is the outer product of two identical vectors.

Notice that you can interchange rows and columns, and the information is the same. It simply becomes harder to read if the values are not in increasing order.

11.4 Visualizing Square Matrix Multiplicationr

11.4.1 Visualizing Determinants

So far, the visualizations have been about *how* a matrix multiply is computed. This section is about visualizing what a matrix multiply actually *does* to data.

The matrix $\begin{pmatrix} 1 & 0 \\ 0 & 1 \end{pmatrix}$ does not do anything when multiplied. It is like multiplying by 1. A matrix with all 1s down the diagonal and 0s everywhere else is the *identity matrix*.

If you multiply a 2-by-2 matrix **A** with a column vector $\begin{pmatrix} x_1 \\ y_1 \end{pmatrix}$ visualized as coordinates of a point **r** in the *x-y* plane, it becomes a different point $\begin{pmatrix} x_2 \\ y_2 \end{pmatrix}$ in general.

Multiplying a matrix times a vector (blue dot) results in a different vector in general.

Imagine applying **A** to all the coordinates of a counterclockwise path around the unit square. It will help the visualization if we assign a different color to each edge:

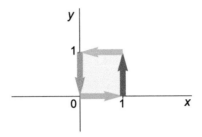

Start with the unit square, traversed in a counterclockwise direction.

Now multiply $A = \begin{pmatrix} 2 & -\frac{1}{2} \\ 1 & 2 \end{pmatrix}$ by every part of that path. It transforms the unit square into a parallelogram in general:

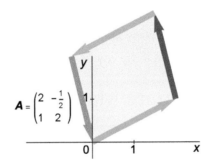

Start with the unit square, traversed in a counterclockwise direction.

The *determinant* of a 2-by-2 matrix $\begin{pmatrix} a & b \\ c & d \end{pmatrix}$ is $a \cdot d - b \cdot c$, and it has a very easy visualization: It is the area of that parallelogram! In this case, $2 \times 2 - \left(-\frac{1}{2} \times 1\right) = 4.5$. But we call it a "negative area" if the traversal changes from counterclockwise to clockwise. Interchanging the rows or interchanging the columns flips the sign of the determinant. Here is what happens for those interchanges, which produce an area of –4.5:

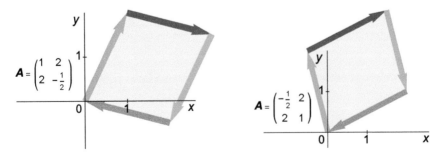

Interchange rows (left example) or interchange columns (right example)

It can happen that the unit square gets flattened into a line segment. For example:

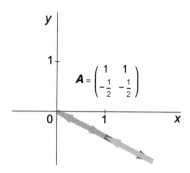

If rows or columns are proportional, the unit square collapses and becomes one-dimensional.

This happens when one row is a multiple of the other. Or equivalently, one column is a multiple of the other. Such a matrix is said to be *singular*. If the area of the parallelogram (the determinant) is nonzero, the 2-by-2 matrix is said to have *rank 2* or to be of *full rank*. If it squashes to a line (which has no area), it is said to have *rank 1*. It could happen that all the entries in **A** are zero, in which case the unit square turns into the point at the origin. That has no dimensions, so that would mean **A** is *rank 0*, which is another way the matrix can be singular.

Exercise for the Reader: What does multiplying by $A = \begin{pmatrix} 0 & 1 \\ 1 & 0 \end{pmatrix}$ do to the unit square?

11.4.2 Shear Transformations

Matrices of the form $\begin{pmatrix} 1 & c \\ 0 & 1 \end{pmatrix}$ shear the unit square in the x direction, to the right if c is positive and to the left if c is negative:

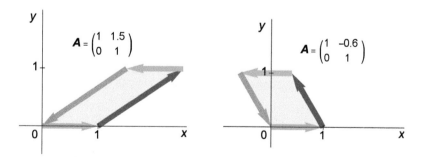

Shearing to the right by 1.5 and to the left by 0.6

Similarly, matrices of the form $\begin{pmatrix} 1 & 0 \\ c & 1 \end{pmatrix}$ shear the unit square in the y direction, upwards if c is positive and downwards if c is negative.

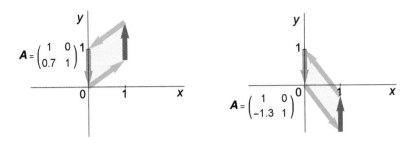

Shearing up by 0.7 and down by 1.3

Notice that shearing does not change the area of the unit square; the determinant remains 1. The traversal also remains counterclockwise.

11.4.3 Rotation and Scaling Matrices

Continuing to add tools to the toolkit: A *rotation matrix* has the following form:

$$A = \begin{pmatrix} \cos(t) & -\sin(t) \\ \sin(t) & \cos(t) \end{pmatrix}$$

Multiplying by it will rotate everything by the angle t, counterclockwise if $t > 0$ and clockwise if $t < 0$. Here are unit square examples for 30° ($\pi/6$ radians) and −45° ($-\pi/4$ radians), for which the sines and cosines can be expressed with square roots:

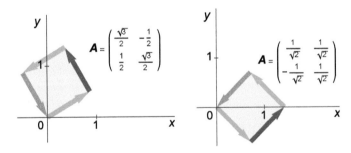

Rotate left by 30° (cosine is $\sqrt{3}/2$ and sine is $1/2$); rotate right by 45° (cosine and sine are $1/\sqrt{2}$).

The determinant is $\cos(t)^2 + \sin(t)^2 = 1$, so it rotates without changing the area.

A *scaling transform* by a factor s can be done separately for the x and y directions. If the scale factor is negative, the rectangle is flipped to the other side of the axis.

$$A = \begin{pmatrix} s & 0 \\ 0 & 1 \end{pmatrix} \text{ scales } x \text{ by } s, \text{ and } A = \begin{pmatrix} 1 & 0 \\ 0 & s \end{pmatrix} \text{ scales } y \text{ by } s.$$

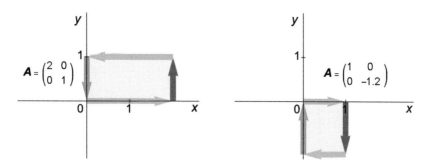

Scale *x* by a factor of 2; scale *y* by a factor of –1.2, which flips it over and gives it negative area.

Either one scales the area by a factor of *s*, of course. Scaling by $\left(\begin{smallmatrix} s & 0 \\ 0 & s \end{smallmatrix}\right)$ scales both dimensions by the same amount and preserves the square shape, but with area s^2.

11.4.4 Composing Transformations: Be Careful

This brings us to an important fact that may seem strange, but the visualization tools will make it clear why it is true:

> Matrix multiplication is *not* commutative in general. That is, $\boldsymbol{A}\cdot\boldsymbol{B} \neq \boldsymbol{B}\cdot\boldsymbol{A}$ in general.

It is, however, associative. So if you apply a matrix transformation $\boldsymbol{B}$ to a point $\boldsymbol{r}$, then apply a matrix transformation $\boldsymbol{A}$, you could write that as $\boldsymbol{A}\cdot(\boldsymbol{B}\cdot\boldsymbol{r})$, but it is valid to group it as $(\boldsymbol{A}\cdot\boldsymbol{B})\cdot\boldsymbol{r}$ because that does the same thing. You can compute $\boldsymbol{C} = \boldsymbol{A}\cdot\boldsymbol{B}$ as a new transformation $\boldsymbol{C}\cdot\boldsymbol{r}$ that composes the transformations $\boldsymbol{A}$ and $\boldsymbol{B}$. But it will not, in general, be the same as applying $\boldsymbol{A}$ to $\boldsymbol{r}$ and then applying $\boldsymbol{B}$, written $\boldsymbol{B}\cdot\boldsymbol{A}\cdot\boldsymbol{r}$.

Visualization makes it clear why this is. Suppose we rotate by $-45° = -\pi/4$, followed by a scaling of the *y* dimension by 2. First apply $\boldsymbol{B} = \left(\begin{smallmatrix} 1/\sqrt{2} & 1/\sqrt{2} \\ -1/\sqrt{2} & 1/\sqrt{2} \end{smallmatrix}\right)$ to rotate; then apply $\boldsymbol{A} = \left(\begin{smallmatrix} 1 & 0 \\ 0 & 2 \end{smallmatrix}\right)$ to scale in the *y* direction by 2. The product $\boldsymbol{A}\cdot\boldsymbol{B}$ is $\left(\begin{smallmatrix} 1/\sqrt{2} & 1/\sqrt{2} \\ -\sqrt{2} & \sqrt{2} \end{smallmatrix}\right)$:

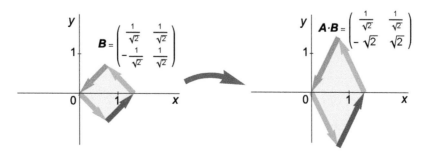

Rotate right (clockwise) by 45°, then scale in the *y* direction by a factor of 2.

If we instead scale the *y* direction by 2 first, and *then* rotate clockwise by 45°, the result is quite different. The matrix product $B \cdot A$ is $\begin{pmatrix} 1/\sqrt{2} & 1/\sqrt{2} \\ -\sqrt{2} & \sqrt{2} \end{pmatrix}$, different from $A \cdot B$:

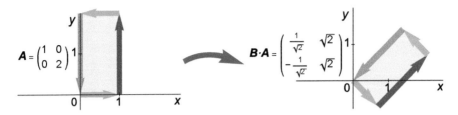

Reverse the order: scale in the *y* direction by a factor of 2, then rotate right by 45°.

There are certain types of matrices for which $A \cdot B = B \cdot A$. A *symmetric* matrix is one that is unchanged if you turn the rows into columns (or vice versa). Symmetric matrices are commutative for multiplication. The 2D rotation matrices are not symmetric, but they also are commutative since it does not matter which angle you rotate first; the angles add. But in the general case, matrix-matrix multiplication is *not* commutative and the above example shows why, graphically.

11.4.5 The Translation Trick

Notice that whatever happens to the square, the corner at the origin remains pinned to that corner. That has to be, because multiplying by a matrix is a *linear* operation, that is, $A \cdot (m \cdot r) = m \cdot (A \cdot r)$, so the point at the origin is unchanged. What if we want to *translate* the unit square such that the bottom left corner is instead at a point $\begin{pmatrix} u \\ v \end{pmatrix}$?

Of course that simply means adding $\begin{pmatrix} u \\ v \end{pmatrix}$ to every point as a second step, but that is not something you can do with 2-by-2 matrix multiplication.

Computer graphics people developed a trick for doing this that still looks like a matrix multiply, but you have to raise the dimension of the multiplication to 3-by-3, by padding it like this:

$$\begin{pmatrix} a & b & u \\ c & d & v \\ 0 & 0 & 1 \end{pmatrix} \cdot \begin{pmatrix} x \\ y \\ 1 \end{pmatrix}$$

That has the effect of computing $\begin{pmatrix} a & b \\ c & d \end{pmatrix} \cdot \begin{pmatrix} x \\ y \end{pmatrix} + \begin{pmatrix} u \\ v \end{pmatrix}$ for the 2-by-2 upper left corner, which you can still then visualize as a 2D transformation.

The identity matrix leaves the unit square unchanged, but the 3-by-3 trick then translates it, like this:

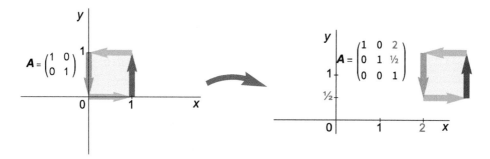

In plotting, the padded last coordinate is ignored when using the translation trick.

All the previous transformations can be thought of as the $\begin{pmatrix} a & b \\ c & d \end{pmatrix}$ upper left-hand part of this 3-by-3 transformation. Here is one that rotates by $60°$, doubles it in size in both dimensions, and translates it by 2 horizontally and $\frac{1}{2}$ vertically like the above example.

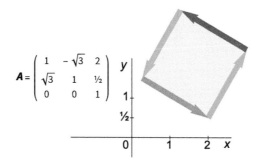

The upper-left 2-by-2 submatrix of **A** is the product of a 2× scaling matrix and a $60°$ rotation matrix.

11.4.6 Matrix Multiply Visualizations in 3D

The visualization extends to three dimensions as well. Instead of the unit square, we start with the unit cube. In three dimensions, the determinant is the volume of the parallelepiped produced by multiplying the matrix by the unit cube.

The determinant of a 3-by-3 matrix is

$$\mathrm{Det}\begin{pmatrix} a & b & c \\ d & e & f \\ g & h & i \end{pmatrix} = a \times \mathrm{Det}\begin{pmatrix} e & f \\ h & i \end{pmatrix} + b \times \mathrm{Det}\begin{pmatrix} f & d \\ i & g \end{pmatrix} + c \times \mathrm{Det}\begin{pmatrix} d & e \\ g & h \end{pmatrix}$$

$$= a \cdot e \cdot i - a \cdot f \cdot h + b \cdot f \cdot g - b \cdot d \cdot i + c \cdot d \cdot h - c \cdot e \cdot g.$$

Start with the unit cube, again coloring edges for clarity:

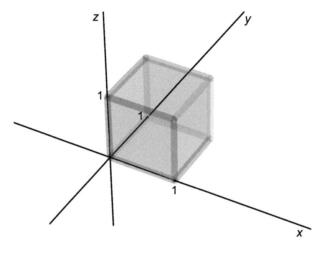

The unit cube

If the matrix takes the unit volume to a nonzero volume, the determinant is nonzero and the matrix has "full rank," rank 3. Here is an example where the determinant is $\frac{5}{8}$:

$$A = \begin{pmatrix} 1 & \frac{1}{4} & \frac{1}{4} \\ \frac{1}{4} & \frac{1}{2} & 0 \\ \frac{1}{4} & 0 & \frac{3}{2} \end{pmatrix}$$

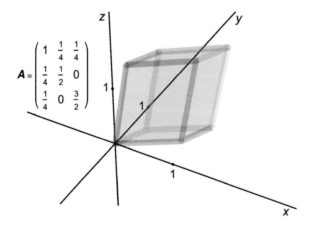

Unless **A** is singular, multiplication turns the unit cube into a parallelepiped.

If we swap the first two columns of that particular **A**, notice how the edges now look like they are in a mirrored order from those of the unit cube, and the determinant is now $-\frac{5}{8}$:

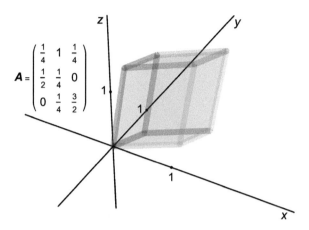

This parallelepiped is the mirror image of the previous one, so the volume is negative.

If the matrix squashes the unit cube flat (which can happen in many planes), then it is singular and has rank 2. The determinant is zero and the volume is zero. For example, making the last row of **A** all 0 entries collapses the parallelepiped onto the *x-y* plane:

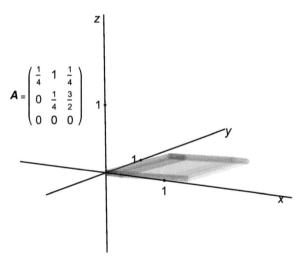

Rank 2 matrices turn the unit cube into parallelograms with one corner at the origin.

That may *look* sort of 3-dimensional, but that is an illusion. The shape is a parallelogram with all of its edges in the *x-y* plane.

And it could be that the matrix has rank 1, so the unit cube turns into a line segment.

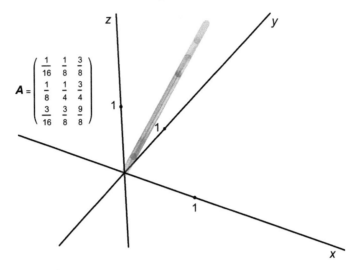

$$A = \begin{pmatrix} \frac{1}{16} & \frac{1}{8} & \frac{3}{8} \\ \frac{1}{8} & \frac{1}{4} & \frac{3}{4} \\ \frac{3}{16} & \frac{3}{8} & \frac{9}{8} \end{pmatrix}$$

This happens when all three rows (or all three columns) are scalings of the same vector.

There are 3D transformations for shear, rotations, scaling, and translations like those for 2D. To do translations, the matrices have to be promoted to 4-by-4, where the translation is in the last column. For example, this matrix revolves the unit cube by 45° ($\pi/4$ radians) around the y axis, then scales by $\frac{1}{4}$ in the x direction, then translates it by $\{2, -1, 3\}$ in $\{x, y, z\}$. Note: the multiplies should be read from right to left to show the order in which they are applied to a vector:

Translate by $\{2, -1, 3\}$		Scale by $\frac{1}{4}$ in x direction		Revolve by $\pi/4$ about y – axis		Composite transformation	

$$\begin{pmatrix} 1 & 0 & 0 & 2 \\ 0 & 1 & 0 & -1 \\ 0 & 0 & 1 & 3 \\ 0 & 0 & 0 & 1 \end{pmatrix} \cdot \begin{pmatrix} \frac{1}{4} & 0 & 0 & 0 \\ 0 & 1 & 0 & 0 \\ 0 & 0 & 1 & 0 \\ 0 & 0 & 0 & 1 \end{pmatrix} \cdot \begin{pmatrix} \frac{1}{\sqrt{2}} & 0 & -\frac{1}{\sqrt{2}} & 0 \\ 0 & 1 & 0 & 0 \\ \frac{1}{\sqrt{2}} & 0 & \frac{1}{\sqrt{2}} & 0 \\ 0 & 0 & 0 & 1 \end{pmatrix} = \begin{pmatrix} \frac{1}{4\sqrt{2}} & 0 & -\frac{1}{4\sqrt{2}} & 2 \\ 0 & 1 & 0 & -1 \\ \frac{1}{\sqrt{2}} & 0 & \frac{1}{\sqrt{2}} & 3 \\ 0 & 0 & 0 & 1 \end{pmatrix}$$

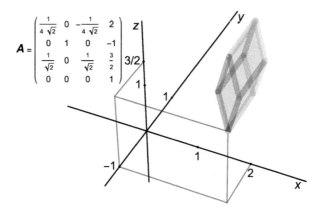

As with 2D visualization, the last coordinate is ignored when showing the effect of 4-by-4 matrix multiply.

As the opening image of this chapter showed, doing fast 4-by-4 matrix multiplies was the key to making video gaming fully 3D, where the camera angle could be changed and every polygon in the scene would transform correctly in 3D space. It was the job of a separate hardware unit to turn the 3D positions into the graphics frame buffer for display, and that unit took care of correct perspective.

11.5 Quires Make Matrix Multiplication Accurate

If you round after each multiply-add when computing the dot products in a matrix multiply, there is no bound on the relative error you can suffer. Error can be infinite.

The rounding error is often magnified when you compute a sum of the form large + small − large from left to right. Suppose the matrices are only 3-by-3, and one of the row-column dot products is $\{64, \frac{1}{64}, -64\} \cdot \{64, \frac{1}{64}, 64\}$. The first and third products are 4096 and −4096 and they cancel out, so the correct answer is the second product, $\frac{1}{64} \cdot \frac{1}{64} = \frac{1}{4096}$. If we use IEEE Std 754 32-bit floats, all of those numbers can be expressed exactly, and so can all of the products before they are summed. If the products are added from left to right, the first add does you in:

$$4096 + \frac{1}{4096} = 4096.$$

Then the addition of −4096 gives you 0 instead of $\frac{1}{4096}$. That is an infinite relative error, per Chapter 2. The next-larger float after 4096 is $4096 + \frac{1}{2048}$. The sum $4096 + \frac{1}{4096}$ is not representable exactly and rounds back to 4096. Then, adding −4096 cancels it back to 0 and the $\frac{1}{4096}$ is completely lost.

Posits are subject to the same unbounded hazard (though a 32-bit standard posit will compute the above example without error). In any number system that rounds, it is always possible to create an example like this where the amount of relative error is infinite. This is why the quire is essential for matrix multiplication. As long as the matrix is less than 2^{31} (about two billion) elements on a side, the quire will compute the dot product perfectly with no possibility of overflow; it can then be rounded back to the nearest posit, with a bound on the amount of relative error.

Most of this chapter has been about small matrices, but very large scale matrices that are thousands or millions of elements on a side abound in technical computing. For example, in designing a stealth aircraft, the aircraft is modeled with as many millions of elements as the computer can store. The question is then, what happens when the aircraft is hit with a radar pulse? Every element of the model resonates to the radar pulse and re-emits a reflection of that, which interacts with all the other elements. For solving such models, a single matrix multiply could involve trillions of data points for the matrices, and quintillions of multiply-add operations.

For a traditional *n*-by-*n* matrix-matrix multiply, the code looks something like this:

```
For[i = 1, i ≤ n, i++,
  For[j = 1, j ≤ n, j++,
    C[i, j] = 0;
    For[k = 1, k ≤ n, k++,
      C[i, j] = C[i, j] + A[i, k] * B[k, j]
      (* Accumulate dot product *)
    ]
  ]
]
```

It is a triply-nested loop with each loop going from 1 to *n*, so the number of multiply-adds that have to be done in the innermost loop is n^3. The loops can be nested in any order, so there are 3 != 6 ways to express the matrix-matrix multiply; if **k** is the index of the innermost loop, the dot product becomes obvious in the origami way shown in Section 1. If that dot product is done with traditional floating-point arithmetic, the step `C[i,j]=C[i,j]+A[i,k]*B[k,j]` will round every time it executes, and every entry in the result matrix will have gone through *n* rounding errors. It will be 2*n* rounding errors if the multiply-adds are not fused. If you use posits, then the quire can give the result of every dot product with just *one* rounding error, that of converting the dot product from quire format to posit format.

11.6 Why This Chapter Is Here

- Solving systems of linear equations, the next chapter, is the inverse operation of matrix multiply. Therefore, matrix multiply needed to be presented first.
- The visual method is not the usual way matrix operations are taught, but I believe visualizations make the concepts much more approachable.
- Many computer programmers use square arrays and think that matrices are the same as two-dimensional lists, and that memory is organized as a one-dimensional address space. Those are fallacies.

I wish someone had shown me, back when I was learning this type of math, the geometric meaning of matrix operations in two and three dimensions as illustrated here. Textbooks jump right into *n*-by-*n* general matrices for which we have no mental image, and we have to rely on our skill at pushing symbolic expressions around. Much of matrix algebra was developed by physicists who needed it to describe things like electromagnetic field behavior in 3D. Once the mathematicians generalized everything to be any number of dimensions, linear algebra became a subject sufficiently confusing that it is not required for engineering majors or even some math programs in college. It is really not that much more difficult than analytic geometry, which is taught in K-12 education. In fact, it is actually rather beautiful.

Referring to Dominic Walliman's amazing Map of Mathematics on the next page, much of this book has been about the gold central circle region. With this and the next chapter, the "You Are Here" pointer would be just to the left of that gold circle.

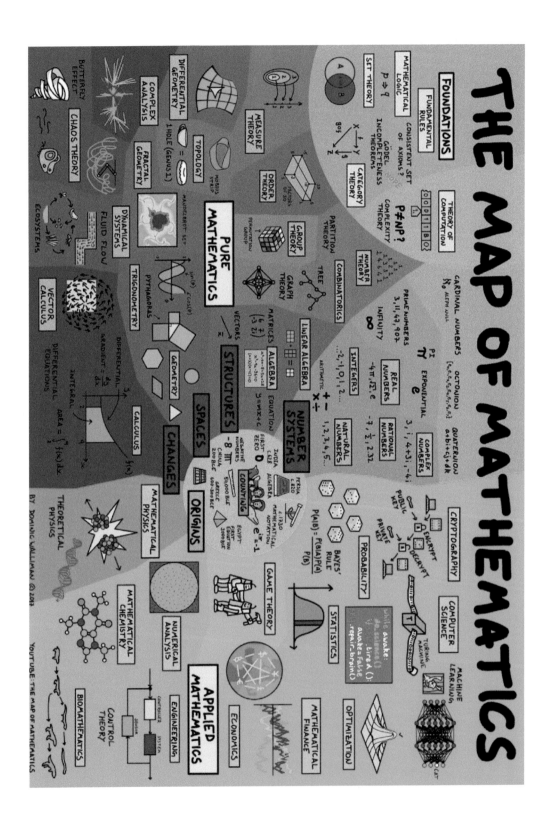

12 Linear Solvers that Suck Less

"This method usually works" is one of the most
exasperating things numerical analysts say.

12.1 Inverse Operations and Double-Struck Letters

Asking inverse questions leads to new fields of mathematics. Chapter 10 explained matrix multiplication. Its inverse problem is "If $A \cdot x = u$, given matrix A and vector u, what is x?" The answer is the subject of this chapter.

Addition is repeated counting. Multiplication is repeated addition. Exponentiation (power function) is repeated multiplication. All are closed under the set of counting numbers 1, 2, 3, ... for which mathematicians use the symbol $\mathbb{N}$. Double-struck letters have been used since the 1930s to represent sets of a certain type of number.

When you start asking inverse questions like, "What number x, when operated on by a, equals b?" you create new functions and new kinds of numbers. For addition, the inverse function needed to solve "$a + x = b$; what is x?" is subtraction, and that leads to the possibility of negative numbers. Mathematicians use $\mathbb{Z}$ for the set of all integers: positive, negative, and zero. Addition, subtraction, and multiplication are closed in $\mathbb{Z}$. Exponentiation is not closed, but we will get to that. Why a double-struck Z, you wonder? It comes from the German *Zahlen*, which means "numbers."

Once you ask, "Given a and b, $a \times x = b$; what is x?" you create the inverse of multiplication: division. That leads to the set of rational numbers, usually symbolized by $\mathbb{Q}$ (a double-struck Q for *quotient*). Division also introduces very simple-looking arithmetic problems that have *no solution*. If you ask, "If 0 times x equals 3, what is x?" the only truthful answer is, "That problem has no solution."

Some non-mathematicians go down the route of saying, "It is three divided by zero, which is infinity," but that is glib and incorrect because zero times infinity is certainly not three. It is indeterminate. If you compute 3.0/0.0 with IEEE Std 754 arithmetic, however, the result will be positive infinity.

By the way, when you were around age 10 and were first taught to divide numbers with more than one digit, it was also the first time grade school arithmetic included *guesswork*. It was as if a line had been crossed, since all the other methods taught up to that point were just pencil-pushing, following the rules. When you see a problem like

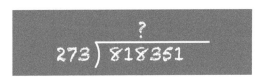

Chalkboard long division

you are supposed to guess what digit to use for "?". If you guess 3 (which is a reasonable guess), you overshoot the mark, since $3 \times 273 = 819$. Then you curse under your breath, erase, and do it over with 2. Many inverse problems are like this; you have to make a trial guess and compute the forward problem to find out what correction is needed. That approach will arise throughout this chapter.

Before leaving the subject of inverse functions, notice that there are two ways to make inverse functions out of exponentiation. We could ask what x satisfies $x^a = b$, which leads to roots $x = \sqrt[a]{b}$ that usually are irrational numbers, as well as complex numbers $u + v i$ where $i^2 = -1$. Those are still algebraic numbers, as defined in Chapter 10, and relatively easy to compute with correct rounding. But you can also ask what x satisfies $a^x = b$, and that creates logarithms: $x = \log_a(b)$. The good news is that once you allow for the set of all complex numbers $\mathbb{C}$ and a "not a number" for impossible problems, you are done with complicating the number set. All the elementary operations and their inverses will stay within that set.

For linear solvers, be prepared for a lot of guessing just like you do for long division.

12.2 Linear Equations Before Computers

12.2.1 A 2-by-2 Problem from Babylonia

There is a Babylonian tablet from around 300 BC that describes what appears to be an exercise:

> There are two fields whose total area is 1800 square yards. One produces grain at the rate of two-thirds of a bushel per square yard while the other produces grain at the rate of half a bushel per square yard. If the total yield is 1100 bushels, what is the size of each field?

Jennings, A., "Matrices, Ancient and Modern," *Bull. Inst. Math. Appl.*, 13 (5) 1977, pages 117–123.

Whoever translated it to English also took the liberty of changing the ancient units to yards and bushels, which is fine for readability. The Iowa boy in me wants to know how they were able to produce many times as much grain per square yard as prime farmland can using modern agricultural technology. Half a bushel per square yard? Amazing. That is like four gallon containers of corn from a square yard of dirt.

If x and y are the sizes of the two fields, the word problem becomes the following:

$$x + y = 1800$$
$$\tfrac{2}{3}x + \tfrac{1}{2}y = 1100$$

If all you knew was elementary arithmetic, how might you go about solving this? Perhaps by guessing. Try making the fields equal size, 900 and 900 so that they add up to 1800 for the first equation. But then the second equation is $\tfrac{2}{3} \times 900 + \tfrac{1}{2} \times 900$ which is 1050, so you need more area in the more productive field. Try 1500 and 300; $\tfrac{2}{3} \times 1500 + \tfrac{1}{2} \times 300$ is 1150 which is too large. Home in on it until you discover that using 1200 and 600 solves the second equation. That is an example of an *iterative linear solver*. Not exactly an algorithm, but at least a method of guessing.

Here is the matrix *A* and right-hand side *u* for the Babylonian problem:

$$\begin{pmatrix} 1 & 1 \\ \frac{2}{3} & \frac{1}{2} \end{pmatrix} \cdot \begin{pmatrix} x \\ y \end{pmatrix} = \begin{pmatrix} 1800 \\ 1100 \end{pmatrix}$$

$$\boldsymbol{A} \cdot \boldsymbol{x} = \boldsymbol{u}$$

A linear equation is one of the form $a_1 \cdot x_1 + a_2 \cdot x_2 + \ldots + a_n \cdot x_n = u$, one that sets a dot product of given coefficients $\{a_1, a_2, \ldots, a_n\}$ and unknown vector $x = \{x_1, x_2, \ldots, x_n\}$ to a given coefficient *u*. To avoid subscripts, we can write $\boldsymbol{A} = \begin{pmatrix} a & b \\ c & d \end{pmatrix}$, $\boldsymbol{x} = \begin{pmatrix} x \\ y \end{pmatrix}$, and $\boldsymbol{u} = \begin{pmatrix} u \\ v \end{pmatrix}$ when there are only two equations. Given all the coefficients of $\boldsymbol{A}$ and $\boldsymbol{u}$, a set of *n* linear equations can often determine all the values in $\boldsymbol{x}$. But not always, because several things can go wrong. It is like the risk of dividing by zero but in multiple dimensions. It is even worse than that, because solving linear systems is a multi-step problem with rounding errors along the way.

Perhaps the reader works with matrices and linear algebra all the time. I am explaining the basics partly because I recently discovered that solving systems of equations is no longer part of every grade school curriculum. Also, I have known Caltech seniors (not math majors) who could not answer the question, "How would you solve a system of ten linear equations in ten unknowns?" in a job interview. The other reason is that I have heard misconceptions about linear algebra (sometimes called matrix algebra) from technical professionals that can lead to trouble.

The Babylonian problem can be written three other ways without changing a thing:

$$\begin{pmatrix} 1 & 1 \\ \frac{1}{2} & \frac{2}{3} \end{pmatrix} \begin{pmatrix} y \\ x \end{pmatrix} = \begin{pmatrix} 1800 \\ 1100 \end{pmatrix} \text{ or } \begin{pmatrix} \frac{1}{2} & \frac{2}{3} \\ 1 & 1 \end{pmatrix} \begin{pmatrix} x \\ y \end{pmatrix} = \begin{pmatrix} 1100 \\ 1800 \end{pmatrix} \text{ or } \begin{pmatrix} \frac{2}{3} & \frac{1}{2} \\ 1 & 1 \end{pmatrix} \begin{pmatrix} y \\ x \end{pmatrix} = \begin{pmatrix} 1100 \\ 1800 \end{pmatrix}.$$

That may look wrong, like putting *y* before *x* or using the last equation in the word problem as the first one, but they all produce the same answer for *x* and *y*. This is the point made in the previous chapter: Matrices look ordered, but they really are not. That is why you can interchange rows and columns and express the same problem.

Another way to see the lack of ordering is to solve the equations graphically.

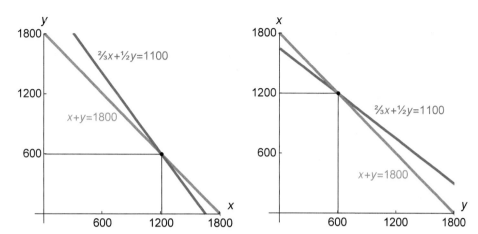

Solving two simultaneous linear equations is exactly like finding where two lines in the plane intersect.

It makes no difference which unknown variable is the horizontal axis and which one is the vertical axis.

12.2.2 Ancient China Took It Up to 6-by-6

It was the Chinese who first specified the procedure we now call Gaussian elimination. Like the Babylonians, the Chinese of that era were also excellent at mathematics. Unlike the ancient Greeks who sought to build up abstract propositions from axioms and logic, the Chinese were more like applied mathematicians who sought ways to solve numerical problems of practical interest.

Starting around 1000 BC, Chinese mathematicians began accumulating mathematical methods in what became the *Jiuzhang Suanshu*, or *Nine Chapters on the Mathematical Art* (九章算术). Generations of scholars maintained its contents, much like a religious document. Each entry was the statement of what we now call a "word problem" to be solved, followed by the method used to solve it, and then the actual calculation. It did not have formal proofs. It recognized both positive and negative numbers, and fractions. They knew what we now call the Pythagorean theorem. They knew how to find the roots of quadratic and cubic equations. And they certainly spelled out how to solve systems of linear equations.

By 200 BC, its Chapter 8 *Fangcheng* (方程) had a description of "calculation by square arrays." In other words: matrix algebra. That chapter gives 17 examples of solving systems of n equations in n unknowns, with n ranging from two to five. There is also an example with six unknowns but only five equations, the first example of an *underdetermined* system. Like the Babylonian example, they were interested in problems involving agricultural yields. If we had only one equation in the Babylonian example, all points on the line satisfy the system of linear equations.

For n equations in n unknowns, the approach is to scale the first equation so the coefficient of the first variable matches that of the second equation, then subtract to eliminate that variable from the second equation. Repeat for all the other equations. Then scale the second equation so the coefficient of the second variable matches that of the third equation, etc. until you are left with just one last variable. That is one equation in one unknown, and that is easy. Then backsolve to find the other unknowns. For instance, for the Babylonian problem, multiply the first equation by $\frac{2}{3}$ so that the first term matches that of the second equation:

$$\frac{2}{3}x + \frac{2}{3}y = 1200$$
$$\frac{2}{3}x + \frac{1}{2}y = 1100$$

Subtract the second equation from the first to get

$$\frac{1}{6}y = 100$$

So this directly finds $y = 600$, which means $x = 1800 - 600 = 1200$, which is the step called *backsolving*.

The methods that crank out the values for x directly are often called *direct solvers* to contrast them with iterative solvers, but if we do those methods using computers that make *rounding errors*, they are actually iterative solvers.

Since the amount of arithmetic needed to solve n equations directly grows as the cube of n, to get all the way up to a 5-by-5 system armed with only an abacus takes considerable patience.

12.2.3 Ada and the Analytical Engine

British poet Lord Byron (1788 1824) was prolific in more ways than one. We really do not know how many children he had, but the only one he had in wedlock was Augusta Ada Byron. He was disappointed; he was hoping for a boy. He left his wife and daughter in London a month after Ada was born and never returned to England.

Ada Byron (1815–1852), at age 7. Portrait by Comte d'Orsay at Sommerville College, Oxford.

Lord Byron, like many creative people, was bipolar. He had been quite a challenge to stay married to. He nicknamed his wife "Princess of Parallelograms" to mock her respect for mathematics. Lady Byron vowed to raise Ada with strict scholarship, in the hope Ada would not turn out as crazy as her father. It turned out that Ada loved math and became brilliant at it. When she was 18, she attended a soirée hosted by mathematician Charles Babbage at Babbage's home in London, at which Babbage proposed to build what he called the Analytical Engine (see Section 4.5), a room-sized decimal computer based on gears and mechanical latches. The Analytical Engine was to be *programmable*, with the program and input data stored on punched cards like those used to control Jacquard looms invented in 1804. The proposed Analytical Engine could do conditional branching and loops, making it general purpose. It was 1833, so it was to be steam-powered. Just imagine, a coal-fired computer. For output, the intent was that the machine could drive a printer for numerical results, plot curves for graphics, and ring a bell.

Ada was fascinated, and she "got it." She realized that a programmable machine could do *anything*, not just calculations, and that programming was a revolutionary concept. She started writing complete programs for the proposed machine, and thus became the world's first computer programmer. There would not be another programmer for 110 years. One of her first programs was a linear equation solver for ten equations in ten unknowns. In creating it, she invented a technique for re-using an instruction for all the numbers in a list, what we now call a *vector instruction*.

She once said that by using loops, the machine could "solve a system of linear equations, no matter how big it was." Had the machine been built, it would have taken about sixty hours to solve ten equations in ten unknowns, and to this day, sixty hours is about how long supercomputer users are willing to wait for an answer to a problem. It would have taken about four minutes per multiply-add operation, with operands that had fifty decimal digits. Before you laugh at how slow that is, compare it with how long it would take you to multiply two fifty-digit numbers using the only alternative available in 1833, pencil and paper.

Two years after meeting Babbage, she married William King-Noel, First Earl of Lovelace, and thus became Ada, Countess of Lovelace. Or Lady Lovelace. She is today often referred to as "Ada Lovelace," but that is not how nobility titles work. If George becomes the Duke of York, that does not make his name George York.

This discussion is in the "Linear Algebra Before Computers" section because Babbage never finished his machine. It would have been quite expensive, and they suspected that the mechanical precision required was beyond the capabilities of that era. Both Ada and Babbage grasped the importance of the possibility that the machine could solve large systems of linear equations, automatically; it would have been the very first supercomputer. In pitching the machine to the British government, they also singled out the immense value of being able to solve sets of linear equations, often called "simultaneous equations." Here is an excerpt from the government committee's report:

> Another important desideratum to which the machine might be adapted... is the solution of simultaneous equations containing many variables. This would include a large part of the calculations involved in the method of least squares... In the absence of a special engine for the purpose, the solution of large sets of simultaneous equations is a most laborious task, and a very expensive process indeed, when it has to be paid for, in the cases in which the result is imperatively needed.

Reprinted in *The Origins of Digital Computers*, B. Randell, ed., 3rd Ed., Springer-Verlag, NY, 1982, page 64.

Babbage was too far ahead of his time, so it was never built. It would be a century later that the first computers capable of solving linear equations would be built. The lasting accomplishment was Ada's: the concept of computer programming of complete applications. She invented software a hundred years before there was hardware that could run it.

12.3 Even 2-by-2 Systems Have Issues

12.3.1 Input Roundings, and Complete Sets of Solutions

What does it mean to "solve equations" with a computer that commits rounding errors? Your first reaction might be, "Find the rounded answer closest to the exact mathematical answer." But, what if the number formats cannot even express the exact problem you want to solve?

To make the issue clear, we can use low-precision floats or posits. With 16-bit standard posits, The Babylonian problem becomes

$$\begin{pmatrix} 1 & 1 \\ \frac{2}{3} & \frac{1}{2} \end{pmatrix} \cdot \begin{pmatrix} x \\ y \end{pmatrix} = \begin{pmatrix} 1800 \\ 1100 \end{pmatrix}$$

$$\downarrow$$

$$\begin{pmatrix} 1 & 1 \\ \frac{2731}{4096} & \frac{1}{2} \end{pmatrix} \cdot \begin{pmatrix} x \\ y \end{pmatrix} = \begin{pmatrix} 1800 \\ 1100 \end{pmatrix}$$

because $\frac{2}{3}$ cannot be expressed exactly by a number of the form $m \times 2^j$ where j and m are integers. The other input numbers can be represented exactly. The closest standard 16-bit posit value to $\frac{2}{3}$ is $\frac{2731}{4096} = 0.667480\,46875$, which makes it a different problem for which the exact answer is no longer $x = 1200$ and $y = 600$. The exact answer to the *rounded* problem becomes, using mixed fractions, $x = 1199\frac{283}{683}$ and $y = 600\frac{400}{683}$.

Fine, you say; round those exact answers to the nearest representable value. If you do that, you get $x = 1200$ but $y = 601$, which clearly does not satisfy the first equation. Or does it? Try adding x and y, rounding to the nearest 16-bit posit:

$$\overline{1200 + 601}$$

1800

There are two very different ideas here. One is to ask what is the exact solution in the Math Layer, then bring it up to the Compute Layer with a rounding. The other is to find *all* the pairs of representable values in the Compute Layer that, when the equations are evaluated in the Math Layer (which can be done using the quire), return to the Compute Layer as the rounded value of the right-hand side.

To illustrate this, expand the first graph at the end of Section 12.2.1 at the intersection point of the orange and blue curves representing the two equations:

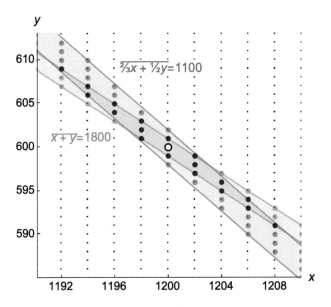

Discrete-point view of the intersection of two lines in the plane that solve each equation

The grid of dots is values representable with standard 16-bit posits near the solution $x = 1200$, $y = 600$. The ULP spacing is 2 for x values and 1 for y values in this region. Any dots that lie within the orange band will satisfy the first equation after rounding, $\overline{x + y} = 1800$. Similarly, any dots that lie within the blue band will satisfy the second equation after rounding, $\overline{\tfrac{2}{3} x + \tfrac{1}{2} y} = 1000$. If they belong to only one band, they are shown with large dots of the corresponding color. If they belong to both bands, they are shown with large **black** dots and an even larger hollow dot for the exact solution at {1200, 600}. There are 19 such large black dots. Going to 32-bit or 64-bit would have little effect on the shape; it would simply make the ULP spacings smaller.

> **Definition**: The *c-set* (for *complete set*) for an inverse problem is the set of all representable solutions that satisfy the forward problem after final rounding.

For systems of linear equations, the c-set is connected, so if you can find one representable solution that works, you can write a program that explores all the neighboring representable points by going up an ULP or down an ULP in each dimension, and adds that point to the set of known solutions if it equals the right-hand side after rounding. Repeat with any new point added.

Eventually, every new neighbor of every point in the set fails the equations, and that is when you have the complete c-set. It sounds like a lot of work, but remember that the *forward* problem is usually a much cleaner and faster calculation than the inverse problem. That is certainly true when solving linear equations, and you can compute the dot product without rounding. The work to compute $A \cdot x$ increases as the square of the number of equations, which is dwarfed by the work to find a guess in the first place, which increases as the cube of the number of equations.

Why would you want the entire c-set? Maybe you do not. But if you want to be safe, obtaining all possible solutions and examining them will inform you if you have an ill-posed problem (when the shape of the c-set is very eccentric, sharp-pointed). Or, like the {1200, 601} example given above, it would allow you to discover that there is a solution right nearby, {1200, 600}, that works even *without* rounding. This is my first tip for writing linear solvers that suck less, if the number of equations is small: Consider looking at ±ULP neighbors of the solution you find to see if they work, too.

12.3.2 Perfect Determinants; Textbook Gaussian Elimination

Chapter 10 mentioned determinants of 2-by-2 and 3-by-3 matrices and their geometric meaning. They have a geometric meaning when doing the inverse problem of matrix multiplication, that is, solving a linear system. If solving 2 equations in 2 unknowns, $A \cdot x = u$ and the determinant of A is zero, then you are trying to find the intersection of parallel lines. There are two possibilities that depend on u:

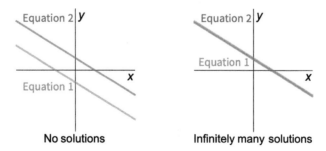

The left example might be a linear system like

$$2x + 3y = 1$$
$$4x + 6y = 0.$$

Multiply the second equation by ½, and the left-hand side is the same for both equations, but one is equal to 1 and the other to 0, a contradiction. No solution.

The geometric interpretation is that distinct parallel lines never meet. Whereas for a system like

$$2x + 3y = 4$$
$$4x + 6y = 8,$$

the *b* values are such that the two equations say the same thing, so the parallel lines are one and the same line. That is actually only *one* equation in two unknowns, $2x + 3y = 4$, so you could solve for *x* in terms of *y* or *y* in terms of *x*, but you cannot state the full solution as a single pair of real numbers. This aspect of "what can go wrong" is part of the standard curriculum on the subject of solving linear systems. Less well-known is what happens when using computer arithmetic and the determinant is very small, but not zero. Hence, this advice:

In solving 2-by-2 systems using posits, first compute the determinant using the quire. If the determinant is 0, the system does not have a real solution {x, y}. If the determinant is small compared to the input data, beware of ill-posed behavior.

This fails for a float environment because there is no quire, and when you compute $\det\begin{pmatrix} a & b \\ c & d \end{pmatrix} = a \cdot d - b \cdot c$, rounding could make it wrong. I will show an example later.

It is also not advisable for systems larger than 2-by-2; the determinant will not be computable exactly in general using the quire. But for 2-by-2, the quire computation is exact, and it returns something that rounds to 0 as a posit if and only if the quire is 0. If it comes back very small, like $\pm minPos$, or small relative to the input values, that provides a warning that the problem may be ill-posed because the lines are *almost* parallel and the slightest change in either side of the equation could change the answer by a huge amount.

Here is a linear system that can be expressed exactly using 16-bit posits, so we can ignore the issue of rounding from decimals to posits:

$$1.5625x + 3.6875y = 5.8125$$
$$3.3125x + 7.8125y = 12.3125$$

This has a very simple exact answer: $x = -1$, $y = 2$.

If we apply Gaussian elimination, we would first eliminate x from the second equation by subtracting the first equation scaled by 3.3125/1.5625, rounded. Or we could use the second equation to eliminate x from the first equation. Or we could eliminate y from either equation using the other one, for a total of four different ways. Try all four possibilities, using fused multiply-adds where possible to reduce the number of roundings. Here is the dataflow graph of a 2-by-2 Gaussian solver for $\begin{pmatrix} a & b \\ c & d \end{pmatrix} \cdot \begin{pmatrix} x \\ y \end{pmatrix} = \begin{pmatrix} u \\ v \end{pmatrix}$:

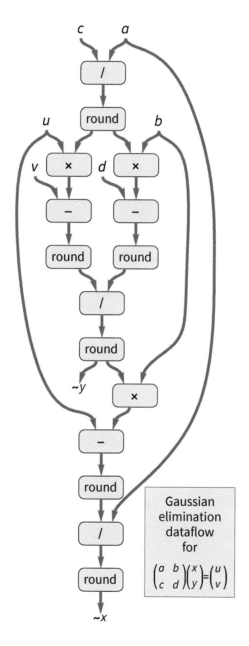

Gaussian elimination dataflow for

$$\begin{pmatrix} a & b \\ c & d \end{pmatrix}\begin{pmatrix} x \\ y \end{pmatrix} = \begin{pmatrix} u \\ v \end{pmatrix}$$

The parallelism in that dataflow is what a compiler might produce using automatic parallelization, or what a microprocessor might exploit by detecting whichever of the 15 operations can execute concurrently. The graph is still 12 stages high.

Here is code for a linear solver with no rounding, and no guards against divide-by-0:

```
linearSolve2By2[{{a_, b_}, {c_, d_}}, {u_, v_}] := Module[{t, x, y},
    t = c / a;
    y = (v - t * u) / (d - t * b);
    x = (u - b * y) / a;
    {x, y}
]
```

That is the way Gauss himself would have done it, carrying as many decimals as needed for each step. Besides being the leading mathematician in the world, he was one of those "lightning calculators" who could do arithmetic quickly in his head.

Carl Friedrich Gauss (1777–1855). Painting by Gottlieb Biermann.

But in computers with fixed-size number formats, rounding is generally done without such flexible precision. Here is a version that shows the rounding steps as overbars. All three multiply-adds are fused, which helps the accuracy a little:

```
linearSolve2By2r[{{a_, b_}, {c_, d_}}, {u_, v_}] :=
 Module[{t, x, y},
    t = c / a;
    y = (v - t * u) / (d - t * b);
    x = (u - b * y) / a;
    N[{x, y}]
]
```

Of the four possible ways to select the entry in **A** we first use to eliminate another term, the technique of choosing the right one is called *pivoting*. *Partial pivoting* is the most commonly used, which is to choose the entry in the first column that has the largest magnitude. To quote Wikipedia's **Pivot element** entry, "Partial pivoting is *generally* sufficient to *adequately* reduce round-off error" (emphasis added).. The problem as stated has the smallest magnitude element in the top-left corner, so start by trying that:

```
a = {{1.5625, 3.6875}, {3.3125, 7.8125}};
u = {5.8125, 12.3125};
Print["Exact solution:    ", linearSolve2By2[a, u]]
Print["Computed solution: ", linearSolve2By2r[a, u]]
```

```
Exact solution:     {-1., 2.}
Computed solution: {-0.9206542969, 1.9663088594}
```

The first answer, {–1., 2.}, confirms that `linearSolve2By2` works. The second answer is off. Now you know why the Wikipedia quote had some caveats.

Textbook numerical analysis says partial pivoting should produce a better result than using the smaller element in the column. Usually. Swap rows in the system to put the larger column element, 3.3125, into the top left corner:

```
a = {{3.3125, 7.8125}, {1.5625, 3.6875}};
u = {12.3125, 5.8125};
Print["Exact solution:    ", linearSolve2By2[a, u]]
Print["Computed solution: ", linearSolve2By2r[a, u]]
```

```
Exact solution:     {-1., 2.}
Computed solution: {-0.9426269531, 1.9755885938}
```

It made little difference in this case. The term *full pivoting* means picking the largest magnitude entry in all of the unsolved part of **A** to use to eliminate the next term. Textbooks assure us that full pivoting is the best kind of pivoting of all, though it is much more work than partial pivoting since you have to find the maximum of an entire square of values instead of just one column. The largest magnitude entry in **A** is 7.8125, so swap columns in **A**. That reverses the roles of *x* and *y*, so the correct answer is now {2., –1.}:

```
a = {{7.8125, 3.3125}, {3.6875, 1.5625}};
u = {12.3125, 5.8125};
Print["Exact solution:    ", linearSolve2By2[a, u]]
Print["Computed solution: ", linearSolve2By2r[a, u]]
```

```
Exact solution:    {2., -1.}
Computed solution: {1.73583 9844, -0.37744 14063}
```

If you look for a proof that full pivoting is better than partial pivoting, you will not find it. That is because, as the example demonstrates, it is *not* always better. It gave the most inaccurate answer so far. The main thing partial pivoting does is prevent dividing by zero and overflow, not necessarily minimize relative error.

The one remaining case is if we use 3.6875 as the pivot:

```
a = {{3.6875, 1.5625}, {7.8125, 3.3125}};
u = {5.8125, 12.3125};
Print["Exact solution:    ", linearSolve2By2[a, u]]
Print["Computed solution: ", linearSolve2By2r[a, u]]
```

```
Exact solution:    {2., -1.}
Computed solution: {1.47705 0781, 0.23406 98242}
```

That produces the worst inaccuracy of any of the four pivots. It did not even get the *sign* right for the second variable.

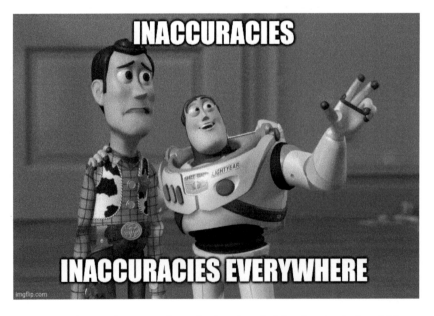

All four answers show why Gaussian elimination (with pivoting) should be regarded not as a "direct method" so much as a method that *usually* gets close as a first guess.

Was that a contrived problem? Not really. Such ill-posed systems arise all the time in one of the most common uses for 2-by-2 solvers: Least-squares fits, also called linear regression. Fitting a line to a set of statistical value pairs requires solving 2 equations in 2 unknowns. If there is a significant correlation between two statistical values, the system will be well-posed. But if there is not, the resulting fit will be garbage. In the right-hand diagram below, the scatter plot is completely random. That does not stop a linear system solver from finding a line that minimizes the squared distance to the points. The slope and intercept of that line vary greatly with tiny changes in the positions of the data points, which is the definition of ill-posed.

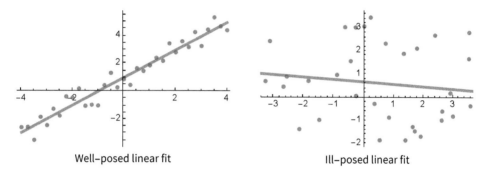

Well–posed linear fit   Ill–posed linear fit

Curve-fitting is often the first place a novice discovers just how disastrous it can be to use computers to solve systems of equations.

12.3.3 Constructive Reverse Error Analysis

As Douglas Adams would say, the answer is 42. But... what was the question?

The idea of reverse error analysis is this: "Maybe we cannot bound the error, but we can show that the answer is the exact solution to a problem that was close to the one you asked." Textbooks and papers on numerical methods often fall back on this argument, but they do not actually construct what problem was solved. With posits, we can do just that. Take the example of the last pivot we tried:

```
a = {{3.6875, 1.5625}, {7.8125, 3.3125}};
u = {5.8125, 12.3125};
Print["Exact solution:    ", linearSolve2By2[a, u]]
Print["Computed solution: ", linearSolve2By2r[a, u]]
```

```
Exact solution:    {2., -1.}
Computed solution: {1.477050781, 0.234069 8242}
```

Those are the decimal approximations to 16-bit posit values $\frac{3025}{2048}$ and $\frac{3835}{16384}$. So, for what right-hand side are those values the exact answer?

N[ā.{3025 / 2048, 3835 / 16 384}, 20]

{5.81235 88562 01171 8750, 12.31481 55212 40234 375}

The quire allows construction of the exact right-hand side for which the computed value is the answer. The user can then examine it and decide if that is sufficiently close to the problem that was posed in the first place. But again, it is critical to check the determinant when solving 2-by-2 systems, as a quick check for a nearly-parallel (or worse, perfectly parallel) pair of lines expressed by the two-equation system.

12.3.4 An Argument for Cramer's Rule

Suppose we solve the system by *cross-multiplying* to make the first terms match:

$$a x + b y = u$$
$$c x + d y = v$$

⬇

$(a \cdot c) x + (b \cdot c) y = u \cdot c$ by multiplying the first equation by c,

$(a \cdot c) x + (a \cdot d) y = a \cdot v$ by multiplying the second equation by a,

so $(a \cdot d - b \cdot c) y = (a \cdot v - u \cdot c)$ by subtracting the first equation from the second.

You will recognize $D = (a \cdot d - b \cdot c)$ on the bottom left as the determinant of **A**. One way to write $(a \cdot v - u \cdot c)$ is $\det\left(\begin{smallmatrix} a & u \\ c & v \end{smallmatrix}\right)$, replacing the second column of **A** with **u** and taking the determinant. So $y = (a \cdot v - c \cdot u)/D$.

If you do that reversing of the roles of x and y, you get $x = (u \cdot d - b \cdot v)/D$ and notice that $u \cdot d - b \cdot v = \det\left(\begin{smallmatrix} u & b \\ v & d \end{smallmatrix}\right)$. That is *Cramer's rule* for a 2-by-2 system:

To solve $\left(\begin{smallmatrix} a & b \\ c & d \end{smallmatrix}\right) \cdot \left(\begin{smallmatrix} x \\ y \end{smallmatrix}\right) = \left(\begin{smallmatrix} u \\ v \end{smallmatrix}\right)$ for x and y, first compute the determinants

$D = \det\left(\begin{smallmatrix} a & b \\ c & d \end{smallmatrix}\right) = a \cdot d - b \cdot c$, $D_x = \det\left(\begin{smallmatrix} u & b \\ v & d \end{smallmatrix}\right) = u \cdot d - b \cdot v$, $D_y = \det\left(\begin{smallmatrix} a & u \\ c & v \end{smallmatrix}\right) = a \cdot v - u \cdot c$.

If $D = 0$, the solution is singular. Otherwise, $x = D_x/D$ and $y = D_y/D$.

With posits, all three determinants can be evaluated exactly using the quire, then rounded back to the nearest posit. If the system is singular, the quire will *always* detect that, and if the system is solvable, the quire will always get that right, too. We can test whether Cramer's rule does any better on the example than did the four ways of doing Gaussian elimination with pivoting.

Here is the first form of the equations:

$$1.5625\,x + 3.6875\,y = 5.8125$$
$$3.3125\,x + 7.8125\,y = 12.3125$$

Once the data is assigned to variables, Cramer's rule only takes two lines of code:

```
{{a, b}, {c, d}} = {{1.5625, 3.6875}, {3.3125, 7.8125}};
{u, v} = {5.8125, 12.3125};

{det, detX, detY} = {a×d-b×c, detX = u×d-b×v, detY = a×v-u×c};
{x, y} = {detX/det, detY/det}
```

{-1, 2}

Cramer's rule nailed it, and we did not have to make decisions about pivots. Each result experienced three roundings: the conversion of two exact quire results to posit format and then the rounding after dividing.

With floats, Cramer's rule is numerically dangerous because the determinants all require *two* rounding steps, like $a \times d - b \times c$ if we use fused multiply-add. To see why, suppose we use 16-bit IEEE floats and set all four values a, b, c, d to the same number. The determinant obviously should then be zero. Set them all to 0.3 (the closest float to 0.3), and compute the determinant with two roundings:

```
setFloatEnv[{16, 5}] (* IEEE Std 754 half-precision *)
a = b = c = d = 0.3;
Print["The closest float to 0.3 is ", N[a, 12]]
Print["Determinant with two roundings is ", N[a*d-b*c, 4]]
```

The closest float to 0.3 is 0.300048828125

Determinant with two roundings is 2.444×10^{-6}

Instead of detecting that the system is singular, floats would barrel ahead and compute a wildly misleading result. Do not use Cramer's rule without a quire.

Now contrast the Cramer's rule dataflow with that for Gaussian elimination:

Each input value is used twice, for six multiplies that can take place in parallel. Then three subtractions can take place in parallel to produce the three determinants needed. The divides can also execute in parallel, so the total number of operation delays is only 5, versus 12 for Gaussian elimination. The result is often exact. In Gaussian elimination, the initial rounding from c/a propagates to every other operation, whereas for Cramer's rule, the roundings are almost independent of one another. For 2-by-2 systems, Cramer's rule *looks* like a win. However, we can be more scientific in assessing different methods, which is the topic of the next section.

12.3.5 If *Consumer Reports* Rated 2-by-2 Solvers

Ratings *Compact refrigerators*   ● Excellent ◐ Very good ○ Good ◑ Fair ● Poor

Brand & model	Price	Overall score	Refrigerator performance	Freezer performance	Energy efficiency	Noise
		0 100 P F G VG E				
MEDIUM SIZED MODELS *These cost and weigh a bit less and are easier to carry.*						
Kenmore 9275(9)	120	48	◐	●	○	○
Haier HAS01A	120	48	◐	●	○	○
Sanyo SR2560M	130	47	◐	●	○	◐
Frigidaire FRL03DQ2E(L)	130	15	◑	●	○	○

This is not real data; it is an example of how *Consumer Reports* displays product ratings.

The product testers at Consumer Union, publisher of *Consumer Reports,* are quite proficient at objective experimental science. Perhaps we could emulate the way they rate consumer products like refrigerators and paper towels in looking for a better 2-by-2 linear solver. First, declare the criteria and the testing method, then run the tests and summarize the results. I propose these criteria:

- If the exact answer is expressible in the number format, the solver should find it.
- If the exact answer is not expressible in the number format, the answer should be the pair of numbers closest to the exact answer.
- If the problem is singular, that should be detected and reported.
- The solver should work for a representative set of real numbers.
- If the problem is ill-posed, a warning should be provided along with the answer.
- It should be inexpensive in terms of the size of its dataflow graph.

We should test the *method,* not the number format. Obviously, a 64-bit format can hide errors resulting from a bad method better than a 16-bit format. The representative real set **RRSet** from Chapter 7 can generate 2-by-2 linear systems for which the answer is expressible exactly, and then we can test how well the numbers get crunched. We can also test cases where the answer cannot be expressible exactly.

I tested 10 000 2-by-2 systems using standard 32-bit posits, with entries drawn from **RRSet**. The offerings tested were Gaussian elimination, Gaussian elimination with partial (column) pivoting, Cramer's rule, and Cramer's rule with residual correction like that shown in Section 6.4.

Ratings *2–by–2 linear solvers*

Excellent	Very good	Good	Fair	Poor
◉	◓	○	◒	●

Brand & model	Price	Overall score		Finds exact answer	Finds closest answer	Worst-case error	Average error
		0 100					
		P F G VG E					
32–BIT POSIT FORMAT *These cost less than legacy high–precision solvers.*							
Gauss No pivoting*	15	3		○	●	●	●
Gauss Partial pivoting	23	21		○	○	●	●
Cramer Base model	18	63		◒	●	○	◒
Cramer Refined	37	99		◉	◉	◉	◉

*We rate this brand *not recommended*. Use can result in severe injury to a calculation.

The code for this testing is in Appendix E.

The error was measured in the total difference in ULPs between the computed answer and the correctly-rounded answer. The price is the estimated number of functional units on a dataflow machine that would be needed for the method. All but the first method check the determinant and quit with an error message if the system is singular. The first method can pass the determinant test yet divide by zero if the top left entry of **A** is zero, hence the footnote warning.

When given a problem with an exactly representable answer, Gaussian elimination without pivoting was able to find it about half of the time. Straight Cramer's rule found the exact answer less often, 40% of the time. With a refinement step using the quire, the success rate was 100%, but it is possible to contrive cases for which the refinement will still miss the exact answer, so the Overall Score is less than 100.

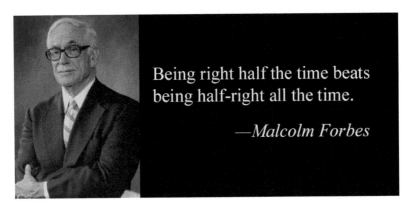

Being right half the time beats being half-right all the time.

—*Malcolm Forbes*

The more likely situation is that the exact answer is not exactly representable in the format. With no pivoting, Gaussian elimination found the correctly-rounded answer 27% of the time; using partial pivoting raised that score to 40%. Cramer's rule also scored 27% as a first guess, but the average error was far less: an average of 5.5 ULPs for Cramer's rule, compared to 118 000 ULPs for Gaussian elimination with pivoting! With refinement, Cramer's rule achieved correct rounding for all 10 000 test cases, but there exist rare situations where it will not do so.

Cramer's rule is not practical for larger systems because the cost of finding all the determinants for n equations grows as $n!$, the factorial, and the precision needed also becomes intractable. Gaussian solvers with partial pivoting scale tolerably, with a cost that grows as n^3, like the cost of n-by-n matrix-matrix multiplication.

12.4 Linear Equations After Computers

The Atanasoff-Berry Computer (ABC) was mentioned in Section 4.5. John Atanasoff had, since the early 1930s, sought a way to automate the process of solving systems of linear equations. From his paper describing the ABC after it was finished:

> Now it is easy to see that the principal term in the amount of labor needed to solve a system of equations is kN^3 in which N is the number of unknowns and k is a constant. Since an expert [human] computer takes about eight hours to solve a full set of eight equations in eight unknowns, k is about 1/64. To solve twenty equations in twenty unknowns should thus require 125 hours...The solution of general systems of linear equations with a number of unknowns greater than ten is not often attempted.

Reprinted in *The Origins of Digital Computers*, B . Randell, ed., 3rd edition, 1982, page 316.

The term "computer" originally meant a person, not a machine, and Atanasoff's estimate is clearly based on someone with access to a mechanical desktop calculator. He also stated the applications that needed such an automated solver:

> 1. Multiple correlation
> 2. Curve fitting
> 3. Method of least squares
> 4. Vibration problems including the vibrational Raman effect
> 5. Electrical circuit analysis
> 6. Analysis of elastic structures
> 7. Approximate solution of many problems of elasticity
> 8. Approximate solution of problems of quantum mechanics
> 9. Perturbation theories of mechanics, astronomy, and quantum theory

Reprinted in *The Origins of Digital Computers*, B . Randell, ed., 3rd edition, 1982, page 315.

He wrote that list in 1940, but it is still current today as an excellent summary of high-performance computing. The ABC launched a decade of intense work developing automatic computers, and by the late 1940s there were stored-program electronic computers that are easily recognizable as ancestors of the ones we use today.

Another fellow who recognized the need for linear solvers was Franz Alt at Bell Laboratories. Instead of using vacuum tubes, he opted for the much less expensive electromechanical relays which were already heavily used in the telephone industry. After a series of smaller systems, by 1948 they attempted a system that used 9000 relays, occupied about 1000 square feet, and weighed ten tons. It was based on decimal arithmetic, not binary. Alt was unaware that Konrad Zuse's relay-based Z3 had implemented floating-point years earlier, and claimed the Bell Labs machine to be the first computer to use floating-point arithmetic.

Alt's floats had seven significant digits and an exponent that ranged from 10^{-19} to 10^{19}; each float required 62 relays. It could do an add in 0.3 second, a multiply in 1 second, and a divide or square root in about 5 seconds; the speed depended on the digits in the inputs. Alt said the device was about five times faster at solving problems than a human equipped with a good desk calculator. It is always described as "A Bell Laboratories Computing Machine," or sometimes "Mark V."

So after a thorough description of the machine's design, what was the first *application* he describes in his 1948 paper? Linear solvers.

> (1) *Systems of linear equations.* Several methods of solution have been tried out on the machine. In all of them, elimination is used to arrive at a first approximation of the solution, and if this is not close enough (because of the cumulative effects of rounding errors), iterations are used to find better approximations. …The computing time for the elimination process for 13 equations is about 3½ hours…. The time required for systems of higher order varies approximately as the cube of the order. A more stringent limitation is set by the seven-digit accuracy of the machine. It is believed that this will limit the process used, even if used iteratively, to about 20 or 30 unknowns.

Reprinted in *The Origins of Digital Computers*, B . Randell, ed., 3rd edition, 1982, page 289

If you read between the lines, you can tell that they tried Gaussian elimination with no pivoting and found out the hard way that the calculation can very easily blow up. This is not so obvious if you are limited to, say, six equations in six unknowns and do the calculation by hand. But when the systems gets larger, you discover that Gaussian elimination (without pivoting) is *unstable*. Which brings us to this guy:

John von Neumann (1903–1957) and his EDVAC computer, circa 1952

John von Neumann is known for so many accomplishments that it is hard to know where to start. He may have been the most brilliant mathematician since Gauss (and he shared Gauss's lightning calculator ability). He was a key physicist on the Manhattan Project. In 1945, he remarked to his second wife, "What we are creating now is a monster whose influence is going to change history, provided there is any history left." But he was not talking about the atomic bomb. He was talking about the growing power of computing machines.

Von Neumann was no fan of floating-point arithmetic. He thought that letting someone use floats for solving linear equations on an automatic computer was like handing a five-year-old a loaded gun to play with. He believed the format should be fixed-point, with cognizant programmers managing the position of the radix point and being careful with rounding error. He knew very well that naive use of Gaussian elimination on ill-posed systems could produce wildly wrong results; when someone suggested it could be made safe by using very high-precision arithmetic, he replied, "If it is not worth doing, it is not worth doing well." Section 12.3.4 showed how difficult it is to build a bombproof solver even for systems as small as two unknowns.

The breakthrough for linear solvers that suck less came from James Wilkinson in 1965: he published *The Algebraic Eigenvalue Problem*, a paper proving that with pivoting, under certain conditions, linear solvers could be stable and could succeed using floats of modest precision. Thus began in impressive academic lineage. Cleve Moler was Wilkinson's student. Jack Dongarra was Moler's student. Moler would later found The Mathworks and create MATLAB, one of the most commercially successful pieces of mathematical software ever written.

It was in the late 1970s when Moler was a professor at University of New Mexico that he and his students produced LINPACK, an extraordinarily high-quality and portable software library for analyzing and solving linear equations. These were the early days of email, and they invented something of equally high impact: free, electronically-distributed software. You could send an email to a particular address, and it automatically replied with the source code for LINPACK. I believe it was Dongarra who had the idea of including a short list of computers and their execution times for solving a system of 100 equations, just for fun, and inviting users to email timings for other computers to him. That was the beginning of the best-known technical computing benchmark test of all time.

From left to right, Dongarra, Moler, Bunch, and Stewart, and the User's Guide
to their solvers that sucked less. Notice the vanity plate on the car.

To this day, many people hear "LINPACK" and think of the benchmark, forgetting that it was originally a well-designed software library for getting work done, not for comparing computers. Because of Moore's law doubling the transistor densities every couple of years, it was not long before 100 equations was too small to use as a test problem; the fastest computers could run the benchmark in less than 1/100 of a second, faster than the display could refresh. There was also a growing problem with "benchmark rot," the industry term for designing systems specifically to be good at a benchmark that is used to make purchasing decisions, when the system was actually not that fast in general. A ground rule of the benchmark was that the source code (Fortran) could not be altered in any way. At least one company created a "LINPACK recognizer" in their Fortran compiler that would test if the source code was the LINPACK benchmark, and if it was, it emitted an executable binary that had been hand-tuned at the machine language level. Someone tried misspelling "DONGARRA" in a comment line in the source code, and the performance of that computer mysteriously dropped by a factor of seven. This kind of clever cheating is sometimes called "marketeering." I will not name names here.

So Dongarra raised the benchmark problem size to 300-by-300, to make it about $3^3 = 27$ times more operations and $3^2 = 9$ times as much storage. It was not long before that, also, became a too-small problem. A third version was released that was 1000-by-1000 so that the system took eight megabytes to store and almost a billion float operations to solve, and the "no changes to source code" rule was relaxed so that some of the more sophisticated parallel and vector computer systems could exploit features that compilers could not yet handle automatically. The innermost kernel operation, performed on the order of n^3 times and the bulk of the task of the solver, looked something like this:

```
A (I, J) = A (I, J) - A (I, K) * A (K, J)
```

That looks a lot like a matrix-matrix multiply, but instead of the product being a separate matrix, it is writing back to the input matrix. That spooks a compiler because if you try to do things in parallel for higher speed, it looks like it could overwrite an input value before it is fetched. A human studying the algorithm can see that it is safe, but the compilers could not. More on this in the next section.

In 1988, I wrote a very short (two-page) paper, "Reevaluating Amdahl's law" on the need to scale computer workloads to the power available, not measure time reduction. To my surprise, people thought that was a profound new idea and started calling it "Gustafson's law." I suggested to Dongarra that as custodian of the LINPACK benchmark (actually three different benchmarks by that point), he could make the test scalable to whatever size is appropriate for the computing capability, and then the figure of merit would be the speed in (64-bit) float operations per second. That made the benchmark immune to Moore's law, and the scalable version of the LINPACK benchmark (now called HPL) that came out in 1992 is still in use today, 32 years later. It is still used to create the "TOP500" list of the world's fastest supercomputers. The original LINPACK library was modernized over and over to accommodate parallel computer architectures. The linear systems solved by super-computers for the benchmark have tens of millions of simultaneous equations, at the time of this writing.

12.5 Visualizing Gaussian Elimination

Remember the visualization of the outer product as a matrix-matrix multiply in Section 11.3? We can treat a step of Gaussian elimination as swapping rows to put the largest magnitude element on the diagonal, the scaling of that row by dividing by the pivot, then taking the outer product of the row and column to the right of and below the pivot. Remember that this is the same as forming a multiplication table. That table is then subtracted from the matrix. Here is the visualization:

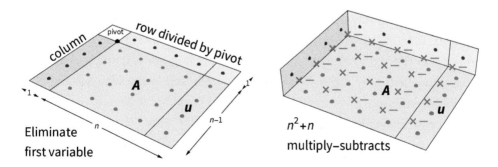

Gaussian elimination is repeated subtraction of a row-column outer product from the linear system.

This figure shows why it is safe to rearrange that line of code mentioned before:

```
A (I, J) = A (I, J) - A (I, K) * A (K, J)
```

There is no way the row or column inputs (the red and blue vectors in the figure) will be affected by rearranging the code for higher performance, since only the green part of the matrix is changed.

The process is repeated on the smaller system (the part shown in green) to eliminate the second variable from the remaining equations, and so on, to complete the forward elimination stage of the solver. This can be visualized as a stack of the boxes shown on the right in the above figure:

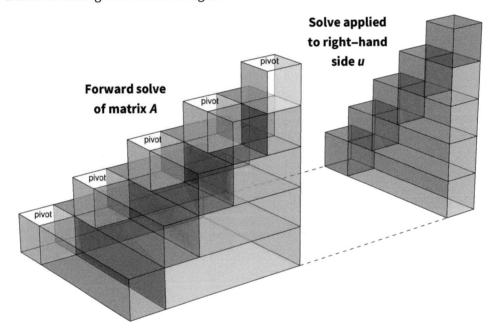

Gaussian elimination visualized; iterations work their way upward to the top of the pyramid.

The number of multiply-subtract pairs is the volume of the pyramid for **A** and stair step for **u**. The volume of a pyramid is ⅓ the area of its base times its height, or $\frac{1}{3}N^3$ in this case. The right-hand side and the backsolving are just triangular stair step shapes that have only order N^2 volume. So now you know why solving N linear equations takes order N^3 time. Backsolving and checking the answer (computing the residual or the defect in the answer) take only order N^2 time.

There is another reason for showing this visualization. Computer programs perform the forward elimination with a set of nested loops over the row number, the column number, and the iteration number (the height of the pyramid).

Here is something that is not obvious about the three nested loops, but quite useful:

> The nesting of the loops can be in *any order*. If you arrange the loops so that the loop over the iteration number (the vertical direction in the visualization) is the innermost loop, that innermost loop will be in the form of a dot product.

The dot product is economical because the running sum is kept in a register so there are fewer memory references per multiply-subtract. One of the vectors can be re-used for the next dot product, so it can be kept in a cache or scratch storage close to the processor. Hence, that raises performance for computers in general.

But for posits, it has the extra benefit that doing the dot products with the quire reduces the number of rounding steps in the forward solver from order N^3 to order N^2, which is a huge accuracy improvement if N is large.

12.6 The LINPACK Benchmark and Posits

The LINPACK benchmark works like this: Populate a matrix A with pseudo-random numbers (64-bit float values) evenly distributed between 0 and 1. If x is a vector of all 1.0 values, then $A \cdot x = u$ will make each element of u the sum of the corresponding row of A, so create that u. Apply a Gaussian solver with partial pivoting to see if it can compute something close to an x with all 1.0 entries.

> There is a subtle flaw in the LINPACK setup. The row sums in u are *rounded* to the nearest 64-bit float value, so the correct solution is *not* a vector of all 1.0 values.

Even a problem as small as 4-by-4 can demonstrate this:

```
n = 4;
SeedRandom[3141]; (* Make the sequence reproducible. *)
setFloatEnv[{64, 11}] (* IEEE Std 754 double precision. *)
A = Table[RandomReal[{0, 1}], {i, 1, n}, {j, 1, n}];
u = Total[Transpose[A]]; (* u = row sums of A. *)
Print["u ≈ ", N[MatrixForm[u], 17],
  "  computed u = ", N[MatrixForm[u], 17]]
```

$$u \approx \begin{pmatrix} 1.7523208033791653 \\ 1.5988911631741087 \\ 2.0962940328478370 \\ 1.4655469235079583 \end{pmatrix} \quad \text{computed } u = \begin{pmatrix} 1.7523208033791653 \\ 1.5988911631741087 \\ 2.0962940328478368 \\ 1.4655469235079583 \end{pmatrix}$$

My apologies for the long decimals, but these are 64-bit floats as required by the benchmark, to 17 decimals. The third entry of **u** is not computed exactly; it had to be rounded. Here is the (rounded) solution with an exact **u** versus the rounded **u**:

```
LinearSolve[A, u]  (* Mathematica's built-in exact solver *)
LinearSolve[A, u]  (* Solve exactly with rounded u, then round *)
```

$$\{1, 1, 1, 1\}$$

$$\left\{1, \frac{4\,503\,599\,627\,370\,497}{4\,503\,599\,627\,370\,496}, 1, \frac{9\,007\,199\,254\,740\,989}{9\,007\,199\,254\,740\,992}\right\}$$

Oops. You see the problem. Just as we could not express the ⅔ coefficient in the Babylonian 2-by-2 problem in Section 12.2.1, we cannot express the desired right-hand side the way the LINPACK benchmark generates it. Hence, it is not fair to compare the computed answer with a list of all 1.0 values, because all 1.0 values is *not* the float answer closest to the exact answer.

This is easy to fix. Tweak the **A** values in their lowest-order bits so that **u** is expressible exactly. Then you have a fair problem to solve. One can program automatic ways to create expressible versions of LINPACK-type problems, but for now, simply adjust one of the elements of **A** by scraping off its least significant **1** bit.

```
A[3,1] = fToX[BitShiftLeft[BitShiftRight[xToF[A[3,1]], 4], 4]];
u = Total[Transpose[A]]; (* u = row sums of A. *)
Print["u ≈ ", N[MatrixForm[u], 17],
  "  computed u = ", N[MatrixForm[u], 17]]
```

$$u \approx \begin{pmatrix} 1.75232\,08033\,79165\,3 \\ 1.59889\,11631\,74108\,7 \\ 2.09629\,40328\,47836\,8 \\ 1.46554\,69235\,07958\,3 \end{pmatrix} \qquad \text{computed } u = \begin{pmatrix} 1.75232\,08033\,79165\,3 \\ 1.59889\,11631\,74108\,7 \\ 2.09629\,40328\,47836\,8 \\ 1.46554\,69235\,07958\,3 \end{pmatrix}$$

Appendix F has the source code for two Gaussian solvers with pivoting, one for floats, `linearSolveFloat`, and one for posits, `linearSolvePosit`. While the test matrix is well-posed, a Gaussian solver manages to miss by just a little on all four elements of the {1, 1, 1, 1} answer:

```
N[linearSolveFloat[A, u], 17]
```

$$\{0.99999\,99999\,99999\,8, 1.00000\,00000\,00000\,2,$$
$$1.00000\,00000\,00000\,4, 0.99999\,99999\,99999\,4\}$$

Those answers are off by –2, 1, 2, and –5 ULPs. Now try it with posits. We do not need 64-bit posits to express the problem and answer exactly. I realize it is a peculiar size, but 57-bit standard posits can solve the same problem as 64-bit floats.

```
setPositEnv[{57, 2}];
x = linearSolvePosit[A, u];
N[x, 17]
```

{0.99999 99999 999998, 1.00000 00000 000002,
 1.00000 00000 000002, 0.99999 99999 999997}

The 57-bit posit answer is *slightly* better than the 64-bit floats, with errors of –2, 1, 1, and –3 ULPs. But the main thing posits can do that floats cannot is to compute the defect in the answer. This is very much like the long division problem presented at the beginning of this chapter: Guess the value that works, check it with an exact multiplication, then correct it. We strive for $A \cdot x$ to equal u, but it missed. Compute the difference, called the *residual*, $t = u - A \cdot x$, and round it to posit format:

```
t = u - A.x;
N[t, 6]
```

{−5.20376 × 10⁻¹⁷, −5.09119 × 10⁻¹⁸, 8.88853 × 10⁻¹⁷, 3.26817 × 10⁻¹⁷}

Finally, the punchline: Solve the same system with the residual as the right-hand side, add that as a correction to x, and round to the nearest posit:

```
x + linearSolvePosit[A, t]
```

{1, 1, 1, 1}

Exact. This is how a posit environment with a quire can lead to linear solvers that suck less, and suck even less than the polished LINPACK routines. Posits allow defect correction in approximate solutions, no matter what solver method is used.

12.7 Inverses and the Dreaded Hilbert Matrix Test

Influential mathematician David Hilbert, 1862–1943

The identity matrix, usually written with the letter I, is 1s down the diagonal and 0s everywhere else. We saw it in Chapter 10 as the matrix that leaves things unchanged when you multiply by it. Given a matrix A, finding a matrix X such that $A \cdot X = I$ is the matrix equivalent of taking the reciprocal of a number. It is very much like Gaussian elimination for $A \cdot x = u$, but with the right-hand side being the identity matrix. The inverse of A is usually written A^{-1}. The routines in the previous section work fine to find A^{-1}; just use I as the right-hand side value, u.

There is a type of matrix that is sometimes used as the "acid test" for the quality of a linear solver, and it arises in trying to curve-fit polynomials to functions. A *Hilbert matrix* is a square matrix H with entries $h_{i,j} = \frac{1}{i+j-1}$. Here is a 4-by-4 example:

$$H = \begin{pmatrix} 1 & \frac{1}{2} & \frac{1}{3} & \frac{1}{4} \\ \frac{1}{2} & \frac{1}{3} & \frac{1}{4} & \frac{1}{5} \\ \frac{1}{3} & \frac{1}{4} & \frac{1}{5} & \frac{1}{6} \\ \frac{1}{4} & \frac{1}{5} & \frac{1}{6} & \frac{1}{7} \end{pmatrix}$$

The inverse matrix is, surprisingly, all integers, and there is a formula for them. So, the exact answer should be expressible with a number format if the integers are not too large.

Here is the inverse of the 4-by-4 Hilbert matrix:

$$H^{-1} = \begin{pmatrix} 16 & -120 & 240 & -140 \\ -120 & 1200 & -2700 & 1680 \\ 240 & -2700 & 6480 & -4200 \\ -140 & 1680 & -4200 & 2800 \end{pmatrix}$$

Hilbert matrices become very ill-posed as the matrix size grows. That fact makes inverting a Hilbert matrix a sensitive test of how well a linear solver (and the number format) can cope with rounding errors.

As always, we want to separate the difficulty of expressing the problem from the difficulty of solving it, and most of the fractions in **H** are not exactly expressible as floats. We can make the problem of finding the inverse expressible by scaling both sides of the equation by $3 \times 5 \times 7 = 105$:

$$\boldsymbol{H \cdot H^{-1} = I} \text{ becomes}$$
$$105\,\boldsymbol{H \cdot H^{-1}} = 105\,\boldsymbol{I}, \text{ or}$$

$$\begin{pmatrix} 105. & 52.5 & 35. & 26.25 \\ 52.5 & 35. & 26.25 & 21. \\ 35. & 26.25 & 21. & 17.5 \\ 26.25 & 21. & 17.5 & 15. \end{pmatrix} \cdot \boldsymbol{H^{-1}} = \begin{pmatrix} 105 & 0 & 0 & 0 \\ 0 & 105 & 0 & 0 \\ 0 & 0 & 105 & 0 \\ 0 & 0 & 0 & 105 \end{pmatrix}$$

All of those inputs are expressible as low-precision IEEE floats or standard posits without rounding. To show how bad things can get, try IEEE Std 754 single-precision and the partial pivoting routine from the last section:

```
H = Table[1 / (i + j - 1), {i, 1, 4}, {j, 1, 4}];
setFloatEnv[{32, 8}]
N[linearSolveFloat[105 × H, 105 × IdentityMatrix[4]], 7]
```

```
{{15.99993, -119.9994, 239.9990, -139.9995},
 {-119.9992, 1199.994, -2699.991, 1679.996},
 {239.9982, -2699.987, 6479.980, -4199.992},
 {-139.9988, 1679.992, -4199.988, 2799.995}}
```

Those numbers are only correct to about five decimals; they should be integers.

A 32-bit standard posit does a little bit better, but still does not land on any integers:

```
setPositEnv[{32, 2}]
X = linearSolvePosit[105 × H, 105 × IdentityMatrix[4]];
N[X, 8]
```

```
{{15.999990, -119.99990, 239.99978, -139.99986},
 {-119.99989, 1199.9990, -2699.9977, 1679.9985},
 {239.99974, -2699.9976, 6479.9946, -4199.9966},
 {-139.99983, 1679.9984, -4199.9966, 2799.9979}}
```

That is about one more decimal of accuracy, but posits can compute the residual and use it to find a correction to the defect, add it to the first guess, and round that:

```
T = 105 × IdentityMatrix[4] - 105 × H.X;
Print["Residual ≈ ", N[MatrixForm[T], 3], "\n"];
X1 = X + linearSolvePosit[105 × H, T];
Print["Corrected result = ", MatrixForm[X1]]
```

$$
\text{Residual} \approx \begin{pmatrix} 6.26 \times 10^{-6} & 0.000200 & -0.0000668 & -0.000267 \\ 6.47 \times 10^{-6} & 0.000340 & -0.000247 & -0.000134 \\ 0.0000104 & 0.000280 & -0.0000935 & -0.000214 \\ 5.86 \times 10^{-6} & 0.000285 & -0.000124 & -0.000174 \end{pmatrix}
$$

$$
\text{Corrected result} = \begin{pmatrix} 16 & -120 & 240 & -140 \\ -120 & 1200 & -2700 & 1680 \\ 240 & -2700 & 6480 & -4200 \\ -140 & 1680 & -4200 & 2800 \end{pmatrix}
$$

All it took was one refinement to get the exact answer. Sometimes it takes more, and sometimes even the refinement method fails.

There are many other solvers for linear systems, like conjugate gradient, GMRES, Cholesky, all striving to suck less. They all have difficulties for some cases.

> The best way to make any linear solver suck less is to check the solution using an exact dot product (the quire), then use the defect to make a correction.

Incidentally, I am being lazy here by calling the linear solver routine twice. That is not necessary. The hard part of partial pivoting that takes order n^3 operations only needs to be done once for an n-equation system. Computing the exact residual also only takes order n^2 operations. If n is even as large as 100, this residual-correction approach is about 100 times faster than the Gaussian elimination part. So for about 1% more work, you can get *far* more accurate answers in most (though not all) cases.

Besides his matrix, many concepts are named after Hilbert. He created a generalized kind of topological space that John von Neumann started calling a "Hilbert space." Hilbert was once attending a lecture in the 1930s given by von Neumann in Hilbert's honor. By this point Hilbert was regarded as one of the Grand Men of Mathematics. Many minutes into the lecture, a puzzled Hilbert finally raised his hand and asked,

"Dr. von Neumann... what is a 'Hilbert space'?"

> **Exercise for the Reader**: If the 2-by-2 matrix $\begin{pmatrix} a & b \\ c & d \end{pmatrix}$ has an inverse, what is it? Under what conditions will it not have an inverse? If computed with posits, how many rounding errors will each entry in the inverse experience?

12.8 Spinoffs of the Quest for Linear Solvers

The quest for a way to solve the ancient problem of simultaneous linear equations has led to quite a few spinoff innovations:

- The invention of matrices, linear algebra, and the method of eliminating one variable at a time from a system
- Vector arithmetic hardware
- The first electronic digital computer
- Seminal and enabling works about numerical analysis
- The first free, electronically disseminated software
- The international benchmark standard for ranking computers by performance

It seems interesting that two of the linear solver pioneers, Gauss and von Neumann, were gifted with the ability to do arithmetic in their head at what seemed like super-human speed. "Lightning calculators" are sometimes people with little under-standing of higher mathematics, and who are sometimes even autistic. But Gauss and von Neumann were brilliant mathematicians, and their arithmetic skill informed them of what to watch out for when solving linear systems.

I will leave you with an example of someone who should *not* be doing arithmetic in his head: the author of this comic book panel. For international readers, there are 16 ounces in a pound. Check his answer:

Superman's Girlfriend, Lois Lane, #10, July 1959.

13 The Future of Posit Computing

Posit adoption is actually well underway.

13.1 Knuth and The Morris Floats

In Donald Knuth's *The Art of Computer Programming* series, there are exercises for the reader, and Knuth provides at least a sketch of the answers. He rates the difficulty of each exercise on a scale from [00] to [50], where [00] is so easy that you can do it in your head, [40] is "Quite a difficult or lengthy problem that would be suitable for a term project ... the solution is not trivial," and [50] is an unsolved research problem. My eyes widened when I saw this Exercise in his section on floating-point arithmetic (*Seminumerical Algorithms*, 2nd Edition, 1981, page 212):

> [40] (John Cocke) Explore the idea of extending the range of floating point numbers by defining a single-word representation in which the precision of the fraction decreases as the magnitude of the exponent increases.

John Cocke (1925–2002) was an IBM Fellow who was one of the first proponents of Reduced Instruction Set Computing (RISC) and of the fused multiply-add. So people were thinking about tapered relative accuracy even as the IEEE Std 754 was being hotly debated. I quickly turned to the ANSWERS TO EXERCISES section to read what Knuth had to say. Had Knuth anticipated posit arithmetic by several decades?

Here is Knuth's answer to the Exercise on page 570 of that Volume:

> See Robert Morris, IEEE Transactions C-20 (1971), 1578–1579. Error
> analysis is more difficult with such systems, so interval arithmetic is
> correspondingly more desirable.

That paper was "Tapered Floating Point: A new Floating-Point Representation" (*IEEE Transactions on Computers*, 1 December 1971). Morris used a bit-field mentality in his design, and it was such a mess that it was not influential, although there was an attempt to build hardware for it in 1989. Knuth suggests that the drawback was more difficult error analysis, but a closer study shows serious flaws that doomed it.

Morris suggests an additional field, the *G* field, that describes how many bits there are in the exponent using an unsigned integer with an implicit offset. With a 2-bit *G* field you could have a range of $2^2 = 4$ possible exponent field sizes. With offset 1, it might be 1, 2, 3, or 4 exponent bits; or with offset 3, it might mean 3, 4, 5, or 6 exponent bits. Here is how he laid out the bit fields for an *n*-bit Morris float, with sign-magnitude for both the exponent and the fraction fields:

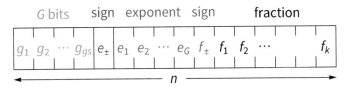

The Morris float bit field format

As was mentioned earlier, one of the quickest ways to find the flaws in a proposed number format is to try it with very low precision. Assume the *G*-field offset is 1. Here is a quick prototype of a Morris environment with precision **nMBit** and *G*-field size **gS**, and the creation of one with 8-bit precision and a *G* field size of 2 bits:

```
setMorrisEnv[{n_Integer /; n ≥ 4, gSize_Integer /; gSize ≥ 1}] :=
  {nMBit, gS} = {n, gSize};
{minMorris, maxMorris} =
```
$$\left\{2^{-\left(2^{2^{gS}}-1\right)},\ 2^{\left(2^{2^{gS}}-1\right)} \times \left(1 + \frac{2^{nMBit-2^{gS}-gS-2}-1}{2^{nMB-2^{gS}-gS-2}}\right)\right\};$$
```
setMorrisEnv[{nMBit = 8, gS = 2}];
```

With that setting, the exponent field can be from 1 to 4 bits long. If it has 1 bit, that leaves 3 bits for the fraction. Next, create a decoder **mToX** that turns a Morris float into the real number it represents. An exception is made for all bits **0** meaning zero. As usual, there is an implicit 1 added to the fraction.

```
mToX[m_Integer /; 0 ≤ m < 2^nMBit] :=
 Module[{eMask, eSign, exp, fMask, frac, fSign, fS, g},
  g = BitShiftRight[m, nMBit - gS]; (* Extract G field *)
  eSign = BitAnd[1, BitShiftRight[m, nMBit - gS - 1]];
  eMask = FromDigits[Table[1, g + 1], 2]; (* Value of G field *)
  fS = nMBit - gS - g - 3; (* Bits remaining for fraction field *)
  fSign = BitAnd[1, BitShiftRight[m, fS]];
  exp = BitAnd[BitShiftRight[m, fS + 1], eMask];
  frac = BitAnd[FromDigits[Table[1, fS], 2], m];
  If[m == 0, 0, (-1)^fSign × 2^((-1)^eSign × exp) × (1 + frac / 2^fS)]] (* Float *)
```

We have everything we need to discover the (nonzero) 8-bit Morris float vocabulary, from $00000001 = \frac{9}{8}$ to $11111111 = -\frac{1}{32768}$. Save it in **morris8**.

```
morris8 = Table[mToX[i], {i, 1, 2^nMBit - 1}];
```

The dynamic range is **minMorris** $= \frac{1}{32768}$ to **maxMorris** $= 32768$. Plot the $\log_{10}$ of the absolute values of the Morris floats from smallest binary to largest binary representation, and a problem becomes obvious:

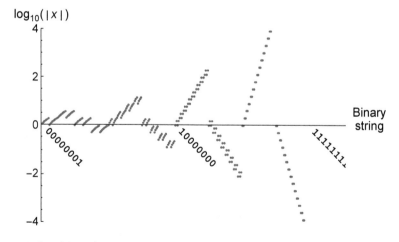

Log base 10 of positive values represented by the 8-bit Morris floats as a function of their bit string

This is the kind of chaos that results from simply carving out a bit field that describes the exponent size instead of thinking about the mapping. The Morris system fails the the second format design criterion from Section 4.8:

---

2. There should be no redundant bit patterns to mean the same thing; every bit counts.

The **Union** command removes multiple occurrences of the same number in a list. Here is how many unique numbers there are in the **morris8** set:

```
morris8 = Union[morris8];
Print["morris8 has ", Length[morris8], " unique numbers."]
```

```
morris8 has 144 unique numbers.
```

An 8-bit format should be able to express $2^8 = 256$ different numbers (or exception values), but the Morris definition wastes almost half of them.

> **Exercise for the Reader**: How many different ways can the 8-bit Morris floats express the value 1.0? What are the bit patterns that do so?

13.2 The Broader Definition of a Posit

Chapter 1 introduced the simplest kind of posit, with no exponent bits. Chapter 3 introduced exponent bits to provide a larger dynamic range, without losing any of the elegant properties of posit format. Chapter 7 added the maximum regime size parameter, which meant losing only the property that *minPos* and *maxPos* are perfect reciprocals. Then Chapter 8 presented the option of using logarithmic binades instead of linear binades, which adds the feature of perfect reciprocals and easier multiplication at the cost of perfect math in the quire and scalability to any precision. Just how far can the rules be stretched and something still be called a posit? Here are the ground rules that any posit format should follow:

- **Monotonicity**: The bit strings, treated as signed (2's complement) integers, must map monotonically to the signed reals.
- **Unique zero**: An all-**0** bit string represents zero.
- **Unique Not-a-Real**: A **1** bit followed by an all-**0** bit string represents Not-a-Real (NaR), and there are no other exceptions supported. Any operation with a NaR input produces a NaR output and it compares as $-\infty$, less than all real numbers.
- **Reciprocation**: Reciprocals of posits representing integer powers of 2 are exact and are formed by negating all but the sign bit. Note that this means **01000···000** always represents 1.0 and **11000···000** always represents −1.0.
- **Binade choice**: The binades may be equal-spaced values (linear posits) or logarithmically spaced (log posits). Linear posits must support the quire for exact dot products up to length 2^{32}; log posits do not need to support a quire. The word "posit" without the "log" qualifier implies a linear posit.
- **Stability**: An n-bit posit ending in a **0** bit represents the same value as does an $(n-1)$-bit posit without that last **0** bit.

The fundamental framework allows considerable flexibility. Notice that there is no requirement that posits have tapered relative accuracy. The *sane float* format introduced in Chapter 7 by restricting the regime to 2 bits provides flat relative accuracy (except for wobble within a binade) for its entire dynamic range. By the above broad definition, a sane float is a posit.

A *standard posit*, at least as of the time of writing, is a posit with a linear binade, a two-bit exponent size ($eS = 2$), and no restrictions on the regime size rS. Chapter 7 showed we could actually improve the fidelity, at least for HPC applications, by making choices for eS and rS that differ from that of standard posits, so in the future there could be a sweeping change to *Posit™ Standard (2022)*. A later section in this chapter will present a very promising variant recently proposed by Laslo Hunhold, but which has not yet undergone the extensive and independent testing needed to show superiority over standard posits. Time will tell.

There is no need to obey a standard if there is no interchange of data with other computing systems. A graphics processor, for example, simply has to get the right pixels onto the screen, and it can use any internal number representation that serves that purpose well. Similarly, neural networks that classify sounds and images can use whatever provides an acceptably-accurate classification. The intent of this book is to supply the tools needed to design and customize formats for real numbers, and future posits may be quite different from those presently in use.

13.3 Twenty Questions and A-to-D Convertors

The "Twenty Questions" parlor game originated in the 1800s in the US
and became a radio game show in the 1940s (Wikimedia Commons photo).

If you get truthful answers to twenty yes-no questions asked of some person thinking of an object, in principle you could distinguish between $2^{20} = 1\,048\,576$ different things, from Winston Churchill's cigar to the pencil a person is holding.

If you restrict the game to "I am thinking of an integer between one and a million" as the starting set, then there is an obvious way the questioner can determine the integer every time, using divide-and-conquer. "Is it greater than 500 000?" as the first question, followed by either "Is it greater than 250 000?" or "Is it greater than 750 000?" depending on the answer to the first question. And so on. Since a million is less than $2^{20} = 1\,048\,576$, you are guaranteed to have it narrowed down to a single integer after the twentieth question.

> Posit format is a system for playing Twenty Questions (or actually *n* questions, where *n* is the precision) when the starting set is "I am thinking of a real number."

Of course you are very unlikely to pin down the exact number in only twenty questions, but what should be your strategy for getting as close as possible?

The obvious first question should be, "Is it negative?" and if Yes, you know the first bit of its posit representation is a 1; if No, then the first bit must be 0.

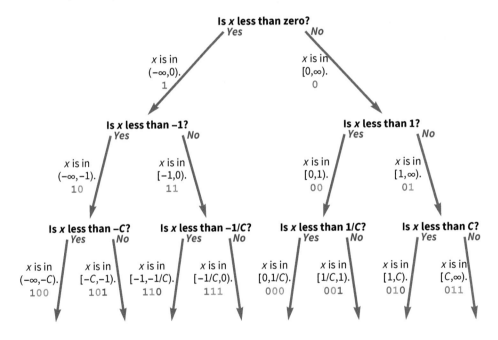

Strategy for guessing a real number, first three questions

What you choose for the constant *C* is up to you, but it must be greater than 1. Up to now, we have been using *uSeed*. There are also many ways to ask the fourth question, which leads to the various types of posit.

Notice how different reading bits from left-to-right is from bit-field mentality (Section 4.6). Posits use a mathematical way to define real numbers called *Dedekind cuts*. The posit framework does not even restrict you to powers of 2 and binades, though that is the most hardware-friendly choice for moderate precision and higher.

It also informs how to build a potentially important bit of electronic hardware: a successive-approximation analog-to-digital convertor (ADC) where the digital form is in posit format, not fixed-point as is commonly done today. Consider the enormous dynamic range of human vision and hearing:

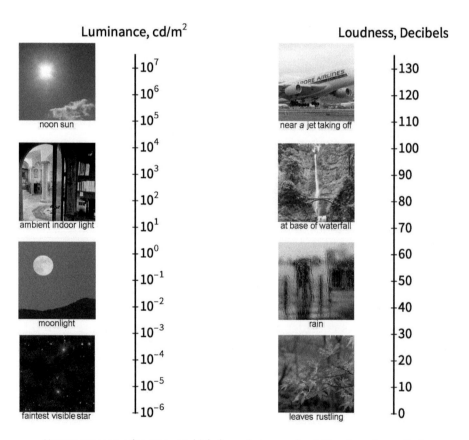

Human senses work over a very high dynamic range, about 14 orders of magnitude

You use an ADC every time you speak into your mobile phone or take a digital photo with it, for example. But those convertors give perhaps 12 bits of fixed-point precision, about three orders of magnitude. They work by converting sound or light into a voltage that can then be progressively compared against a built-in voltage to emit a **0** or **1**, then subdividing those ranges, just like Twenty Questions. But the results are equispaced points. The posit format points to a way to make ADCs behave much more like our eyes and ears in dealing with large dynamic ranges.

Before we leave the topic of ADCs, there is an important trick to know that I found with the help of circuit genius Akshat Ramachandran. Suppose we have a five-bit ADC that generates values between −1 and just below 1, which in binary signed fixed-point is $1.0000 = -1$ to $0.1111 = \frac{15}{16}$ in steps of $\frac{1}{16}$. Suppose we want to convert those values to posit format, 6-bit precision or higher.

Take a look at the bottom half of the posit ring plot for posit(6,0):

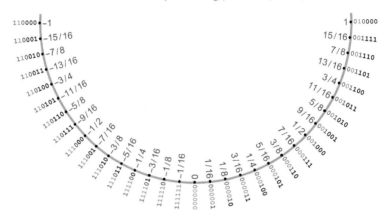

Bottom half of a posit ring (values ranging from −1 to 1) for 6-bit posits with *eS* = 0.

These are the posits presented in Chapter 1, with no exponent bits. If you set *eS* = 0, then the posits between −1 and 1 are *equispaced*. Which means there is nothing to do to the fixed-point values but pad them on the left with a copy of the sign bit. Posit arithmetic can then handle the calculations that take the numbers outside the [−1, 1) range, like for signal processing. Trying to convert signed fixed point to a normal float format is considerably more difficult. To get to more than 6-bit posit precision, simply append **0** bits. Either with high dynamic-range ADC design, or with conventional ADC design and posits with *eS* = 0, it looks like posits and ADCs are going to play very well together in the future.

13.4 The Hunhold Variation: Takum Format

13.4.1 The Extreme Magnitude Constants Argument

Some of the defenders of legacy floating-point format have pointed out that posits lose accuracy too quickly as values go farther from the fovea. Also, 32-bit posits and smaller do a poor job of representing physical constants that have extreme magnitudes, which are sometimes needed for technical computing. That is a fair criticism, especially if we are trying to replace 64-bit IEEE floats with 32-bit standard posits.

Here are some physical constants and their representation with 32-bit standard posits, with incorrect digits shown in **red**. The units have been omitted but all of the entries are in meter-kilogram-seconds international standard units. The posits values are shown using just enough decimals to specify the bit pattern of the posit.

Extreme magnitude constant	Known value	32-bit posit form
Speed of light (physics)	2.99792458×10^8	2.997924×10^8
Charge of an electron (electromagnetism)	$1.602176634 \times 10^{-19}$	1.6022×10^{-19}
Avogadro's number (chemistry)	$6.02214076 \times 10^{23}$	6.021×10^{23}
Boltzmann constant (thermodynamics)	$8.31446262 \times 10^{-23}$	8.313×10^{-23}
Planck's constant (quantum mechanics)	$6.62607015 \times 10^{-34}$	$7. \times 10^{-34}$
Cosmological constant (general relativity)	1.1056×10^{-52}	$7. \times 10^{-34}$
Mass of known universe (cosmology)	1.5×10^{53}	1.3×10^{36}

Physical constants in increasing order of how extreme their magnitude is

The last two constants are probably not used much in computer programs, but serve to illustrate an upper bound on the dynamic range that a number system might be called upon to represent. This is one of the few areas where IEEE Std 754 32-bit floats do better than the same precision standard posit, since the floats have about 7 decimals of accuracy even in these extreme magnitude ranges.

The term "Golden Zone" was proposed by Florent de Dinechin of the CITI Laboratory (*Posits: The good, the bad, and the ugly*, CoNGA 2019):

> **Definition**: The *Golden Zone* of a real number format is the dynamic range where its relative accuracy is as good as or better than that for IEEE Std 754 floats of the same precision.

This reminds us of the *fovea*, the region that was introduced in Section 2.5 where the number of significant bits is at a maximum. The Golden Zone is usually a much broader dynamic range than the fovea. The following accuracy plot shows us where the Golden Zone is for 32-bit standard posits:

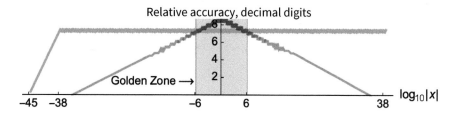

Accuracy plots for IEEE 32-bit floats and 32-bit standard posits, showing the Golden Zone

The Golden Zone is $2^{-20} \leq |x| \leq 2^{20}$, that is, values with magnitude between about one-millionth (10^{-6}) and one million (10^6). That may seem like a robust range, but there is a way to do much better that was recently (December 2023) invented by University of Cologne mathematician Laslo Hunhold.

13.4.2 A Different Kind of Regime Field

The regime field, as defined so far, is compact only when numbers it represents are not far from zero. Positional notation is more compact once the represented numbers get outside the range –3 to 3. Hunhold's observation is that we only need the first bit after the sign to indicate the direction away from 1 (or –1) that the binary represents. Then, a small fixed-size regime field can specify *eS* with positional notation. What is not obvious is that this preserves the monotonicity property of posits as well as all their other elegant features; it also does not lead to redundant representations of values the way the Morris floats do. Hunhold's variation *might* be better than standard posit format for some applications; we will know after we amass a body of experiments with application performance and with hardware design.

As we always do, turn the precision far down to see how this works. Suppose we have eight-bit precision. We can color-code the direction specifier with a **purple** bit for contrast with the adjacent colors. Start in the fovea. The number 1.0 is the same bits as all posit formats, but it gets there by a new route:

Hunhold: 01000000=1.0

Standard posit: 01000000=1.0

For the Hunhold case, the sign bit indicates the value is positive, and the direction bit 1 means the number is in the range [1, *maxPos*]. The next three bits specify the number of explicit exponent bits that follow as an unsigned integer, so 000 means no exponent bits. Prepend an implicit 1 bit to the bits (there are none in this case), treat that as an unsigned integer, and subtract 1. That is the power-of-two to scale by. Compared to standard posits, Hunhold posits have as much accuracy in the fovea but the fovea is narrower: from $\frac{1}{2} \leq |x| \leq 2$ instead of $\frac{1}{16} \leq |x| \leq 16$. However, there is an enormous benefit to sacrificing the accuracy just outside the fovea. Look at the maximum values representable, with ghost bits shown after the 8 explicit bits:

Maximum Hunhold posit: 01111111|00000 = $2^{239} \approx 8.8 \times 10^{71}$

Maximum standard posit: 01111111|00000 = $2^{24} \approx 1.7 \times 10^7$

To clarify where 2^{239} comes from, here are the steps:

- The three regime bits `111` mean 7 as an unsigned integer, so there are seven exponent bits. The direction is `1` so there is no need to negate the regime bits.
- There are only three explicit bits left, `111`, so use the next four ghost bits to get to a total of seven exponent bits: `111|0000`.
- Prepend the exponent bits with an implicit `1` bit: `1111|0000`.
- Find the value of all eight exponent bits, treated as an unsigned integer: `11110000` = 240.
- Subtract 1 from the result to get 239.

The dynamic range of Hunhold's variant is jaw-dropping for an 8-bit posit type; it is enough to cover all the scientific constants, from the cosmological constant to the mass of the known universe. Yet it has more than a decimal of accuracy for $|x|$ near 1. The use of regime bits for positional notation instead of the run-length encoding is so different from other posit types that Hunhold gives this format a new name: *takum* format. The word takum comes from two Icelandic words, '**tak**markað **um**fang,' meaning 'limited range.' Floats use a bit field to express the scaling exponent. Posits have (up to now) used two bit fields. The takum variant uses *three* fields for the direction, regime, and exponent.

Recall that Leonardo Torres Quevedo (Section 4.5) said computing needs a maximum power of 10 that is more than 10^9 but less than 10^{99}. The first handheld scientific calculators supported a dynamic range of 10^{-99} to 10^{99}. All the scientific constants fit in that range. The takum system uses three regime bits because that means the maximum expressible power of 2 is `011111111111` = $2^{254} \approx 3 \times 10^{76}$. If we had only two takum regime bits, the maximum expressible power of 2 would be far too small, only `0111111` = $2^{14} \approx 1.6 \times 10^4$. If we had four takum regime bits, the maximum expressible power of 2 would be overkill and result in many bit patterns that no one would ever use: `011111111111111111111` = $2^{65534} \approx 5 \times 10^{19727}$. By a fortunate mathematical coincidence, the use of three regime bits in takum format lands us perfectly in the range of real numbers of the most interest to humans for practical calculations.

If we can figure out what maximum dynamic range is really needed for technical computing, the design aesthetic of takums is to get there as quickly as possible as the precision increases. Once the dynamic range is covered, use all further increases in precision to lengthen the fraction to get more significant digits.

Hunhold has a good way to visualize this: plot the dynamic range extremes as a function of the number of bits.

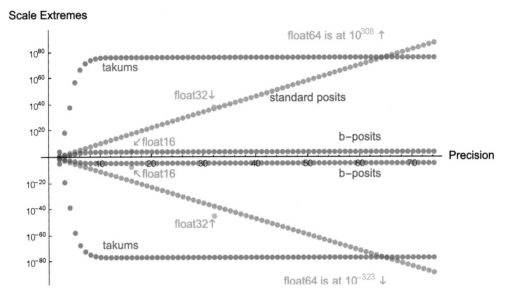

Scale Extremes

Log base 10 of the smallest and largest magnitude positive real representable by various formats

The problem with standard posits and floats is that their dynamic ranges keep growing far beyond what scientific computing needs. That means a large fraction of the possible bit patterns are almost never used. Takums prioritize coverage of the necessary dynamic range and then use the remaining bits to improve the relative accuracy. The same is accomplished by b-posits that limit rS as shown in Chapter 7, although b-posits are poor at expressing the extreme magnitude physical constants.

If r is the integer expressed by the three regime bits in a takum, the expressible scaling powers range from $2^r - 1$ to $2^{r+1} - 2$. There are only eight possible values for r, suggesting that the use of a lookup table in hardware could be very efficient. Also, in shifting the takum fraction bits to align the radix point, the shift is at most 7 bits, so the shift register only needs a 3-bit input. That is a hardware advantage over standard posits, where the shifts can be almost the size of the entire bit string. The same advantage is shared by b-posits.

Here are the eight possible regime values and the corresponding possible powers each regime can represent, for values $1 \leq x < \infty$:

Regime value, r	Representable power, 2^r-1 to $2^{r+1}-2$
0	0
1	1, 2
2	3, 4, 5, 6
3	7 to 14
4	15 to 30
5	31 to 62
6	63 to 126
7	127 to 254

Takum regime values and the integer power ranges possible within that regime

There is a different way to think about takums versus standard posits that Hunhold uses: *For a particular power (the integer power of the base), how many bits are left for expressing the fraction?* It resembles a relative accuracy plot, but without the wobble. We can graph that for standard posits and for takums, for 16-bit precision:

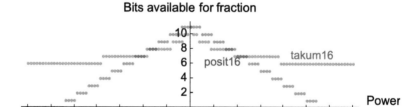

Relative accuracy plot for 16-bit takums and standard posits, over the dynamic range that posits cover

However, that graph only shows the dynamic range covered by posit16, which is 2^{-56} to 2^{56}. The takum format keeps going, all the way from 2^{-254} to 2^{254}:

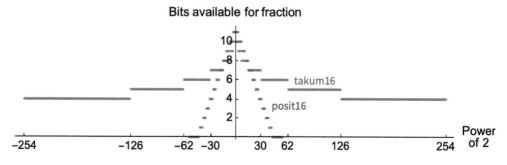

Relative accuracy plot for 16-bit takums and standard posits, over the dynamic range that takums cover

Posits usually taper accuracy linearly with magnitude, but takums taper logarithmically. Linear tapering is more accurate for the small magnitudes but has less range.

The number of bits needed to express the power is as low as 4 in the takum fovea, to 11 for the extreme magnitudes. Hence, there is still tapered accuracy but any precision with more than 12 bits will always have explicit fraction bits and be able to express every possible integer power. That contrasts with standard posits that can lose all fraction bits and skip over integer power values for extreme magnitudes.

If we draw a ring plot for standard posits with radial lines where the regime increments in value, the radial lines crowd near 0 at the bottom of the ring and near NaR at the top; see Section 2.4. But with takums, the radial lines are *equispaced*. Here is the northeast quadrant of a ring plot for 8-bit takums.

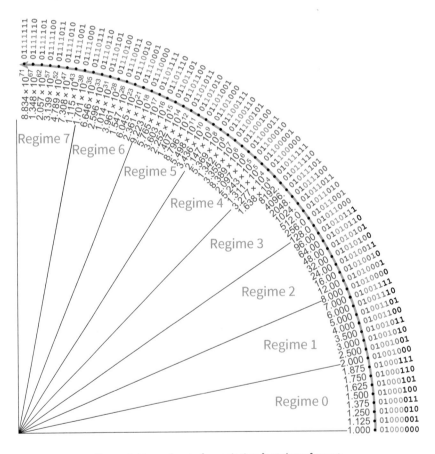

Upper right quadrant of a posit ring for takum format

In playing Twenty Questions, we know what the takum C value is for the third question: $2^{15} = 32\,768$. That aggressively finds out early if the person is thinking of an extreme magnitude number, which might be a good strategy.

With both standard posits and floats, when reducing precision you have to worry not only about losing relative accuracy but also about insufficient dynamic range. With the takum approach, the worry about dynamic range disappears. Even an 8-bit takum can express numbers from 2^{-239} to 2^{239}. And because the system for decoding the sign-direction-exponent is the same for all precisions and only takes a maximum of 12 bits, takum format may be more hardware-friendly than standard posits. But remember that we already have a way of limiting the set of bits that express the scale factor, so we should do an experiment to compare with that.

13.4.3 A Test of Takum Fidelity

Remember that in Section 7.3, experiments with **RRSet** showed that the limited-length regime approach produced the best results for $eS = 3$, $rS = 6$, the b-posit. That means the sign-power bits only take from 6 to 10 bits no matter what the precision is. That compares with the 4 to 12 bits for takums, but b-posits have a smaller dynamic range of about $10^{\pm 14}$. Also similarly, that eS-rS combination results in simple hardware to find the scale factor that does not change as precision varies.

We can test the fidelity of takums using the same method. We can write routines **xToT** and **tTox** to convert reals to takum format and takum format to reals. Then assign the over-vector notation $\overrightarrow{\dots}$ to mean "round ... to nearest takum." The code for this is in Appendix G.

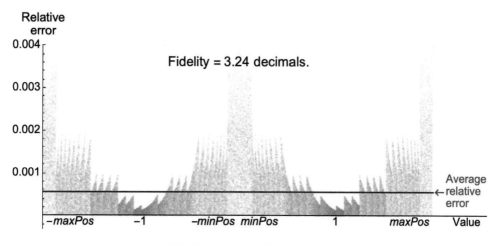

Fidelity plot for 32-bit takum format

Surprisingly, they did not do as well as standard posits or b-posits, which both have a fidelity of 3.52 decimals. Put another way, the b-posits were accurate to one part in 3300 on average; this type of takum is accurate to one part in 1700.

13.4.4 Logarithmic Takums

Hunhold prefers the log version of takum format, and instead of base 2, he uses base $\sqrt{e} \approx 1.65$. Once you decide to use a logarithmic system, you can use any number larger than 1 as the base. Because $\sqrt{e}$ is less than 2, using $\sqrt{e}$ makes the vocabulary more closely spaced and raises relative accuracy, giving takums slightly higher accuracy in their fovea than standard posits. It shrinks the dynamic range from about $10^{\pm 77}$ to about $10^{\pm 55}$, which is still enough to cover all the scientific constants.

As explained in Chapter 8, logarithmic systems have greatly simplified (and more often exact) multiplication, division, square root, and x^y operations, but addition and subtraction operations depend on tables and are difficult to scale higher than 32-bit precision. The choice of e or $\sqrt{e}$ as the base makes it easier to build the tables needed for addition and subtraction. Here are accuracy plots for 16-bit standard posits, log base $\sqrt{e}$ takums, and linear base 2 takums:

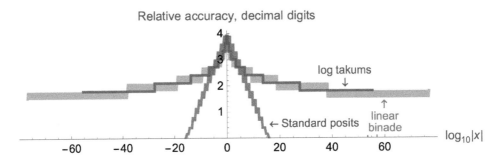

Accuracy plots for standard posits, log takums, and linear takums (all 32-bit precision)

Log takums have higher fidelity, since represented values are more closely spaced:

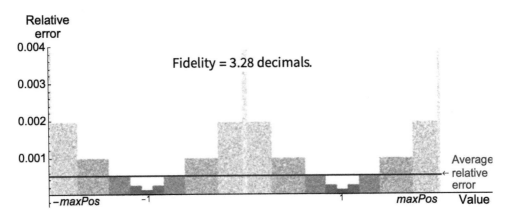

Fidelity plot for 32-bit log takums, with $\sqrt{e}$ as the base of the logarithm

We can also compare the sorted errors for linear and log takums:

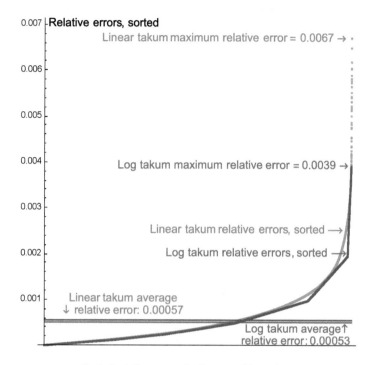

Sorted relative errors for linear and log takums

The 16-bit b-posits have an average relative error of 0.0003, about half that of either kind of takum. The maximum relative error of 0.0067 is the same for 16-bit linear takums and 16-bit b-posits, but the log takums dramatically reduce the maximum relative error to 0.0039.

13.4.5 An Idea: Human-Friendly *and* Hardware-Friendly?

If you bite the bullet and decide to use a logarithmic system where every nonzero number is $\pm base^{power + fraction}$, with all the tradeoffs described in Chapter 8, here is an idea: Use a log takum with a base of $b = 10^{1/16}$. This gives us something we have never had with binary floating point: *The ability to express integer powers of ten exactly.* Usually, the anthropomorphic goal of making computers use base ten is hardware-unfriendly, as Chapter 4 pointed out. But log takums express powers from –254 to 254 of any base, so the dynamic range would be $\left(10^{1/16}\right)^{-254}$ to $\left(10^{1/16}\right)^{254}$, or about 10^{-16} to 10^{16}. That range is nearly ideal for covering **RRSet**.

The original motivation for tapered accuracy is to make the more-common numbers more accurate, and more-common numbers seem to have magnitudes near 1. But when people do need to express very large or very small numbers, the common way to do it is with words like "billion" or "trillionth," or metric prefixes like "giga-" or "pico-". The number one-thousandth, "0.001" in the Human Layer would be

$$00111010100000110001001001101111 = \frac{85\,859\,935}{8\,589\,934\,592}$$
$$= 0.00999\,54119\,65064\,70441\,81823\,73046\,875$$

in a Compute Layer using 32-bit IEEE floats. Linear posits have the same issue. But with 16-bit log takums using $10^{1/16}$ as the base, look what can be expressed in the Compute Layer without rounding error:

Prefix	Value	Decimal	16–bit base $10^{1/16}$ takum
peta	10^{15}	1 000 000 000 000 000	0111000011000010
	10^{14}	100 000 000 000 000	0111000000101111
	10^{13}	10 000 000 000 000	0110111100111000
tera	10^{12}	1 000 000 000 000	0110111000010001
	10^{11}	100 000 000 000	0110110011101010
	10^{10}	10 000 000 000	0110101111000011
giga	10^{9}	1 000 000 000	0110101010011101
	10^{8}	100 000 000	0110100101110110
	10^{7}	10 000 000	0110100001001111
mega	10^{6}	1 000 000	0110011001010001
	10^{5}	100 000	0110010000000011
	10^{4}	10 000	0110000110110110
kilo	10^{3}	1000	0101111011010001
hecto	10^{2}	100	0101101000110110
deca	10^{1}	10	0101001100110110
	10^{0}	1	0100000000000000
deci	10^{-1}	0.1	0010110011001010
centi	10^{-2}	0.01	0010010111001010
milli	10^{-3}	0.001	0010000100101111
	10^{-4}	0.0001	0001111001001010
	10^{-5}	0.00001	0001101111111101
micro	10^{-6}	0.000001	0001100110101111
	10^{-7}	0.0000001	0001111001001010
	10^{-8}	0.00000001	0001101111111101
nano	10^{-9}	0.000000001	0001100110101111
	10^{-10}	0.0000000001	0001011110110001
	10^{-11}	0.00000000001	0001011010001010
pico	10^{-12}	0.000000000001	0001010101100011
	10^{-13}	0.0000000000001	0001010000111101
	10^{-14}	0.00000000000001	0001001100010110
femto	10^{-15}	0.000000000000001	0001000111101111

Powers of ten can be expressed exactly with 16-bit log takums using base $10^{1/16}$.

Because $10^{1/16} \approx 1.15$, values between 1 and $10^{1/16}$ are much more closely spaced than values in a binade from 1 to 2. As a result, the base $10^{1/16}$ takum has the highest (but narrowest) fovea accuracy of any 16-bit format so far, 4.5 decimals, as shown in this zoomed-in accuracy plot for values in the range of $\frac{1}{1000}$ to 1000:

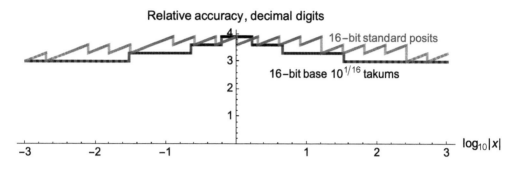

Accuracy plot of 16-bit standard posits versus 16-bit log base $10^{1/16}$ takums

For some application areas, having higher accuracy near 1 and –1 can be important. Probabilities range from 0 to 1 with 0 being impossible and 1 a certainty, but when you compute something to be "almost certain" and money is involved, a probability that is a little less than 1 had better be accurate. Another area is anything involving trig functions. For small angles t, the value of $\cos(t)$ becomes so close to 1 that it rounds up to 1, which means important information is lost. When signal processing programs do analysis of the frequency spectrum, they typically use blocks of 1024 points, and then need the value of t as low as $\pi/256$ for that task. Here is a comparison of 16-bit formats struggling to express $\cos(\pi/256)$ accurately:

Cosine of $\pi/256$ to six decimals	0.999924⋯
16–bit IEEE float	1.
16–bit standard posit	1.
16–bit linear takum	1.
16–bit base $\sqrt{e}$ takum	1.
16–bit base $10^{1/16}$ takum	0.999929

Approximating the cosine with 1 provides a relative accuracy of 4.1 decimals. The next smaller number in each 16-bit set is farther from the correct value than 1.0, except for the base $10^{1/16}$ takums; the value they can represent is correct to 5.4 decimals relative accuracy.

Further study is needed to discover which applications are best served by takum format. For linear binades, changing the number of exponent bits has always allowed coarse-grained control over the tradeoff between dynamic range and relative accuracy. When using a logarithmic system, we have yet another dial to adjust, which is the value of the base. Since that can be any real number greater than 1, that dial offers very fine-grained control.

13.5 A Summary of Fovea and Golden Zone Widths

I have presented quite a few variations on the fundamental posit idea and often have used 16 bits as the example precision since that makes more obvious what the differences are. For general technical computing, we should really be looking at 32-bit precision, at least. The following table shows properties for several 32-bit versions of formats mentioned so far.

Format	Described in	Dynamic range	Fovea relative accuracy, decimals	Fovea width	Golden Zone
Standard floats	IEEE Std 754	10^{-45} to 3×10^{38}	7.2–7.5	10^{-38} to 3×10^{38}	—
Simplest posits	Chapter 1	10^{-17} to 3×10^{16}	9.0–9.3	1/2 to 2	$\frac{1}{128}$ to 128
Standard posits	Chapter 5	8×10^{-37} to 10^{36}	8.4–8.7	1/16 to 16	10^{-6} to 10^{6}
Log posits	Chapter 8	8×10^{-37} to 10^{36}	8.6	1/16 to 16	10^{-6} to 10^{6}
b–posits	Chapter 7	4×10^{-15} to 3×10^{14}	8.1–8.4	1/256 to 256	2×10^{-10} to 4×10^{9}
Sane float, $eS{=}5$	Chapter 7	2×10^{-20} to 5×10^{19}	7.5–7.8	2×10^{-20} to 5×10^{19}	2×10^{-20} to 5×10^{19}
Takums, linear	Chapter 13	2×10^{-77} to 6×10^{76}	8.4–8.7	1/2 to 2	5×10^{-10} to 2×10^{9}
Takums, log base $\sqrt{e}$	Chapter 13	4×10^{-56} to 2×10^{55}	8.7	0.6 to 1.6	2×10^{-7} to 5×10^{6}
Takums, log base $10^{1/16}$	Chapter 13	10^{-16} to 9×10^{15}	9.0	0.7 to 1.3	3×10^{-4} to 4×10^{4}

Dynamic range and accuracy comparisons for floats and the posit variants

If it is important to be able to reproduce behavior perfectly from one system to another, then standard posits are the way to go. Otherwise, we have a rich set of choices that can be matched to the distribution of numbers needed by a workload. The next section helps visualize how to turn the dials and push the buttons.

13.6 Visualizing Posit Parameter Effects

We seek a real number system that has a distribution similar to that needed to run an application. To put it another way, the density (accuracy) of the number vocabulary should track the density of all the real numbers that arise if a calculation were performed with exact arithmetic. The relative accuracy plots provide a quick way to see the density of numbers as a function of the magnitude. Here is a visual summary of what happens to the relative accuracy plot when you tweak parameters.

Increase the exponent size, *eS* (Chapter 3):

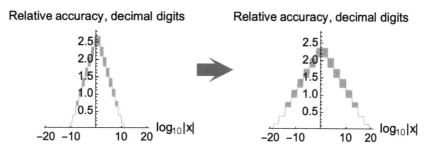

Increasing the number of exponent bits by 1 doubles the posit dynamic range.

Each increment of *eS* in a posit doubles the width of the dynamic range, but reduces the maximum relative accuracy by about 0.3 decimals. The above example starts with a 12-bit posit (*eS* = 2) and raises the number of exponent bits to *eS* = 3.

Blunt the tapering by limiting regime size *rS* (Chapter 7):

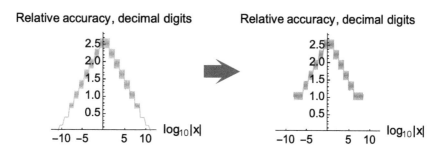

Limiting the regime size blunts tapering and raises accuracy at the magnitude extremes.

On the left, a standard 12-bit posit. On the right, the regime is restricted to a maximum size of *rS* = 8 bits, which removes the outermost three binades on either extreme and doubles the width of the outermost remaining tapering level.

If you know there is no chance that numbers will go outside the restricted dynamic range, then the bits are better spent maintaining accuracy at the extremes of magnitude.

Eliminate accuracy wobble by using logarithmic binades (Chapter 8):

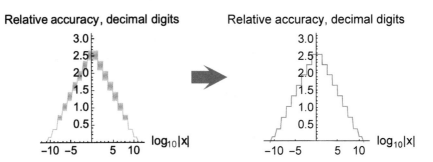

Logarithmic binades eliminate relative accuracy wobble within each binade.

Removing the wobble raises the worst-case relative accuracy in each binade by about 0.15 decimal. Remember the tradeoffs: The use of log binades gives you perfect reciprocals, easy (and often exact) multiplies and divides, but makes adds and subtracts complicated and difficult to scale to high precision. The quire also becomes impractical to support, so you sacrifice that powerful tool.

As mentioned in Section 13.4, with a logarithmic system you are free to change $2^{\text{power+fraction}}$ to $b^{\text{power+fraction}}$ where b is any real number greater than 1. That lets you fine-tune the tradeoff of relative accuracy and dynamic range the way changing eS gives you coarse-grain tuning. The power + fraction (characteristic + mantissa, to use logarithm jargon) still works like a base-2 fixed-point number even when you have a $b \neq 2$. The main thing you have to change when you change b is the tables for addition and subtraction.

Switch to takum format to express the power (Chapter 13):

First, a view restricted to the dynamic range of the standard posit to make the shape of the area near the fovea easier to see:

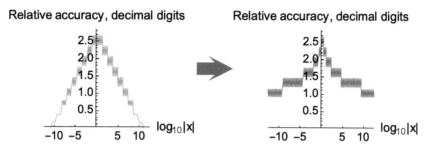

The takum variant reaches its maximum dynamic range quickly, leaving more bits for the significand

The linear tapering has higher accuracy than takums for many numbers in this range. Now zoom out to show the full dynamic range possible with the three-bit regime used as positional notation for the number of exponent bits:

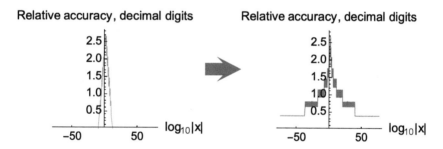

This comparison shows the full dynamic range of takum format for both graphs.

This is the linear binade form of takums. This approach is a good one if you know you have a valid need for a very wide dynamic range and still need decent accuracy at the extreme magnitudes.

All of these tweaks can be applied independently of one another. The above list is not intended to be ordered like a recipe; they are simply all tools in the tool set for fitting a number vocabulary to what is needed.

13.7 Two Other Ways to Bend the Rules

Training for AI takes an immense amount of energy and time, and this has led to considerable research in ways to reduce the number of bits needed to get acceptable results. Here is a typical histogram of the weights used in a neural net (Himeshi De Silva PhD dissertation, 2020), where 32-bit IEEE floats were used:

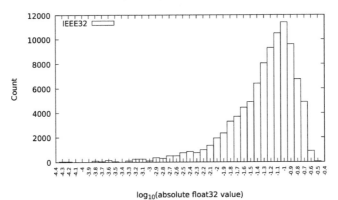

$\log_{10}$(absolute float32 value)

https://scholarbank.nus.edu.sg/handle/10635/176929

The range of magnitudes is really not that large, spanning fewer than five decades, so the use of a float with an 8-bit exponent like the IEEE 32-bit float or Google's bfloat16 is obviously wasteful. If we start to use the parameters in Section 13.6 to fit the needed distribution efficiently, two things are immediately apparent:

■ The peak of the distribution is not centered at $10^0 = 1$, but off to the left around 10^{-1}.

■ The distribution is lopsided; there is more need for small-magnitude numbers than larger-magnitude numbers.

Besides all the variations on posit format presented so far, there are two other ways to bend the rules:

■ Skew the power by a fixed amount, call it *ebias*, to put the maximum of the distribution where it should be.

■ Use different encoding rules for $|x| \geq 1$ than for $|x| < 1$ to match the lopsided distribution. Testing the first regime bit says which posit variant to use.

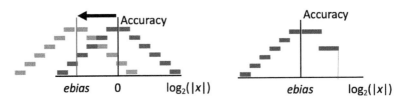

Biasing the exponent and allowing asymmetry in the tapering can allow better customization.

Both of these break the rule that reciprocation of integer powers of 2 is exact, but all the other merits of posits are preserved, like perfect negation, monotonic ordering, no redundant bit patterns, and deterministic behavior. Neural networks really do not need to take the reciprocals of weights, so that is a small thing to give up for a much better match to the needed distribution.

At NUS, we were able to make adjustments to standard posit format as described above and get very successful AI training results with posit precision as low as 8 bits, sometimes as low as 5 bits. Classification of images was 97.76% correct, compared to 98.7% for IEEE 32-bit floats.

There is a lot of effort currently to hang on to the idea of float format and somehow make it work for AI at 8-bit precision. Researchers notice that sometimes they need 4 exponent bits in the 8-bit float, and sometimes 5, so they support both. They do not realize that when you have two different exponent sizes, the computer needs a bit to know which size exponent to use, which means they have actually created a 9-bit format. An IEEE committee presently studying ways to rescue floats for AI has even explored letting the number of exponent bits in an 8-bit float range from 2 to 7. And the number of exponent bits is recorded ... where? Perhaps in a dedicated bit field? Does this sound familiar?

It is time we heard from this man again:

"It is difficult to free fools from the chains they revere." — Voltaire

Floats have a flat accuracy profile and thus are a mismatch to the distribution of numbers needed in AI, so many different scalings are adjusted for different parts of the calculation to compensate. It is like fitting a square peg into a bell-curve-shaped hole. Also, this kludge does not admit that *scalings* consume bits. The approach looks like a throwback to the late 1940s when programmers had to manage the position of the radix point of every operation. At NUS, we tried 8-bit floats and found float8 classification accuracy unacceptably low, 86.42% with 4 exponent bits and only 69.56% with 5 exponent bits.

One can imagine a set of 8-bit floats with 4-bit exponent, scaled by 2^{-6}, bolted onto a a set of 8-bit floats with 5-bit exponent, scaled by 2^{-31}, which would constitute a 9-bit system with an accuracy plot that looks like this:

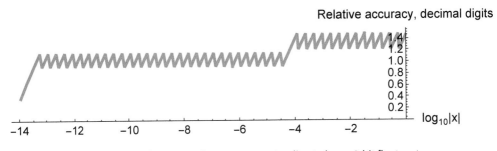

A 9-bit choice-of exponent float system pretending to be an 8-bit float system

That would resemble the distribution of AI data, if you can somehow manage to clean up the issues of redundant NaN representation, whether to represent infinities, signed zero (four different kinds!), non-monotone mapping to integers, and gradual underflow. The people clinging to the legacy float formats are gradually backing into the concept of tapered precision, but they are doing it the hard way. A clean-slate design is needed.

13.8 A Custom Number Set for AI

Facebook's Jeff Johnson was the first to drive precision for AI inference down to five-bit quantities, using log posits. Pushing it down to four bits resulted in too much classification inaccuracy from the neural network.

What if we hand-picked a minimal set of numbers for AI, so low in precision that the exact log hardware approach described in Chapter 9 eliminates the need to "decode" the bits?

Here is a strawman suggestion for such a set:

$$x = \left\{ \frac{1}{8192}, \frac{1}{4096}, \frac{1}{2048}, \frac{1}{1024}, \frac{1}{512}, \frac{1}{256}, \frac{3}{512}, \frac{1}{128}, \right.$$
$$\frac{3}{256}, \frac{7}{512}, \frac{1}{64}, \frac{5}{256}, \frac{3}{128}, \frac{7}{256}, \frac{1}{32}, \frac{9}{256}, \frac{5}{128}, \frac{3}{64}, \frac{7}{128},$$
$$\left. \frac{1}{16}, \frac{9}{128}, \frac{5}{64}, \frac{3}{32}, \frac{7}{64}, \frac{1}{8}, \frac{9}{64}, \frac{3}{16}, \frac{1}{4}, \frac{3}{8}, \frac{1}{2}, 1, 2 \right\};$$

The idea is to use logarithmically-spaced values for the smallest numbers, transitioning to evenly-spaced binades for the numbers close to 1/16 near the peak of the histogram shown at the beginning of Section 13.7. Here is a histogram of **x**:

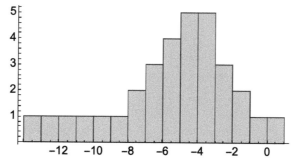

Values were chosen to approximate the distribution needed for AI.

The relative decimal accuracy plot shows that it covers the needed dynamic range and has similar emphasis on expressing numbers near 10^{-1} with about one decimal digit of accuracy:

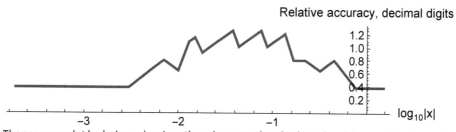

The accuracy plot looks irregular since the values were hand-selected and the precision is very low.

The values were hand-picked to make the above graph fit the required distribution. The multiplication table for **x** has only 164 distinct entries, ranging from 2^{-26} to 2^2. That means we can use a 64-bit fixed-point accumulator and have enough carry protection bits that dot products are guaranteed exact for vectors up to two billion elements long, more than enough for even the largest AI inference tasks.

We can multiply without pulling apart bit fields (decoding) to figure out what $m \times 2^j$ is represented by a bit string. We simply map the values in **x** to the following set of 10-bit unsigned integers, **L**:

L = {0, 41, 82, 123, 164, 205, 229, 246, 270, 279,
287, 300, 311, 320, 328, 335, 341, 352, 361, 369, 376,
382, 393, 402, 410, 417, 434, 451, 475, 492, 533, 574};

Those integers were derived by the methods presented in Chapter 9. Here is the rank plot, whether multiplying real values in **x** or adding integers in **L**:

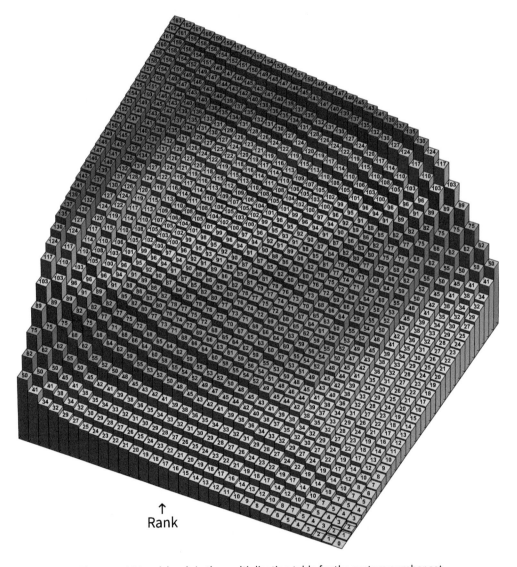

↑
Rank

There are 164 rank levels in the multiplication table for the custom number set.

The sign bit is easily computed when multiplying two signed numbers; it is simply the XOR (exclusive OR) of the two input sign bits. The five-bit magnitudes are mapped to the 10-bit integers in **L** with the type of simple circuit shown in Subsection 9.5.4. Since there can be a carry bit, an 11-bit adder is needed to sum the integers. That produces one of 164 possible pointer values.

Each pointer value is mapped to the m and j value of the exact product, $m \times 2^j$. Each m integer is between 1 and 81 and thus takes seven bits to store. The j value can be treated as the amount to shift m, a range covering 28 integers that can be stored in a five-bit integer, before adding m to a 64-bit signed integer accumulator.

The last step is to 1) apply what is called an *activation function* to the accumulator and 2) convert it back to a sign bit and the five-bit custom number set bit identifier, **00000** to **11111**. But *those two steps can be combined*. After determining the sign, five successive tests suffice to map to a bit string. Just like Twenty Questions. The dividing points can be anywhere you want, as long as the function is monotone.

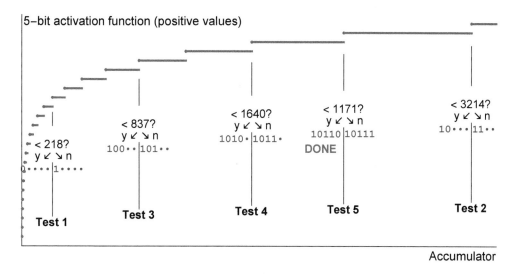

Combined activation function and recoding into the low-precision posit format

Comparison operations are quick, and the application of the activation function and conversion back to 5-bit format only occur after many multiply-accumulates, so it should not require a large fraction of the time and energy consumed.

This is a simplification of what modern neural nets do, especially when training. Inference is more like what is described here: computing dot products and putting the accumulations through nonlinear functions like the one shown above.

A final remark, just for fun. The very first neural networks were called *perceptrons*, invented in 1943 by Warren McCulloch and Walter Pitts. The first use of posit format in machine learning was done much more recently by a team at Rochester Institute of Technology, and they coined the term "positron" for their version of a perceptron. If posits do indeed prove to be the format of choice for artificial intelligence, it is possible that the "electronic brains" of future robots and AI computers could be *positronic*. Isaac Asimov would be pleased.

13.9 The Nature of Technological Revolutions

In the 1980s, technical computing was done with the Windows proprietary operating system, the x86 proprietary processors of Intel and AMD, and the proprietary Intel real number format that became IEEE Std 754. Forty years later, community-supported standards have displaced or are in the process of displacing all of these: Linux as the operating system, RISC-V as the processor design, and posits as the real number format.

In an email discussion with RISC-V developers, I remarked that I did not expect to see much hardware support for IEEE Std 754 ten years from now. That provoked some outrage from one Berkeley professor, who wrote back to the discussion:

> I will bet you one million dollars, or whatever amount you are willing to name, that IEEE floating-point will be in use in all computers ten years from now.

I replied that I was surprised at his position, since hardware support for the full IEEE Std 754 is almost extinct already. Intel, AMD, and ARM only support normal floats in hardware and treat the rest of the standard as a software option, one that most users turn off. The GPU builders flush subnormals to zero, so there is no guarantee that a bit string means the same thing on different machines. And a veritable zoo of new formats is springing up, including posits, Google's bfloat16, IBM's DLFLoat, 8-bit microfloats that discard the IEEE exceptions, and so on.

His anger reminded me of a famous quotation from physicist Max Planck:

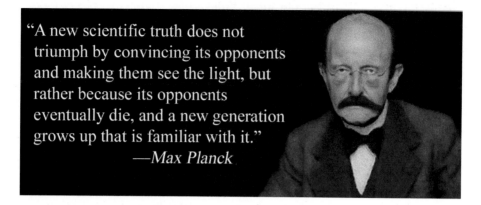

Or as it is often more bluntly paraphrased,

Planck worked to introduce quantum mechanical concepts to a community that insisted Newtonian mechanics was the only true physics, and that all this stuff about relativity and quantum theory was utter nonsense. I experienced similar fury four decades ago when I argued that the future of high-performance computing was in parallel computing, and was told that vector mainframes like the Cray-1 would reign supreme forever.

This revolution feels different. Introducing parallel computing was like rolling a boulder up a hill. Posit computing is more like watching the boulder roll downhill. Posit computing has caught fire in a computing community that has had a burgeoning desire for some better way to perform computer arithmetic with real numbers.

Posits were quickly adopted by a 5G network provider. Posit arithmetic is the basis for the accelerator part of the European Processor Initiative. Over 400 papers have been written about their application, many of them on their advantages for AI. Bosch built posits into an intelligent video camera that automates motion detection, since posits solved the problem of NaN generation they experienced when they used floats. The University of Oxford has shown that 16-bit posits can simulate weather and climate as well as 64-bit floats.

Perhaps most significantly, Calligo Tech (Bangalore, India) has introduced an 8-core RISC-V system with native posit arithmetic in VLSI, and they plan a 64-core follow-on using leading-edge chip resolutions. They have compilers, libraries, applications, and a complete hardware/software stack commercially available. That would seem to indicate that the idea has crossed the chasm between academic early adopters and real mainstream use. Such changes always start with the smaller companies who have nothing to lose and everything to gain by introducing something new; the larger companies have long-range plans, and it takes them years to change direction.

It always seems to take about ten years for a new technological idea to become mainstream, even when it has proved to be uniformly better. I type this seven years after the introduction of posit computing, and everything seems to be right on schedule. □

"IT STARTED WITH A SIMPLE CASE OF PEER-REVIEW."

Glossary

$\overline{}$: The "$\overline{}$" overbar symbol, when placed over a number or mathematical expression, indicates that what is under the bar is rounded to the nearest posit. Section 3.5, pages 57–58.

$\underline{}$: The "$\underline{}$" underbar symbol, when placed under a number or mathematical expression, indicates that what is above the bar is rounded to the nearest float using banker's rounding (the default rounding for IEEE floats). Section 6.1, page 146.

accuracy: A measure of the closeness of a computed number $x_{computed}$ to the correct result x_{exact}. The accuracy varies inversely with the error. Section 2.3, pages 19–20.

ADC: An analog-to-digital convertor, also called A-to-D convertor. A circuit that turns physical phenomena into voltages and then converts the voltage level to a binary value. Section 13.3, pages 325–328.

arithmetic mean: For two real values x and y, the *arithmetic mean* is their average, $(x + y)/2$. For example, the arithmetic mean of 1.0 and 4.0 is 2.5. On a linear scale, the arithmetic mean is at the midpoint between x and y. Compare **geometric mean**. Section 2.4, page 22.

b-posit: A particular bounded-range posit format with $rS = 6$ and $eS = 3$ that, with sufficient precision, appears well-suited to HPC workloads. Section 7.4, page 189 and Section 13.4.2, page 332.

banker's rounding: For binary representations of real numbers, *banker's rounding* means rounding to the nearest value, where nearest is defined by which side of the tie point between two adjacent binary representations the value is. The tie point is the in-between value that could be represented if there was one more bit of precision. If the value lands exactly on that tie point, the rounding goes to the binary representation that ends in a **0** bit. Also called *round-to-nearest, ties-to-even*. It is the only rounding mode used by posits. Section 3.4.6, pages 51–52, and Section 5.5, pages 124–125.

basic block: In computer science, a *basic block* is a sequence of operations with no branching or looping. It has a single entry point and a single exit point. Section 6.7, page 164.

BCD or **Binary-Coded Decimal**: The use of four-bit binary numbers to represent decimal digits 0 to 9. Section 9.5, page 230.

binade: A *binade* is the range of real numbers between two successive integer powers of two. That is, x where $2^j \le x < 2^{j+1}$. Section 2.7, page 29 and Section 8.1, pages 195–196.

bit field: A contiguous set of bits in a format with a defined meaning. Section 5.3, page 107.

characteristic: In an LNS, the *characteristic* is the integer part of the fixed-point number in the exponent. Section 8.1, pages 196–197.

c-set: A *c-set* to a problem involving real numbers is the smallest complete set of input values possible in the number format that satisfy the equations after final rounding. Section 12.3.1, pages 293–295.

closure plot: For an operation with two inputs, the *closure plot* is the square plot of all possible input combinations and a color-coded dot for whether the result of the operation is exact, rounded, overflows, or underflows. Section 8.4, pages 203–205.

Compute Layer: The *Compute Layer* is where numbers are represented with finite-precision bit strings. Section 2.8, pages 30–33 and Section 6.3, page 152.

decade: In the context of number systems, a *decade* is the range of real numbers between two successive integer powers of ten. That is, x where $10^j \le x < 10^{j+1}$. Section 2.7, page 28.

decimal accuracy (decAcc): The number of decimals of correctness in a computed real value; it can have a fractional component. The decAcc function computes the decimal accuracy of two inputs x_{exact} and $x_{computed}$. Section 2.3, pages 19–20.

determinant: For n-by-n square matrices, the *determinant* is the volume of the shape you get after applying the matrix to the unit cube in n-dimensional space (negative volume, if the shape is mirror-reversed). Section 11.4.1, pages 271–281 and Section 12.3.4, pages 302–306.

discriminant: In finding the roots of a quadratic equation $a \cdot x^2 + b \cdot x + c = 0$, the discriminant is $b^2 - 4 \cdot a \cdot c$, the sign of which tells you if there are no real roots, two identical roots, or two distinct roots. It discriminates between those three cases. Section 6.8, page 166.

dot product: For two vectors $a = \{a_1, a_2, ..., a_m\}$ and $b = \{b_1, b_2, ..., b_m\}$, the dot product is the sum of pairwise products, $s = a_1 \cdot b_1 + a_2 \cdot b_2 + ... + a_m \cdot b_m$. Often written as $\boldsymbol{a} \cdot \boldsymbol{b}$ with a bold "·" dot. Section 6.1, page 145.

double-precision: A *double-precision* representation of a real number is one that uses 64 bits. Historically, it once meant a number occupying two words of memory so that the number of bits depended on how many bits there were in a word for a particular computer, but double-precision has come to mean 64-bit precision regardless of the computer architecture. Section 4.7, page 82.

error: A measure of the distance between an exact real number x_{exact} and a number $x_{computed}$ produced by a method that is inexact for some reason (usually rounding). There are several ways to define error mathematically. Chapter 2, pages 15–40.

eS: Short for *exponent size*, the number of bits used to express the exponent field. Section 3.1, pages 43–45.

exception: A special case in the interpretation of format bit fields. Section 5.3, page 108.

explicit bits: The **0** or **1** bits expressed in the bit string of a number format, as contrasted with the *ghost bits* implied after the least significant explicit bit (an infinite string of **0** bits). Section 3.2, page 46.

faithful rounding: Rounding to one of the two representable numbers that flank the exact number, not necessarily the closest one. Section 10.3.3, page 250.

fidelity: The *fidelity* of a format is a measure of how well the format vocabulary can express the ideal (exact) values needed in a calculation. It can be thought of as the average relative decimal accuracy in expressing the ideal set of values. Section 7.3, Page 181.

float: Short for floating-point number, a *float* is expressed in a format that imitates scientific notation by using a fixed-size set of digits to represent the significant digits and a fixed-size set of digits to express the integer power of a base b that scales the significant digits. The rules of the format call out certain bit strings to be exceptions to the meanings of the bit fields, and instead express ideas other than real numbers such as an infinite or an indeterminate quantity. Section 1.3, page 9, and Section 4.5, pages 74–76.

format: A set of bit fields and the definition of their meaning. Section 5.3, page 110.

fovea: For posit formats (where the number of significant bits changes with magnitude), the *fovea* is the region where the number of significant bits is at its maximum. Section 2.5, page 25.

fused: An expression with more than one operation that is rounded to the same value that an exact evaluation would round. That is, the expression is evaluated entirely in the Math Layer, with a single rounding to bring it back to the Compute Layer. Section 5.3, page 111.

geometric mean: For two positive real values x and y, the geometric mean is the square root of their product, $\sqrt{x \cdot y}$. For example, the geometric mean of 1.0 and 4.0 is 2.0. On a logarithmic scale, the geometric mean is at the midpoint between x and y. Compare **arithmetic mean**. Section 2.4, page 22.

> What do we mean by the mean,
> This number that's halfway between?
> For 32, 8,
> You can have a debate:
> Is it 20, or only 16?
> —*David Morin, Senior Lecturer on Physics, Harvard University*

ghost bits: The bits beyond the least significant explicit bit in a number. While explicit bits can be **0** or **1**, ghost bits are always **0**. The ghost bits of a posit can serve as regime, exponent, or fraction bits. Section 3.2, pages 45–46, and Section 5.3, page 107.

Golden Zone: The part of the dynamic range of a format where relative accuracy is as good as or better than that of IEEE floats of the same precision. Section 13.4.1, pages 329–330 and Section 13.5, page 340.

heisenbug: A bug in a computer program that disappears when an effort is made to find it. Section 4.7.5, page 94.

HPC: High-Performance Computing. Historically called *supercomputing*, it refers to computing where architectural techniques (such as parallel computing) are used to achieve high speeds. It is usually associated with large, complex calculations involving real numbers. AI has come to be thought of as part of HPC. Section 7.2, pages 179–180

Human Layer: The *Human Layer* is where numbers (and exception quantities) are represented in a form understandable to humans. The Human Layer is not limited to character strings that represent numbers; Human Layer quantities can be images, sound, mechanical motion, or anything else that is physically perceivable by a human or can be supplied by a human as input to a computer system. Section 2.8, pages 30–33 and Section 6.3, page 152.

ill-posed: A numerical computing problem is *ill-posed* if a small change in the input values results in a large change in the answer. Section 6.4, page 154.

implicit value: In interpreting posits, the *implicit value* is the value added to the fraction based on the sign; it is 1 for positive posits and −2 for negative posits. (There is an implicit value for base-2 floats as well, but it is usually called the "hidden bit.") Section 5.3, page 111.

inner product: See **dot product**. Section 6.1, page 145 and Section 11.3, pages 268–269.

KAA: Karlsruhe Accurate Arithmetic, a computing environment that uses exact dot products and interval arithmetic to make (most) calculations as accurate as if they had been computed entirely in the Math Layer and rounded once. Section 6.2, Pages 148–149.

kludge: A quick-and-dirty engineering design, as opposed to a design that is carefully planned from first principles. Section 4.7.4, page 88.

LNS or Logarithmic Number System: An *LNS* format expresses the nonzero real values as $\pm b^r$ where r is a signed fixed-point binary number and b is a constant greater than 1. Usually, $b = 2$. Chapter 8, pages 195–214.

LSB: The least significant bit of a format or a bit field within a format. Section 5.3, page 112.

Math Layer: The *Math Layer* is the scratchpad part of a computing system where results are computed with sufficient accuracy that the result after rounding is the same as the exact result after rounding. Section 2.8, pages 30–33 and Section 6.3, page 152.

mantissa: Mathematically, the *mantissa* is $L - \text{floor}(L)$ where L is a logarithm. It is a denigrated term for referring to the fraction or significand of a floating-point number. It can still be applied to the part of a logarithmic number format beyond the radix point. Section 8.1, pages 196–197.

maxPos: The largest positive real value representable by a posit. The posit that represents *maxPos* is always a **0** bit followed by all **1** bits. Section 3.4, page 48.

minPos: The smallest positive real value representable by a posit. The posit that represents *minPos* is always all **0** bits followed by a **1** bit. Section 3.4, page 48.

Morris floats: The earliest attempt to create a real-number format that tapers the length of the significand bit field as the absolute magnitude increases. Section 13.1, pages 321–324.

MSB: The most significant bit of a format or a bit field within a format. Section 1.1, page 2, and Section 5.3, page 112.

NaN: Short for "Not-a-Number," NaN is an IEEE float misnomer for imaginary numbers, complex numbers, and some entities that are not a number. It does not include $-\infty$ and ∞, which incorrectly implies that $-\infty$ and ∞ are numbers. Section 4.7.4, page 89.

NaR: Short for "Not-a-Real," meaning anything that is not mathematically definable as a unique real number. It can even be a non-mathematical quantity. Section 1.1, page 3 and Section 5.3, page 112.

NUS: National University of Singapore. In Asia, the abbreviation is as well-known as MIT is in the US, but most US people do not seem to know about NUS. Section 5.1, page 100.

orthogonal: At a right angle. Two nonzero vectors $\boldsymbol{a}$ and $\boldsymbol{b}$ are *orthogonal* if their dot product is zero, $\boldsymbol{a} \cdot \boldsymbol{b} = 0$. Section 6.1, page 146.

outer product: For two vectors $a = \{a_1, a_2, ..., a_m\}$ and $b = \{b_1, b_2, ..., b_m\}$, the *outer product* is the two-dimensional array of all possible products $a_j \cdot b_k$. In other words, it is a multiplication table. Section 11.3, pages 268–271.

panacea: A remedy that claims to cure every problem, but actually falls short of doing so. Section 5.4, page 119.

pIntMax: The largest consecutive integer-valued posit value. Section 5.3, page 113.

positional: The most common way we write numbers, where each digit has an implicit multiplier of a base to a power depending on its position in the character string. For example, the decimal number 23.47 means 2 times 10^1 plus 3 times 10^0 plus 4 times 10^{-1} plus 7 times 10^{-2}. A binary number **10.01** means **1** times 2^1 plus **0** times 2^0 plus **0** times 2^{-1} plus **1** times 2^{-2}. Sections 4.1–4.4, pages 62–74.

precision: The total number of bits for expressing any number format. It is not to be confused with *accuracy*. Section 5.3, page 114.

pToX: Routine that converts a posit p, treated as a signed binary integer, to a real number x. Section 3.1, page 44.

quire: A format with sufficient precision to compute vector dot products without rounding, up to some specified vector length. Section 5.3, page 114, and Chapter 6, pages 145–172.

quire sum limit: The minimum number of additions of posit values that can overflow the quire format. It is generally a very large number; even for 8-bit posits, it is about 3.6×10^{16}, making it possible to add up extremely long lists of posits without any possibility of accumulating rounding error. Section 5.3, page 115.

quire value: Either a real number that can be represented exactly in quire format, or NaR. Section 5.3, page 114.

regime: In posit format, the *regime* is the set of bits immediately after the sign bit that, together with the exponent bits, determine the power j in the scaling factor 2^j. Section 1.1, pages 12–13 and Section 3.1, pages 41–47.

ring plot: A circular plot based on both the projective reals and signed (2's complement) computer integers. The color codings of the bit strings are shown, as well as the real number represented by each bit string. Section 1.1, page 4.

rounded: Converted from a real number to a number representable in a format, according to the rules of the format. Section 5.3, page 115, and Section 5.5, pages 124–125.

RRSet: A proposed Representative Real Set for testing how well a number format expresses the values needed in a complete calculation. It should be computed to much higher precision than the format being tested, so that almost every number will not be representable without rounding. The one proposed here for HPC applications is a random set with a normal distribution of magnitudes centered at 0 and a standard deviation of three decades. Section 7.2, pages 179–181.

rS: Short for *regime size*, the maximum number of bits that can be used to express the regime of a posit type. For standard posits, there is no regime size limit; $rS = \infty$. Section 7.1, pages 173–177.

sane float: A posit with the regime size rS limited to two bits, so there is no tapering of relative accuracy. They have the advantages of posit format but none of the IEEE float format disadvantages such as redundant NaN, subnormals, negative and positive zero, hidden rounding modes, etc. Section 7.1, pages 177–179.

scalar product: See **dot product**. Section 6.1, page 145.

scientific notation: The writing of real numbers as $\pm f \times 10^e$ where f is a number between 1.00...00 and 9.99...99 (a fixed number of digits) and e is a signed integer. As originally proposed in 1914, there was no $\pm$ so only positive values were represented, and the f ranged from 0.100...00 to 0.999...99. Section 4.5, pages 74–76.

sign: The *sign* is 1 for positive numbers, –1 for negative numbers, and 0 for zero. Section 5.3, page 115.

sign bit: For number formats that can represent signed numbers, the *sign bit* is the part of the bit string that indicates if it represents a value less than 0. For all signed formats in this book, the sign bit is always in the MSB position. Section 1.1, page 2.

significand: The value that is scaled by 2^j in a real number format, consisting of the implicit value plus the fraction. For posit format, the significand includes the sign; for float formats it does not. Section 5.3, page 116.

standard posit: A posit that follows *Posit™ Standard (2022)*; the exponent size is $eS = 2$, there are no restrictions on regime size, and the binades are linear (not logarithmic). Section 13.2, page 325, and Chapter 5, pages 99–144.

takum: A real number format that fits the posit framework but uses three fields to express the power instead of two. It is designed to cover a desired dynamic range with as few bits as possible. Section 13.4, pages 328–340.

TMD or **Table-Maker's Dilemma**: The problem of deciding which way to round certain mathematical functions when the function approximation is extremely close to the tie point between two representable values. Chapter 10, pages 245–262.

Twilight Zone: The range of posits where at least one exponent bit is a ghost bit. The rounding rules do not change, but the effect of rounding is to (automatically) use the geometric mean as the tie point between adjacent posits instead of the arithmetic mean. Section 3.2, pages 45–46.

ubit: The ubit is a bit in a unum that is 0 if a number is expressed exactly and 1 if the number represents the open interval between two exactly expressible numbers. That is, a 1 indicates there are more bits after the least-significant bit but no space to store them in the particular format settings. Rhymes with "cubit." Section 1.1, page 3.

ULP: An *ULP* (rhymes with gulp) is the difference between exact values represented by bit strings that differ by one Unit in the Last Place. Some texts use "Unit of Least Precision" as the abbreviated phrase. Section 2.7, page 27.

unum: A *unum* is number format that allows a variable number of bits for the power j and the significant digits m in a number of the form $m \times 2^j$. It also has the option of expressing whether it represents an exact value like "3.14" or a number between two exact values like "3.14⋯" (which means (3.14, 3.15), the open interval). It is pronounced "you-num," since it is short for *universal number*. Section 1.1, page 3.

uSeed: The value $2^{2^{es}}$ that determines the scale factor contributed by each regime bit. The *uSeed* value is always represented by a posit with the first three bits `011` and all other bits **0**. On a ring plot, the *uSeed* value is at the northwest position. Section 3.1, pages 41–43.

vocabulary: When applied to a number format, its *vocabulary* is the set of things the number format can express, both numerical values and exception conditions like NaR. Ideally, an n-bit string can express a vocabulary of 2^n distinct things, but if some representations are redundant by the rules of the format, the vocabulary will be smaller than 2^n. Section 1.1, page 5.

weighted sum: See **dot product**. Section 6.1, page 145.

xToP: Routine that converts a real number x to a posit p, treated as a signed binary integer. Section 3.4, pages 47–55.

I had to work this in somewhere.

Answers to Exercises

Chapter 1

> **Exercise for the Reader**: The standard 16-bit float closest to $\sqrt{17}$ is $\frac{33}{8} = 4.125$.
> Find the rational number represented by 16-bit posit `0111000000111111`.
> Compare the closeness of those approximations to $\sqrt{17} = 4.1231056\cdots$

The sign bit `0` indicates the posit is positive, $s = 0$. The `1110` regime bits have three identical `1` bits in a row, so $k = 3$ and the regime value is $r = k - 1 = 2$. Read the fraction bits as a binary integer from right to left, $1 + 2 + 4 + 8 + 16 + 32 = 63$, and the rest are `0` bits. The fraction has 11 bits total, so the fraction is $f = \frac{63}{2^{11}} = \frac{63}{2048}$. Apply the formula:

$$x = ((1 - 3 \times s) + f) \times 2^{(1 - 2s) \times (r + s)}$$

$$= \left((1 - 3 \times 0) + \frac{63}{2048}\right) \times 2^{(1 - 2 \times 0) \times (2 + 0)}$$

$$= \left(1 + \frac{63}{2048}\right) \times 2^2$$

$$= \frac{2111}{512}$$

$$= 4.1230468750.$$

Use decimals to compare the absolute errors by taking their ratio:

```
Print["Posit error: ", positError = Abs[2111 / 512. - √17]]
Print["Float error: ", floatError = Abs[33 / 8. - √17]];
Print[
  "Ratio of float error to posit error: ", floatError / positError]
```

```
Posit error: 0.00005 87506 1766
Float error: 0.00189 43743 82
Ratio of float error to posit error: 32.24433134
```

The posit approximation is more than 32 times closer to $\sqrt{17}$ than the float approximation. This is an unusually high ratio, but such cases do happen.

Chapter 2

Exercise for the Reader: Why is there a slight asymmetry between the overflow cases and the underflow cases for `posit8` multiplication?

Notice that the multiplication diagram is perfectly symmetrical about the ↗ diagonal axis, because posit multiplication is commutative. However, it is only *approximately* symmetric about the ↘ diagonal. The reason is that the latter case represents flipping the posit ring about the horizontal axis, and reciprocals are only exact for integer powers of 2.

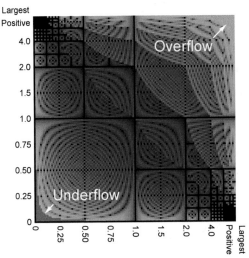

Closure plot for 8-bit posit multiplication

For `posit8`, overflow happens if a product is greater than 128. For example, 12 is a `posit8` value `01111010`, and 12 × 12 = 144 > 128. That creates a light blue point near the top right corner. But what is the corresponding point on the bottom left corner? As shown in Chapter 1, ignoring the sign bit and negating the other bits flips the values about 1.0 and gives an approximate (sometimes exact) reciprocal.

The posit8 value `01111010` flips to `00000110`, representing $\frac{3}{32}$. That fraction is slightly *larger* than $\frac{1}{12}$. Underflow happens if a product is between 0 and $\frac{1}{128}$, but $\frac{3}{32} \times \frac{3}{32} = \frac{9}{1024} > \frac{8}{1024} = \frac{1}{128}$, so that product does *not* underflow and will not create a red point near the lower left-hand corner.

In general, there will be slightly more products that overflow than underflow for this type of posit. There are other types of posit that have perfect reciprocals, and for which the diagram will be perfectly symmetric about both diagonal axes.

Chapter 3

> **Exercise for the Reader**: In going from $eS = 1$ to $eS = 2$, the *uSeed* jumps from 4 to 16. What happens if we try to use 8 as the *uSeed* value?

It can be made to work reasonably well if we alter the rules for inserting a posit ending in a 1 bit. Between 2^k and 2^m where k and m differ by more than one, the system described in the chapter guarantees that $2^{(k+m)/2}$ will always be an integer power of 2. But if *uSeed* is $2^3 = 8$, that rule would put the value between 1 and 8 as $2^{(3+0)/2} = 2^{3/2} = \sqrt{8}$, and then we cannot use linear binades. However we could say to always use the floor or ceiling function of $(k + m)/2$ to get back to an integer. To keep the relative accuracy tapered and also keep the reciprocals exact for powers of 2, we should favor higher density towards $2^0 = 1$. We could replace the fourth rule with this:

- Between 2^k and 2^m where $k < m$ and $m - k > 1$, insert $2^{\lfloor (k+m)/2 \rfloor}$ if $k \geq 0$, and insert $2^{\lceil (k+m)/2 \rceil}$ if $m \leq 0$.

If $(k + m)/2$ is an integer, then the floor and ceiling functions leave it unchanged. For 5-bit versions of such posits with 2 bits of exponent, the ring plot looks like this:

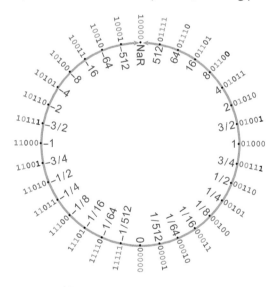

5-bit posits using 8 as the *uSeed* value

The rules for color-coding are unchanged. Compare this with the ring plot on page 43 for ⟨5, 2⟩ posits. The dynamic range is reduced, but now there is better accuracy for $|x|$ near 1, with a single fraction bit showing.

The accuracy plot certainly looks like a posit; the jaggedness at the top corresponds to the binades that have a fraction bit, which creates wobbling relative accuracy.

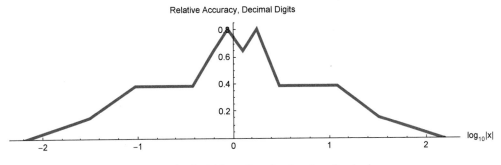

Accuracy plot for 5-bit posits using 8 as the *uSeed* value

> **Exercise for the Reader**: Convert the decimal number 2300 to an 8-bit posit with *eS* = 1. Caution: This number is in the Twilight Zone.

When *eS* = 1, *uSeed* is $2^{2^1} = 4$. The number is positive so the MSB sign bit is 0. The number is greater than 1, so the first regime bit is a 1. Divide 2300 by 4 repeatedly until it is less than uSeed, appending a regime bit each time:

2300	01
575	011
143.75	0111
35.9375	01111
8.984375	011111
2.24609375	0111111

We are up to 7 bits. It is actually 8 bits because we must append the regime termination bit, 0. But for correct rounding, we need to know what it would be to 9 bits. The number is greater than 2, so divide by 2 and append a 1 exponent bit:

2.24609375	01111110
1.123046875	011111101

If we kept going, would there be more nonzero bits? There would, because after the last divide-by-2 the number is greater than 1. So we can append a ubit of 1 to indicate "there are nonzero bits after the last one":

2.24609375	01111110
1.123046875	011111101
1. ···	0111111011

That tells us how to round, and the rounding must be upwards. So 01111110 rounds up to the binary for *maxPos*, 01111111, which for ⟨8, 1⟩ posits is 4096. The next smaller posit is 1024. The log of 2300 is closer to the log of 4096 than to the log of 1024, even though 2300 is closer to 1024 than to 4096. The system of generating an extra bit of precision and then assessing whether there are more bits after that automatically does the right thing, switching from arithmetic rounding to geometric rounding when the bits being rounded are exponent-specifying bits (the Twilight Zone).

Chapter 4

Exercise for the Reader: What would Cistercian numerals look like for base-16 (hexadecimal) notation, if we use the "revised Cistercian" shown above? What would be the range of integers that could be encoded that way?

The glyphs for 1, 2, 4, and 8 each show a **1** bit as a line segment. Their absence shows a **0** bit.

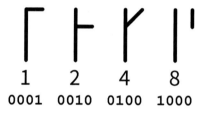

The hexadecimal digits 0–9 and A–F therefore look like this:

The four corners no longer are powers of ten, but powers of 16. Reflect about the vertical axis to get symbols for hex values 0x00, 0x10, 0x20, ..., 0xF0 in the upper left corner. Reflect about the horizontal to get symbols for hex values 0x000, 0x100, ..., 0xF00 in the lower right corner. Reflect about both axes to get symbols for 0x0000 to 0xF000 in the lower left corner. So the range of integers representable is 0x0000 to 0xFFFF, which is 0 to 65 535, one less than 16^4. It is an incredibly compact way to represent a string of 16 bits.

We could therefore use the "revised Cistercian" system to record a 16-bit posit in a single character. Unlike hexadecimal, we can still apply color-coding (for the bits that are 1). For example, the standard 16-bit posit approximating π is 0100 1100 1001 0001 which would be 0x4C91 in hex, but there is no way to color-code the hex. As a base-16 Cistercian, some example posits would look like this:

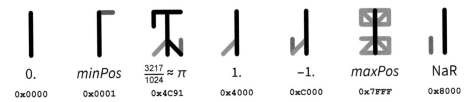

0.	*minPos*	$\frac{3217}{1024} \approx \pi$	1.	−1.	*maxPos*	NaR
0x0000	0x0001	0x4C91	0x4000	0xC000	0x7FFF	0x8000

Examples of 16-bit standard posits written with revised Cistercian, color-coded

We could also use them in a row as positional notation, where each glyph is a power of 65 536. You could write a 64-bit number with just four symbols, and express each bit explicitly.

Exercise for the Reader: An eight-bit format should be able to represent as many as $2^8 = 256$ different values. Suppose we use a format like that of the IBM 704, but with a sign bit, three exponent bits (biased by 4), and four fraction bits; how many distinct real values can be represented?

To help visualize the format:

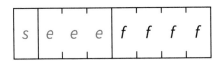

8-bit float that uses the earliest IBM float format

The value of the sign bit s is $S = 0$ or 1, of course, which we can make into +1 or −1 using the expression $(1 - 2S)$. The three exponent bits e form unsigned integers 0 to 7, which with a bias of 4 means the exponent value E ranges from −4 to 3. The four fraction bits f represent $F = \frac{0}{16}, \frac{1}{16}, ..., \frac{15}{16}$. The value represented is $(1 - 2S) \times 2^E \times F$.

Suppose $s = 0$ and the fraction is not all 0 bits. If the last fraction bit is a 1, it cannot be shifted right and the exponent raised by 1 without changing the number, since the 1 bit will be lost. Otherwise, it can. There can be one, two, or three 0 bits in a row ending the fraction. Consider each situation, with the number of redundant values shown in **bold** for clarity:

- If the last two bits of the fraction are **10** (4 cases) and the exponent is less than 3 so that it does not overflow when incremented, the bit string will represent the same number as the fraction shifted right one place and scaled by an incremented exponent. There are 7 exponent values less than 3, creating $4 \times 7 = \mathbf{28}$ redundant values.

- If the last three bits of the fraction are **100** (2 cases) and the exponent is less than 3 (again, 7 cases), it will represent the same number as the fraction shifted right by 1 with an incremented exponent. That is another $2 \times 7 = \mathbf{14}$ redundant values.

- Similarly, if the fraction bits are **1000**, it can be shifted right by one place and with a raised exponent that does not overflow (7 cases), and it will represent the same value. That is **7** redundant values.

That is a total of $28 + 14 + 7 = 49$ redundant positive values. There are also 49 redundant negative values if $s = 1$, for a total of **98** redundancies.

Lastly, we have the case where all the fraction bits are **0**, so the value represented is zero and the other bits do not matter. There are 4 other bits, so that adds $2^4 = 16$ ways to represent zero, of which **15** are redundant. Add those 15 to the 98 for a total of **113** redundant representations out of 256 possible bit patterns. That leaves us with $256 - 113 = 143$ unique values. That is a utilization of only about 56% of the 256 possible bit patterns.

The problem can be solved in two lines of *Mathematica*, by building the table of values, using **Flatten** to make it a simple list, using **Union** to remove redundant values, and then using **Length** to find the length of the list:

```
set = Table[(1 - 2 × S) × 2^E × F, {S, 0, 1}, {E, -4, 3}, {F, 0, 15/16, 1/16}];

Length[Union[Flatten[set]]]
```

143

Chapter 5

> **Exercise for the Reader**: Can a posit with 3 bits of precision have an exponent size of 2 bits? Be careful, because this has the quality of being a trick question.

The exponent size can be any integer 0 or larger and is not restricted by the number of bits of precision. Remember that the *uSeed* value shows up on the posit ring as soon as we have 3 or more bits of precision, and *uSeed* is $2^{2^{es}}$. If $eS = 2$, the set of posits looks like this:

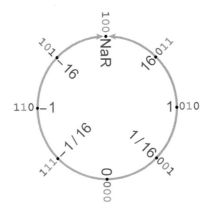

3-bit posit ring with *eS* = 2; exponent bits are always ghost bits in this case, but they are there.

> **Exercise for the Reader**: For standard posits, what percentage of posit bit patterns have $n - 6$ explicit bits for the fraction, just one bit less than in the fovea? What percentage have $n - 7$ explicit bits for the fraction?

There will be $n - 6$ explicit bits for the fraction if there are three bits in the regime, which means the regime bits are either 001 or 110. That is two possibilities out of the three bits that follow the sign bit, $2^3 = 8$ possible bit patterns. So $\frac{2}{8} = \frac{1}{4}$, or 25%. Similarly, $n - 7$ explicit bits means the regime is either 0001 or 1110, two possibilities out of the $2^4 = 16$ possible bit patterns. So $\frac{2}{16} = \frac{1}{8}$, or 12.5%. For each reduction in fraction significance, the number of bit patterns drops by half.

> **Exercise for the Reader**: What is the formula for the *smallest* posit value such that it and all larger posits are guaranteed to be integers?

The *pIntMax* value is the border between an ULP spacing of 1 and an ULP spacing of 2. Above that value, you cannot express an odd integer. What we seek instead is the border between an ULP spacing of $\frac{1}{2}$ and an ULP spacing of 1. Call it *pIntMin*. That is one binade less than the *pIntMax* point, so it should be *pIntMax*/2. The way to make the formula robust all the way down to a precision of $n = 2$ is to subtract 1 from the exponent, shown in blue in the formula below:

$$pIntMin = \left\lceil 2^{\lfloor 4(n-3)/5 - 1 \rfloor} \right\rceil$$

The value of *pIntMin* is 1 for posit precision 2 through 5, and then starts climbing.

Chapter 6

> **Exercise for the Reader:**
>
> *Part 1*: If we have four distinct numbers a, b, c, d, how many different ways are there of adding them if addition is commutative but not associative?
>
> *Part 2*: If the four numbers are 1, 2, 2^{54}, and -2^{54} and we add them using IEEE Std 754 double-precision arithmetic (52 fraction bits) with banker's rounding, what are the possible values for the resulting sum?

Part 1d

There are two distinct dataflow patterns for adding four numbers a, b, c, d. Here is an example of each:

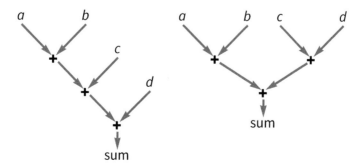

The two possibledataflow patterns for adding four numbers

There are $4! = 4 \times 3 \times 2 \times 1 = 24$ permutations of the four inputs a, b, c, d. In the left diagram, commutativity says we can interchange a and b without affecting the result, and that is true for all the permutations so there are only $24/2 = 12$ distinct ways to compute the sum with that dataflow. Here they are:

$((a+b)+c)+d$	$((a+b)+d)+c$	$((a+c)+b)+d$	$((a+c)+d)+b$
$((a+d)+b)+c$	$((a+d)+c)+b$	$((b+d)+a)+c$	$((b+d)+c)+a$
$((b+c)+a)+d$	$((b+c)+d)+a$	$((c+d)+a)+b$	$((c+d)+b)+a$

The right dataflow pattern (sometimes called a binary sum collapse or binary sum reduction) has more commutativity. You can swap a and b, swap c and d, and swap the a, b pair with the c, d pair without changing the result, each swap cutting in half the number of distinct sum patterns. Cut $4! = 24$ in half three times (divide by 8) and you get only three ways to compute the sum with the right dataflow pattern:

$$(a+b)+(c+d) \qquad (a+c)+(b+d) \qquad (a+d)+(b+c)$$

Hence, there are $12 + 3 = 15$ possible ways to add four distinct numbers if addition is commutative but not associative like it is for computer addition of real numbers.

Part 2

The number 2^{54} is expressible exactly as a 64-bit IEEE float. The fraction is all **0** bits.

```
setFloatEnv[{64, 11}]; colorCodeF[xToF[2^54]]
```

010000110101000000000000000000000000000000000000000000000000000

If you add 1 or 2 to that, the new bit is just past the last bit of the fraction, so it rounds to the nearest even. That is, adding 1 or 2 has no effect.

```
colorCodeF[xToF[2^54 + 2]]
```

010000110101000000000000000000000000000000000000000000000000000

On the other hand, suppose you add -2^{54} to 2. That takes it to the next lower binade where the ULP spacing is smaller, so no rounding is needed to fit the **1** bit in:

```
colorCodeF[xToF[-2^54 + 2]]
```

101111001011000000000000000000000000000000000000000000000000001

However, adding 1 to -2^{54} is just past the last bit, so it rounds back to the original number since that is the one with an even binary string:

`colorCodeF` $\Big[$ `xToF` $\big[-2^{54}+1\big]\Big]$

`1011110010110000000000000000000000000000000000000000000000000000`

Here are the 15 ways to add $a, b, c, d = 1, 2, 2^{54}, -2^{54}$ with float rounding indicated using underscores and the correct answer **3** shown in blue boldface:

$\underline{a+b}+c+d = 4$	$\underline{a+b+c}+d = 4$	$\underline{a+c}+b+d = 0$	$\underline{a+c+d}+b = 2$
$\underline{a+d}+b+c = 2$	$\underline{a+d+c}+b = 2$	$\underline{b+d}+a+c = 4$	$\underline{b+d+c}+a = 3$
$\underline{b+c}+a+d = 0$	$\underline{b+c+d}+a = 1$	$\underline{c+d}+a+b = 3$	$\underline{c+d+b}+a = 3$

$$\underline{a+b+c+d} = 3 \qquad \underline{a+c+b}+d = 2 \qquad \underline{b+c+a}+d = 0$$

The summation of just four IEEE double-precision floats can produce 0, 1, 2, or 4, or the correct answer **3**, depending on what order you sum. The correct sum is produced for only $\frac{2}{9} \approx 22\%$ of the cases. Floats are weapons of math destruction. This is why it is essential for a number format to support exact summation and exact dot products.

Chapter 7

Exercise for the Reader: If the precision is only 8 bits, find the minimal values for *rS* that create a posit format that can cover the dynamic range of **RRSet**, 3×10^{-13} to 3×10^{13}, for *eS* values 3, 4, and 5.

If $eS = 3$, then $uSeed = 2^{2^3} = 2^8 = 256$. If the regime is restricted to a maximum of 6 bits, then *maxPos* will be `01111111` which represents $(2^8)^{6-1} \times 2^4 = 2^{44} \approx 1.8 \times 10^{13}$. That is less than 3×10^{13}, so we need the maximum of 7 bits, which means maximum accuracy tapering. `01111111` represents $(2^8)^{7-1} = 2^{48} \approx 2.8 \times 10^{14}$.

As the relative accuracy plot shows, the relative accuracy is very low outside the magnitude range 10^{-10} to 10^{10}; it actually goes slightly below the *x* axis, which is not plotted. At least we have about a decimal of accuracy in the fovea, $\frac{1}{256} \leq |x| \leq 256$:

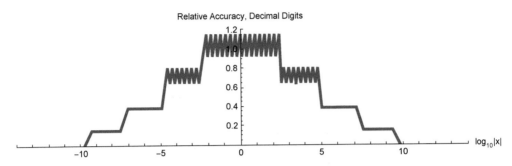

Accuracy plot for posit⟨8,3⟩ with regime size *rS* allowed to go as high as 7 (that is, not restriction)

If *eS* = 4, then *uSeed* = $2^{2^4} = 2^{16} = 65\,536$. If *rS* = 2 (no accuracy tapering, a sane float), *maxPos* will be 01111111, which represents $\left(2^{16}\right)^{2-1} \times 2^{15} \times \left(1 + \frac{1}{2}\right) = 2^{31} \times \frac{3}{2} \approx 3 \times 10^9$, well short of what is needed. But one more factor of *uSeed* should do the trick, so try *rS* = 3. Then *maxPos* is 01111111 which represents $\left(2^{16}\right)^{3-1} \times 2^{15} = 2^{46} \approx 1.4 \times 10^{14}$, and that suffices. We lose one bit of accuracy in the fovea but get much better coverage at the extreme magnitudes:

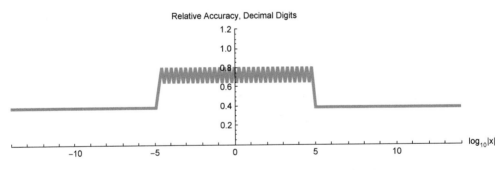

Accuracy plot for posit⟨8,4⟩ and restricted regime size *rS* = 3

For *eS* = 5, we should certainly be able to cover the range using sane floats (*rS* = 2). The *uSeed* value is 2^{32} and the *maxPos* value 01111111 represents a number much larger than needed, $\left(2^{32}\right)^{2-1} \times 2^{31} = 2^{63} \approx 9 \times 10^{18}$. That means wasted bit patterns if we only need the dynamic range specified in the exercise.

Here is what the accuracy looks like, restricted to the dynamic range of the two accuracy graphs on the last page:

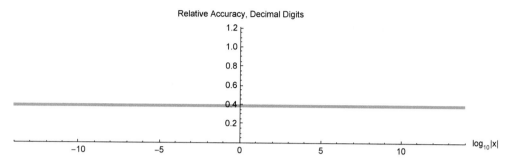

Accuracy plot for posit(8,4) with restricted regime size $rS = 2$ (sane float)

While it is not part of this exercise, the reader may be as curious as I was regarding the *fidelity* of the three possible formats for **RRSet**. So, I ran the numbers.

For the maximum tapering case $eS = 3$, $rS = 7$, fidelity is 1.12 decimals.

For the case $eS = 4$, $rS = 3$, fidelity is 1.00 decimals. The gain in accuracy at the extreme magnitudes was not enough to compensate for the loss of accuracy in fovea.

Not surprisingly, the sane floats do not do well because they waste too many bit patterns representing values outside the range of **RRSet**. Fidelity for $eS = 5$, $rS = 2$ is only 0.75 decimals.

While 8-bit formats with $eS = 3$ seem too small for scientific computations that guided the construction of the **RRSet**, it would be interesting to try them on some HPC applications and find out. It would have to be an application where we only need about one decimal of accuracy in the output.

Chapter 8

Exercise for the Reader: There is one more case of "saturates to *maxPos*" for linear posits than for logarithmic posits in the example above. Find it and explain why it happens.

This is related to the exercise in Chapter 2.

The multiplication closure plot is very close to being symmetric about both diagonal axes, but it fails at just a single point:

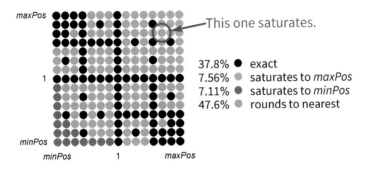

Multiplication closure plot for linear ⟨5,0⟩ posits with the one asymmetrical overflow point called out

The posits values here are the set $\left\{\frac{1}{8}, \frac{1}{4}, \frac{3}{8}, \frac{1}{2}, \frac{5}{8}, \frac{3}{4}, \frac{7}{8}, 1, \frac{5}{4}, \frac{3}{2}, \frac{7}{4}, 2, 3, 4, 8\right\}$, so that point corresponds to 3 times 3. Since the product, 9, is larger than *maxPos* = 8, it saturates. But that does not happen for the corresponding point in the lower left, after reflecting about the diagonal:

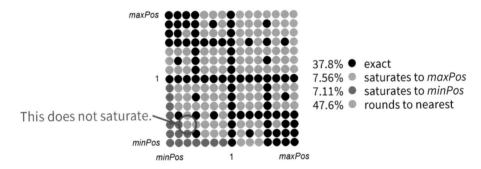

Multiplication closure plot for linear ⟨5,0⟩ posits with the one asymmetrical underflow point called out

That dot corresponds to $\frac{3}{8} \times \frac{3}{8} = \frac{9}{64}$, which is greater than $\frac{8}{64} = \frac{1}{8}$, so it does not fall below *minPos* the way 9 exceeded *maxPos*. While *minPos* and *maxPos* are perfect reciprocals because they are integer powers of 2, the same is not true for values within a binade when the binades are linear. If more precision were used in this example, many other asymmetries would become visible about the top left-to-bottom right diagonal.

Logarithmic posits show perfect reciprocation, so whatever happens for too-large numbers will be symmetrical for too-small numbers when it comes to multiplication.

Chapter 9

> **Exercise for the Reader:** The spacings between integers in L are symmetrical about L_{16}. Why is that? What property of the x list makes that the case in general?

Unlike multiplication, addition is symmetrical about both diagonal lines for its table. The x list is antisymmetric about 0, so using the negative of the table for the positive sums works as a ranking for the negative sums. Anytime this is true of a function, it can be used to simplify the construction of the table as well as the hardware needed to perform the function. It works for multiplication as well if the binades are logarithmic (Chapter 8) instead of linear.

> **Exercise for the Reader:** Suppose x is the set $\{1, \frac{3}{2}, 2\}$. Find the set of all possible products $x_i \times x_j$ and sketch the rank plot. Then find a sequence of increasing integers $L = \{0, L_1, L_2\}$ such that the set of all possible sums $L_i + L_j$ has the same rank and L_1 is as small as possible.

The multiplication table is

$\times$	1	$\frac{3}{2}$	2
1	1	$\frac{3}{2}$	2
$\frac{3}{2}$	$\frac{3}{2}$	$\frac{9}{4}$	3
2	2	3	4

Multiplication table for values $x = \{1, 3/2, 2\}$

There are six ranks, which is the maximum for a commutative function with three inputs. Note that the product in the center, $\frac{9}{4} = 2.25$, is higher rank than 2.

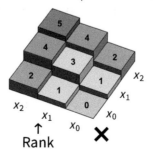

Rank plot for the multiplication table

As usual, assume $L_0 = 0$. Since rank **3** is greater than rank **2**, $L_1 + L_1 > L_0 + L_2 = L_2$. That rules out the sequence $\{0, 1, L_2\}$ for L since the smallest integer that L_2 can be is 2. So try $\{0, 2, 3\}$, which satisfies $2L_1 > L_2$. The possible sums are $\{0, 2, 3, 4, 5, 6\}$, and the addition table shows the same "contour lines" as the multiplication table.

+	**0**	**2**	**3**
0	0	2	3
2	2	4	5
3	3	5	6

Addition table for values $L = \{0, 2, 3\}$

That works, and the integers are as close to 0 as possible.

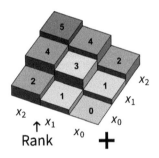

Rank plot for the addition table

This is an example of the "brute force" method of trying every integer sequence starting from $\{0, 1, 2\}$ and in this case it is such a small problem that no computer was needed for the exhaustive search.

Chapter 10

> **Exercise for the Reader**: Complete the proof that z is algebraic for the other cases of the signs of j, k, and n (assuming $m > 0$). Find the polynomial in z with integer coefficients that corresponds to computing $1.25^{0.875}$.

Remember that the case $n = 0$ is easily dispensed with since $x^y = x^0 = 1$.

The possible cases are

> Case 1: $j \geq 0$ $k \geq 0$ $n > 0$ Explained in Section 10.7
> Case 2: $j \geq 0$ $k \geq 0$ $n < 0$
> Case 3: $j \geq 0$ $k < 0$ $n > 0$ Explained in Section 10.7
> Case 4: $j \geq 0$ $k < 0$ $n < 0$
> Case 5: $j < 0$ $k \geq 0$ $n > 0$
> Case 6: $j < 0$ $k \geq 0$ $n < 0$
> Case 7: $j < 0$ $k < 0$ $n > 0$
> Case 8: $j < 0$ $k < 0$ $n < 0$

In general, the approach will be to arrange the equation so that any negative values of j, k, or n appear only as their positive counterparts $-j$, $-k$, or $-n$.

Case 2: $j \geq 0$, $k \geq 0$, $n < 0$

If $n < 0$, $-n \times 2^k$ is a positive integer. Multiply both sides by $\left(m \times 2^j\right)^{-n \times 2^k}$, which is a positive integer and the reciprocal of the right-hand side:

$$z = \left(m \times 2^j\right)^{n \times 2^k}$$

$$\left(m \times 2^j\right)^{-n \times 2^k} \times z = 1.$$

The polynomial is $\left(m \times 2^j\right)^{-n \times 2^k} \times z - 1 = 0$, done.

Case 4: $j \geq 0$, $k < 0$, $n < 0$

If $k < 0$, 2^{-k} is a positive integer. Raise both sides to the power of 2^{-k} to remove the 2^k in the exponent on the right-hand side. Then multiply by the reciprocal of the right-hand side as was done in Case 2 so that the exponents are positive integers:

$$z = \left(m \times 2^j\right)^{n \times 2^k}$$

$$z^{\left(2^{-k}\right)} = \left(m \times 2^j\right)^n$$

$$\left(m \times 2^j\right)^{-n} \times z^{\left(2^{-k}\right)} = 1.$$

The polynomial is $\left(m \times 2^j\right)^{-n} \times z^{\left(2^{-k}\right)} - 1 = 0$, done.

Case 5: $j < 0$, $k \geq 0$, $n > 0$

Multiply both sides of the usual starting equation by the positive integer $\left(2^{-j}\right)^{n \times 2^k}$ to remove the j on the right-hand side:

$$z = \left(m \times 2^j\right)^{n \times 2^k}$$
$$\left(2^{-j}\right)^{n \times 2^k} \times z = m^{n \times 2^k}.$$

The polynomial is $\left(2^{-j}\right)^{\left(n \times 2^k\right)} \times z - m^{\left(n \times 2^k\right)} = 0$, done.

Case 6: $j < 0$, $k \geq 0$, $n < 0$

First multiply by the reciprocal of the right-hand side. Then multiply by the positive integer $\left(2^{-j}\right)^{-n \times 2^k}$:

$$z = \left(m \times 2^j\right)^{n \times 2^k}$$
$$\left(m \times 2^j\right)^{-n \times 2^k} \times z = 1$$
$$m^{-n \times 2^k} \times z = \left(2^{-j}\right)^{-n \times 2^k}.$$

The polynomial is $m^{-n \times 2^k} \times z - \left(2^{-j}\right)^{-n \times 2^k} = 0$, done.

Case 7: $j < 0$, $k < 0$, $n > 0$

Raise both sides to the power of the positive integer 2^{-k}. Then multiply by positive integer $\left(2^{-j}\right)^n$:

$$z = \left(m \times 2^j\right)^{n \times 2^k}$$
$$z^{\left(2^{-k}\right)} = \left(m \times 2^j\right)^n$$
$$\left(2^{-j}\right)^n \times z^{\left(2^{-k}\right)} = m^n.$$

The polynomial is $\left(2^{-j}\right)^n \times z^{\left(2^{-k}\right)} - m^n = 0$, done.

Case 8: $j < 0$, $k < 0$, $n < 0$

All three variables need to appear only with minus signs in front of them.

Start with n, then k, then j, using the techniques shown above for the other cases:

$$z = \left(m \times 2^j\right)^{n \times 2^k}$$
$$\left(m \times 2^j\right)^{-n \times 2^k} \times z = 1$$
$$\left(m \times 2^j\right)^{-n} \times z^{\left(2^{-k}\right)} = 1$$
$$m^{-n} \times z^{\left(2^{-k}\right)} = \left(2^{-j}\right)^{-n}$$

The polynomial is $m^{-n} \times z^{\left(2^{-k}\right)} - \left(2^{-j}\right)^{-n} = 0$, done.

Apply to $1.25^{0.875}$

The case $1.25^{0.875} = \left(\frac{5}{4}\right)^{7/8} = \left(5 \times 2^{-2}\right)^{\left(7 \times 2^{-3}\right)}$. So $m = 5$, $j = -2$, $n = 7$, and $k = -3$. That is Case 7, where $j < 0$, $k < 0$, and $n > 0$.

$$\left(2^{-j}\right)^n \times z^{\left(2^{-k}\right)} - m^n = 0$$
$$\left(2^2\right)^7 \times z^{2^3} - 5^7 = 0$$
$$16\,384 \times z^8 - 78\,125 = 0.$$

This is a constructive proof in that it also shows a way you can compute x^y without using logarithms and exponentials. If we rewrite the last line as $z^8 = \frac{78\,125}{16\,384}$, then z is the 8th root of $\frac{78\,125}{16\,384}$, which you can find by pressing the $\sqrt{x}$ button three times. I tried it just now with a calculator app and got $z \approx 1.21561\,55905\,82591$, which *Mathematica* says is correct in every digit, to 16 decimals.

Chapter 11

Exercise for the Reader: What does multiplying by $\boldsymbol{A} = \begin{pmatrix} 0 & 1 \\ 1 & 0 \end{pmatrix}$ do to the unit square?

Since $\begin{pmatrix} 0 & 1 \\ 1 & 0 \end{pmatrix} \cdot \begin{pmatrix} x \\ y \end{pmatrix} = \begin{pmatrix} y \\ x \end{pmatrix}$, the matrix interchanges the roles of x and y. In other words, the unit square is reflected about the 45° diagonal:

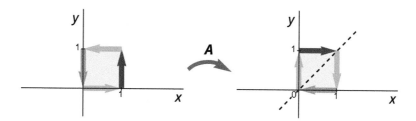

Multiplying the unit square by $\begin{pmatrix} 0 & 1 \\ 1 & 0 \end{pmatrix}$ flips the square about the dashed diagonal line $y = x$.

Chapter 12

Exercise for the Reader: If the 2-by-2 matrix $\begin{pmatrix} a & b \\ c & d \end{pmatrix}$ has an inverse, what is it? Under what conditions will it not have an inverse? If computed with posits, how many rounding errors will each entry in the inverse experience?

We seek a matrix $\begin{pmatrix} w & x \\ y & z \end{pmatrix}$ such that $\begin{pmatrix} a & b \\ c & d \end{pmatrix} \cdot \begin{pmatrix} w & x \\ y & z \end{pmatrix} = \begin{pmatrix} 1 & 0 \\ 0 & 1 \end{pmatrix}$. We can use Cramer's rule on each column of $\begin{pmatrix} w & x \\ y & z \end{pmatrix}$ and each column of the identity matrix. The determinant is $D = a \cdot d - b \cdot c$. The two columns of the identity matrix $\begin{pmatrix} 1 & 0 \\ 0 & 1 \end{pmatrix}$ are $\begin{pmatrix} 1 \\ 0 \end{pmatrix}$ and $\begin{pmatrix} 0 \\ 1 \end{pmatrix}$. By Cramer's rule for solving $\begin{pmatrix} a & b \\ c & d \end{pmatrix} \cdot \begin{pmatrix} w \\ y \end{pmatrix} = \begin{pmatrix} 1 \\ 0 \end{pmatrix}$, we get

$$w = (1 \cdot d - b \cdot 0)/D = \ d/D,$$
$$y = (a \cdot 0 - 1 \cdot c)/D = -c/D.$$

Apply Cramer's rule to the second column, $\begin{pmatrix} a & b \\ c & d \end{pmatrix} \cdot \begin{pmatrix} x \\ z \end{pmatrix} = \begin{pmatrix} 0 \\ 1 \end{pmatrix}$:

$$x = (0 \cdot d - b \cdot 1)/D = -b/D,$$
$$z = (a \cdot 1 - 0 \cdot c)/D = \ a/D.$$

So the inverse matrix, if it exists, is $\begin{pmatrix} d & -b \\ -c & a \end{pmatrix} \big/ D$. It will exist if D is not zero, that is, if the matrix is not singular.

If computed with posits, the determinant is computed exactly in the quire, then rounded to a posit. Each element is then divided by that determinant for a total of two roundings per entry in the inverse. However, the rounding error is often more like a single rounding because D might be exactly expressible in the form of a posit $m \times 2^j$, and each entry in the original matrix is a posit of the form $n \times 2^k$, so the ratio will be $\left(\frac{n}{m}\right) \times 2^{k-j}$, and it frequently happens that 2^{k-j} is expressible exactly with the regime and exponent bits. So, in many cases the only rounding error will be from computing n/m.

Chapter 13

> **Exercise for the Reader**: How many different ways can the 8-bit Morris floats express the value 1.0? What are the bit patterns that do so?

The value 1.0 is expressed when the exponent field represents zero and all the bits in the fraction and fraction sign are **0** bits. The exponent size (the *G* field) and the sign of the exponent do not matter. There are eight ways to express 1.0 with an 8-bit Morris float format:

```
00000000
00100000
01000000
01100000
10000000
10100000
11000000
11100000
```

Like posits, Morris floats can have "ghost bits" past the bits shown explicitly, with all ghost bits having value **0**. For the last two cases, the only fraction bits are ghost bits, so the hidden **1** bit is the only contributor to the significand.

Notice that the numerical meaning of an ULP depends on how many bits there are in the exponent field, and the representations of 1.0 can have four different ULP meanings. That makes numerical analysis of Morris floats a nightmare.

Appendix A Posit Code Listings

The source code of the *Mathematica* routines used in the book is shown here, with explanations. While the reader may know how to read algorithms written in a high-level language, there are keywords and grammar specific to *Mathematica*, so those also are explained as they arise.

This Appendix is intended to assist those who wish to create their own software environment for posit computing, in whatever language they prefer. There are many open source posit software emulators that are easy to find with a web search, written in C, C++, Python, and so on, so there is no need to reinvent the wheel if you can find what you want out there.

A.1 decAcc (Decimal Accuracy)

The **decAcc** function takes two numeric arguments where either one can be regarded as the exact answer x_{exact} and the other as $x_{computed}$. That is, **decAcc[x,y] = decAcc[y,x]**.

```
decAcc[x_, y_] := Which[
  ¬ NumericQ[x] || ¬ NumericQ[y], Indeterminate,
  Im[x] ≠ 0 || Im[y] ≠ 0, Indeterminate,
  x == y, ∞,
  (x < 0 && y > 0) || (x > 0 && y < 0), Indeterminate,
  True, N[-Log10[Abs[Log[x / y]]]]]]
```

Which makes multiple conditional tests on the inputs before applying the usual formula described in Chapter 2. It is important to do the tests in the right order.

The first test uses **NumericQ[x]** to test if **x** is a numeric quantity; if **x** is not numeric (¬ is the logical NOT operator) or **y** is not numeric, return **Indeterminate**. The logical OR operator is the || symbol.

The next line tests if either input is a complex number with a non-zero imaginary part. **Im[x]** extracts the imaginary part of **x**. If either **x** or **y** has nonzero imaginary parts, return **Indeterminate**.

Passing those two tests means that **x** and **y** are real numbers. Doing the test for **x** equal to **y** next (the == operator) neatly takes care of the case when both values are zero, which would otherwise break the final step. If equal, then return ∞ since the values agree to an infinite number of decimals.

As explained in Chapter 2, the function makes no attempt to assess decimal accuracy if **x** and **y** have opposite signs, so that is the fourth test. As in C and C++, a logical AND is the **&&** operator.

The final test is **True** (always satisfied) to make sure all cases are handled. The **N[...]** is necessary to make the code return a numeric answer, not a symbolic one. What if one input is zero? *Mathematica* is able to handle that since it returns ∞ for **Log10**[x/0] and −∞ for **Log10**[0]. The output of **decAcc** will be −∞ for either case. In other computing environments, a different way of handling the case of a zero-nonzero input pair might be needed.

A.2 setPositEnv (Set the Posit Environment)

The **setPositEnv**[n, e] routine must be called before working with posits; it sets the precision **nBitsP** to n bits and the exponent size **eS** to e bits. The routine sets global variables that are used by other routines like **xToP** and **pToX**.

```
setPositEnv[{n_Integer /; n > 1, e_Integer /; e ≥ 0}] :=
  ({nBitsP, eS} = {n, e};
   nPat = BitShiftLeft[1, nBitsP];
   NaR = - (BitShiftRight[nPat]);
   pMask = nPat - 1;
   eSizeP = BitShiftLeft[1, eS];
   uSeed = BitShiftLeft[1, eSizeP];
   {minPos, maxPos} = {uSeed^(-nBitsP+2), uSeed^(nBitsP-2)};
   qSize = Log[2, maxPos^4 * 2^32];)
```

The "/;" applies the condition that follows to the argument and exits if it is not met. So the first line ensures the requested precision n is an integer greater than 1 and the requested number of exponent bits is an integer greater than or equal to zero. If conditions are met, the integer pair {**n,e**} is assigned to {**nBitsP, eS**}. Notice that the restriction on e has nothing to do with n; it is possible for the exponent size to exceed the precision by treating some of the ghost bits as exponent bits.

The number of bit patterns is the integer **nPat**, equal to 2^{nBitsP}. An easy way to create that integer in binary is to shift a **1** bit left by **nBitsP** places. For example, a 3-bit posit would have **1000** = 8 bit patterns. In C and C++, this would need to be an unsigned integer.

The bit pattern for **NaR** is the most negative integer of the same precision. It is easily created from **nPat** by shifting one bit right and making it negative. Like the 3-bit posit **nPat**, **1000** would become **100**, meaning –4 as a signed integer. Because every integer in *Mathematica* is extended precision, there is no MSB to indicate the sign like you would see in C and C++ and other languages.

Subtracting 1 from binary **1000…000** is a trick for creating a string of 1 bits that is **nBitsP** long, **111…111**, which can be used as a mask, **pMask**.

Variable **eSizeP** is 2^{eS}, constructed by shifting a **1** bit left by **eS** places. Since **uSeed** is $2^{2^{\text{eS}}}$, it can be constructed by shifting a **1** bit left by **eSizeP** places.

Each regime bit represents a factor of **uSeed**, and the minimum and maximum values of a positive posit regime are $r = $ **–nBits**+2 and $r = $ **nBits**+2 . For clarity, the assignment of **minPos** and **maxPos** is expressed as powers of those two r values instead of using the bit-shift trick.

Finally, the quire size **qSize** needs to be sufficient to hold the square of **minPos** and the square of **maxPos** and also have 32 bits of carry overflow protection. We can treat the quire as a large integer that is a multiple of the square of **minPos**. Its length in bits is the log base 2 of $\textbf{maxPos}^4 \times 2^{32}$.

A.3 pToX (Posit to X)

The **pToX** function takes a posit bit string and converts it into a real number in the *Mathematica* environment. A routine to convert the bit string to, say, a 64-bit IEEE float would use similar logic.

```
pToX[p_ /; positQ[p]] :=
 Module[{b = BitShiftLeft[1, nBitsP - 2], s, x},
  If[p == NaR, Return[Indeterminate]]; (* Exception values *)
  If[p == 0, Return[0]];
  If[nBitsP == 2, Return[Sign[p]]];
  x = Floor[If[p < 0, -p, p]];
  (* Decode x = |p|. Floor insures the type is Integer *)
  If[BitAnd[b, x] ≠ 0, (* Decode the regime bits of x. *)
   {s, b} = {0, BitShiftRight[b]};
   While[BitAnd[b, x] ≠ 0,
    {x, s} = {BitShiftLeft[x], s + eSizeP}], (*else*)
   {s, b} = {-eSizeP, BitShiftRight[b]};
   While[BitAnd[b, x] == 0, {x, s} = {BitShiftLeft[x], s - eSizeP}]
  ];
  x = BitAnd[x, --b];
  (* x is just the exponent and fraction now. *)
  s += BitShiftRight[x, nBitsP - eS - 3];
  x = BitOr[BitAnd[BitShiftLeft[x, eS], b], ++b];
  (* hidden bit and fraction *)
  x *= 2^(s-nBitsP+3);
  If[p < 0, -x, x]
 ]
```

The `positQ[p]` function is a test that returns True if p is a valid posit and False otherwise. It is a one-liner in *Mathematica*:

```
positQ[p_] := (Floor[p] == p) && (NaR ≤ p ≤ -NaR - 1);
```

To be robust, the routine accepts real numbers like 3.0 if they are integer-valued, and an easy way to check that is if the `Floor[p]` function (sometimes written $\lfloor p \rfloor$) has the same value as `p`. The integer can be anything in the range `NaR` to 1 less than **−NaR**.

This is the first routine that shows the `Module` keyword in *Mathematica*. It allows declaration (and optional pre-assignment) of local scratch variables, in this case `b`, `s`, and `x`, inside the { } braces. Variable `b` is a bit string with just one `1` bit, initially positioned where the first regime bit is by shifting a `1` bit left by `nBitsP`−2.

The *Mathematica* equivalent of **NaR** is **Indeterminate**. That exception and the zero exception are handled first.

The code supports posits as small as **nBitsP** = 2, using

```
If[nBitsP == 2, Return[Sign[p]]];
```

There are only four 2-bit posits, and the cases **10** (NaR) and **00** (zero) were just handled by the exception tests, leaving **01** (+1) and **10** (−1). Some of the logic that follows will not work correctly on such a short bit string.

```
x = Floor[If[p < 0, -p, p]];
```

This line accomplishes three things. First, it takes the absolute value to simplify the logic of decoding the parts of the posit. We haven't lost track of the sign because **p** is unchanged. Second, the **Floor** function accepts both real (floating point) and integer types, but always returns an integer type so we can operate on it as an integer bit string. And third, by having the input value in a temporary **x**, we can make changes to **x**; input values are protected so changes to **p** are not allowed.

A.4 xToP (X to Posit)

The algorithm is explained in detail in Chapter 3, so there is no need to repeat that line-by-line explanation here.

```
xToP[x_] := Module[{i, p, e = BitShiftRight[eSizeP], y},
  Which[
    (x ∈ Reals) =!= True, NaR,
    (* All non-real inputs become NaR. *)
    x == 0, 0, (* the other exception value, zero *)
    y = Abs[x]; (* Otherwise, encode |x|. *)
    True,
    If[y ≥ 1,
      {p, i} = {1, 2}; (* Regime is a run of 1 bits. *)
      While[y ≥ uSeed && i < nBitsP,
        {p, y, i} = {BitOr[BitShiftLeft[p], 1], y / uSeed, i + 1}];
      {p, i} = {BitShiftLeft[p], i + 1},
      (* Always terminate in 0. Handles |x| > maxPos. *)
      {p, i} = {0, 1};
      (* Else regime is a run of 0 bits. *)
      While[y < 1 && i ≤ nBitsP, {y, i} = {y × uSeed, i + 1}];
      {p, i} = If[i ≥ nBitsP, {2, nBitsP + 1}, {1, i + 1}]
      (* Handles |x| < minPos. *)
    ];
    While[e ≠ 0 && i ≤ nBitsP,
      (* Extract exponent bits, MSB to LSB. *)
      p = BitShiftLeft[p];
      If[y ≥ BitShiftLeft[1, e],
        {y, p} = {y / BitShiftLeft[1, e], p + 1}];
      {e, i} = {BitShiftRight[e], i + 1}
    ];
    If[i == nBitsP && y > 3 / 2, p++,
      y--; (* Remove hidden bit. *)
      While[y > 0 && i ≤ nBitsP,
        y += y;
        p = BitShiftLeft[p] + Floor[y];
        {y, i} = {y - Floor[y], i + 1}];
      p = BitShiftLeft[p, nBitsP + 1 - i]; i = BitAnd[p, 1];
      p = BitShiftRight[p];
      (* Sign bit is 0; rounding bit is shifted away. *)
      If[i ≠ 0, If[BitAnd[p, 1] ≠ 0 ||
          (e ≠ 0 && y > BitShiftLeft[1, e]) || y ≠ 0, p++]]
    ];
    If[x < 0, -p, p]
  ]
]
```

A.5 The OverBar Rounding Operator

```
x_ := pToX[xToP[x]];
SetAttributes[pToX, Listable];
SetAttributes[xToP, Listable];
SetAttributes[OverBar, Listable]
```

The **OverBar** is defined as an operator that takes a real input, converts it to the closest posit bit string, then converts that back to a "real" as supported by *Mathematica*. This eliminates the need to build software support for each of the math routines specified in the Posit Standard; we can write normal code but annotate it with overbars to show exactly where an operation is rounded.

SetAttributes is used here to make the three functions *listable*, which means they can apply to entire lists and even nested lists like multidimensional arrays. For example, define a Hilbert matrix:

```
H = Table[1 / (i + j - 1), {i, 1, 3}, {j, 1, 3}];
MatrixForm[H]
```

$$\begin{pmatrix} 1 & \frac{1}{2} & \frac{1}{3} \\ \frac{1}{2} & \frac{1}{3} & \frac{1}{4} \\ \frac{1}{3} & \frac{1}{4} & \frac{1}{5} \end{pmatrix}$$

Suppose we are using a ⟨16,2⟩ posit environment. Instead of having to put an overbar on each entry, the **Listable** attribute lets us do this:

```
setPositEnv[{nBitsP = 16, eS = 2}];
Hp = H̄;
MatrixForm[Hp]
N[MatrixForm[Hp], 5]
```

$$\begin{pmatrix} 1 & \frac{1}{2} & \frac{2731}{8192} \\ \frac{1}{2} & \frac{2731}{8192} & \frac{1}{4} \\ \frac{2731}{8192} & \frac{1}{4} & \frac{3277}{16384} \end{pmatrix}$$

$$\begin{pmatrix} 1.0000 & 0.50000 & 0.33337 \\ 0.50000 & 0.33337 & 0.25000 \\ 0.33337 & 0.25000 & 0.20001 \end{pmatrix}$$

Notice that the output of the overbar operator is always in the form of a rational number. For all nonzero real inputs, the output will be in the form $m \times 2^j$ where m is an odd integer and j is an integer. Sometimes that is the easiest way to read them, and sometimes it is easier to read them as decimals. The **N[...]** function expresses a number in decimal form, with the option of specifying how many digits, and it is used throughout this book. For example:

```
N[π]
N[π, 6]
```

3.141592654

3.14159

As shown in Section 2.8, page 33, and in the *Posit Standard (2022)*, five decimals are sufficient to express a 16-bit standard posit such that it will always convert back to the exact same binary, so the Hilbert matrix used **N[MatrixForm[Hp],5]** to display the posit form as a decimal.

Appendix B Float Code Listings

Throughout this book, there are times when we wish to compare posits with IEEE-type float format. These routines generalize the latter for any precision four bits or greater and also allow a flexible number of bits for the exponent field.

B.1 setFloatEnv (Set the Float Environment)

The **setFloatEnv** [n, e] routine must be called before working with floats; it sets the precision **nBitsF** to n bits and the exponent size **eSizeF** to e bits. The routine sets global variables that are used by routines like **xToF** and **fToX**.

```
setFloatEnv[{n_Integer /; n ≥ 4, e_Integer /; e ≥ 2}] :=
 (
  {nBitsF, eSizeF, fSize} = {n, e, n - e - 1};

  bias = 2^(eSizeF-1) - 1;
  smallSubnormal = 2^(1-bias-fSize);
  smallNormal = 2^(1-bias);
  maxFloat = 2^bias × (2 - 2^(-fSize));
  minRoundable = smallSubnormal / 2;

  maxRoundable = 2^bias × (1 + (2^fSize - 1/2) / 2^fSize);
 )
```

The "$/;$" applies the condition that follows to the argument and exits if it is not met. So the first line ensures the requested precision n is an integer greater than four and the requested number of exponent bits is an integer greater than or equal to two. If conditions are met, the integer pair **{n,e}** is assigned to **{nBitsF, eSizeF}**, and the fraction size is inferred to be **n-e-1** bits. If the exponent is less than two bits, IEEE exception rules create a nonsensical system where all the bit patterns represent exception cases. With fewer than four total bits, there would be no fraction bit and thus no way to distinguish between infinities and NaN.

The way IEEE floats represent negative numbers in the exponent is not with 2's complement but with a *bias*, which creates an asymmetry in the magnitudes that can be represented. The "normal" float magnitudes are centered around 2.0 instead of 1.0, with "subnormal" floats so small that their reciprocal overflows the format.

After computing the bias, $2^{eSizeF-1} - 1$, the routine computes the value of the smallest normal float (**smallNormal**), the smallest subnormal float (**smallSubnormal**), the largest finite real that can be represented (**maxFloat**) and the values that determine when a value underflows to zero or overflows to infinity (**minRoundable**, **maxRoundable**). These are useful for converting real numbers to float format, and for assessing the dynamic range that the float format can represent.

IEEE half precision is set with **setFloatEnv**[{16,5}]. A single precision float environment is set with **setFloatEnv**[{32,8}]. Double precision is set using **setFloatEnv**[{64,11}].

B.2 fToX (float to X)

The **fToX** function takes a float bit string **f** and converts it into a real number in the *Mathematica* environment. The bit string is treated as a signed integer.

```
fToX[f_ /; (Floor[f] == f) && (-2^nBitsF-1 ≤ f < 2^nBitsF-1)] :=
 Module[{exp, fint, frac, sgn},
  fint = Floor[f]; (* Insure f is of Integer type *)
  sgn = If[fint < 0, -1, 1];
  exp = BitShiftRight[BitAnd[BitShiftLeft[1, eSizeF + fSize] -
      BitShiftLeft[1, fSize], fint], fSize];
  frac = BitAnd[BitShiftLeft[1, fSize] - 1, fint];
  If[exp == BitShiftLeft[1, eSizeF] - 1,
   Return[If[frac == 0, sgn × ∞, Indeterminate]]];
  If[exp == 0, Return[sgn × 2^1-bias-fSize × frac]];
  sgn × 2^exp-bias × (1 + frac / 2^fSize)
 ]
```

The code extracts the bit fields of the exponent **exp** and fraction **frac**. Exception cases occur when **exp** is all bits are 0 (subnormal values and zeros) or all bits are 1 (**If[exp==BitShiftLeft[1,eSizeF]-1]**). In the latter case, if **frac** is zero then the number represents signed infinity; otherwise it is a NaN, which in *Mathematica* is equivalent to **Indeterminate**.

B.3 xToF (X to float)

The **xToF** function converts a real number to IEEE float format.

```
xToF[x_] := Module[{e = 0,
   eMask = BitShiftLeft[1, eSizeF + fSize] - BitShiftLeft[1, fSize],
   f, rounding = 0, y},
  If[(x ∈ Reals) =!= True && Abs[x] =!= ∞,
   Return[BitShiftLeft[1, nBitsF - 1] - 1]]; (* NaN cases *)
  y = Abs[x];
  f = Which[
    y ≥ maxRoundable, eMask,   (* -∞ or ∞ *)
    y ≤ smallNormal, Round[y / smallSubnormal],
    True,
    While[y ≥ 2, {y, e} = {y / 2, e + 1}];
    While[y < 1, {y, e} = {y * 2, e - 1}];
    y = (y - 1) * 2^{fSize+1};
    f = Floor[y];
    If[BitAnd[f, 1] ≠ 0,
      If[BitAnd[f, 2] ≠ 0 || y ≠ f, rounding = 1]];
    BitOr[Floor[f / 2],
      BitShiftLeft[bias + e, fSize]] + rounding];
  If[x < 0, f - BitShiftLeft[1, nBitsF - 1], f]
]
```

The first step is to determine if the input **x** is a real number or a signed infinity; if neither, a NaN value is returned (a string of all **1** bits). If it is not NaN, the absolute value of **x** is stored in **y** to simplify the code to find the scaling and significant bits. The **while** loops scale **y** to be in the range [1, 2) and find the exponent e in the scaling 2^e. The fraction bits are then found by removing the hidden bit from **y** (by subtracting 1) and using banker's rounding (the IEEE default rounding. The **BitOr** step assembles the bit fields of biased exponent and rounded fraction into **f**.

The last line checks if the input **x** was negative, and if so, returns a sign-magnitude form of the bit string **f**.

B.4 The UnderBar Operator and Examples

With the **fToX** and **xToF** routines, the **UnderBar** operator can be assigned to round a number or expression to the nearest float value in whatever environment has been set. The **SetAttributes** commands make it possible to apply the routines to arrays of numbers.

```
x_ := fToX[xToF[x]];
SetAttributes[fToX, Listable];
SetAttributes[xToF, Listable];
SetAttributes[UnderBar, Listable]
```

As an example, set the environment to IEEE single precision, and find the number closest to π. Then compute $\pi^2 - \pi^2$ with that approximation to π, using a rounded multiply followed by a fused multiply-add:

```
setFloatEnv[{32, 8}]
pifloat = π
N[pifloat, 23]
pifloat * pifloat - pifloat * pifloat
```

$$\frac{13\,176\,795}{4\,194\,304}$$

$$3.14159\,27410\,12573\,24218\,75$$

$$-\frac{2\,005\,671}{17\,592\,186\,044\,416}$$

If the next step in the calculation was to find the square root, the float environment would incorrectly report that $\sqrt{\pi^2 - \pi^2}$ is NaN instead of 0.

Here is an example of using the underscore to show what happens in IEEE single precision when adding 1 to 100 million (10^8), rounding, then subtracting 100 million and rounding:

$$10^8 + 1 - 10^8$$

0

There is no equivalent of a quire in float arithmetic, so even the sum of three numbers can be disastrously wrong (zero instead of one in this example).

Here is one last example of how these routines can expose hazards of floating-point arithmetic. Suppose we attempt to evaluate $n^2 \big/ \sqrt{n^3 + 1}$ with $n = 42$, using IEEE half-precision arithmetic. This is the example shown in Subsection 6.6.5.

```
setFloatEnv[{16, 5}];
n = 42;
Print["Value correct to ten decimals:    ", n² / √(n³ + 1.)]
Print["Value computed with 16-bit floats: ", (n × n) / √(n × n × n + 1)]
```

```
Value correct to ten decimals:    6.48069 6962
Value computed with 16-bit floats: 0
```

In computing $n \times n \times n$, the too-large result of 74 088 is replaced with infinity by IEEE Std 754 rules. Division by infinity then returns a 0 value instead of a value close to 6.48.

What could possibly go wrong?

Appendix C The **hypot** Function

"EVERYONE'S USING YOUR THEOREM, PYTHAGORAS. I
TOLD YOU YOU SHOULD HAVE PATENTED IT."

The Posit Standard (2022) requires support for the hypotenuse function, **hypot**$(x, y) = \sqrt{x^2 + y^2}$, which means it must be rounded the way an *exact* computation of $\sqrt{x^2 + y^2}$ would round. This can be done via the quire, but the details are not obvious, so they are shown here.

To keep the intermediate values close to the fovea, we use this identity:

$$\sqrt{x^2 + y^2} = |y| \cdot \sqrt{1 - (x/y)^2} \text{ where we sort } x \text{ and } y \text{ so that } |x| \le |y|.$$

As always, check that both inputs are real numbers and return NaR if either one is not. This is a little cryptic in *Mathematica*, but

```
If[(xInput ∈ Reals) =!= True || (yInput ∈ Reals) =!= True,
```

translates to "If the test of the x input being an element of the reals is not identically "True," or the test of the y input being an element of the reals is not identically "True," then we have at least one invalid input.

We have to do the test with "`=!=`" (not identically equal) because if you test directly with "`xInput ∈ Reals`" and `xInput` is something crazy, you will get a result that returns neither `True` nor `False`, like this:

```
Print["The square root of 2 is real: ", √2 ∈ Reals]
Print["The square root of -1 is real: ", √-1 ∈ Reals]
Print["Zebra is real: ", Zebra ∈ Reals]
```

```
The square root of 2 is real: True
The square root of -1 is real: False
Zebra is real: Zebra ∈ ℝ
```

Because *Mathematica* has no idea what Zebra is, it has no choice but to echo back the question instead of evaluating it as true or false.

To show why the **hypot** routine is a challenge, here is a naive implementation where there are four rounding steps that can accumulate error. In many cases the returned result is *not* the same as $\overline{\sqrt{x^2 + y^2}}$ (that is, the exact value after just one rounding to get back to the Compute Layer):

```
hypotNaive[xInput_, yInput_] := Module[{x, y},
  If[(xInput ∈ Reals) =!= True || (yInput ∈ Reals) =!= True,
    Return[Indeterminate]]; (* Return NaR. *)

  {x, y} = {Abs[xInput], Abs[yInput]};
  If[y < x, {x, y} = {y, x}]; (* Insure x ≤ y. *)

  If[y == maxPos || y == 0, Return[y], (* Exception cases *)

  Return[y × √(1 + (x / y)²)]] (* Four roundings *)

]
```

Notice that we take care of exception cases where the sorted largest value *y* is *maxPos* or zero, both of which would cause trouble in the line that follows.

The following test creates the set of all 8-bit standard posits, applies the `hypotNaive` function, then plots the ULP errors:

```
setPositEnv[{8, 2}];
set = Table[pToX[p], {p, NaR, -NaR - 1}];
hypotTable = Table[
    xToP[hypotNaive[set⟦i⟧, set⟦j⟧]] - xToP[√(set⟦i⟧² + set⟦j⟧²)],
    {i, 1, Length[set]}, {j, 1, Length[set]}];
ListPlot3D[hypotTable]
```

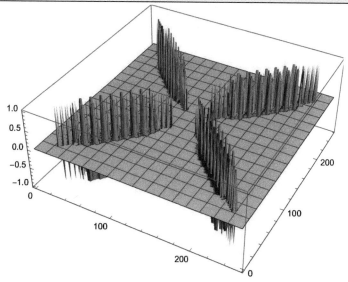

Error plot of naive computation of $\sqrt{x^2 + y^2}$ for standard 8-bit posits

The function is correct in the flat areas, but the spikes indicate ±1 ULP errors near the diagonal where *x* and *y* are close in value. The four roundings accumulate to an error of as much as ±0.75 ULP, and for those outside the ±0.5 ULP range, the rounding will be 1 ULP too high or too low. The error is never more than 1 ULP in either direction.

The first thing the quire can do is find out if the computed value is exact, by computing the square of the computed value minus $(x^2 + y^2)$ in the quire. That has the form of a dot product, and the only way it can then round back to a 0 posit is if it is exactly zero.

If it is inexact, then it lies between two values r_{Lo} and r_{Hi} that are 1 ULP apart. Which side of the tie point are they on? For now, assume a linear binade that is not in the Twilight Zone, so the tie point is the arithmetic average $(r_{Lo} + r_{Hi})/2$.

Let r be the true value of $\sqrt{x^2 + y^2}$. If r is exactly on the tie point, then

$$r = (r_{Lo} + r_{Hi})/2.$$

Multiply both sides by 2 and square:
$$4 \cdot r^2 = 4 \cdot \left(x^2 + y^2\right) = r_{Lo}^2 + 2 \cdot r_{Lo} \cdot r_{Hi} + r_{Hi}^2.$$

Hence, use the quire to compute
$$q = 4 \cdot \left(x^2 + y^2\right) - 2 \cdot r_{Lo} \cdot r_{Hi} - r_{Lo}^2 - r_{Hi}^2.$$

Multiplying by 2 and 4 in the quire is done by adding the quire to itself. The value q then serves as a discriminant; if q is negative, then r_{Lo} is closer to the square root r. If q is positive, then r_{Hi} is closer. If q is zero, then it is a tie point, and by banker's rounding rules, we choose the value r_{Lo} or r_{Hi} for which the posit representation ends in a **0** bit. Converting the quire value q back to a posit preserves the sign or zero value of q. Here is a way to compute the discriminant:

$$q = x \times x + y \times y;$$
$$q = q + q;$$
$$q = q - r_{Lo} \times r_{Hi};$$
$$q = q + q;$$
$$q = q - r_{Lo} \times r_{Lo} - r_{Hi} \times r_{Hi};$$

Here is the complete `hypot` routine that rounds correctly for all inputs:

```
hypot[xInput_, yInput_] := Module[{p, r, rLo, rHi, q, t, x, y},
  If[(xInput ∈ Reals) =!= True || (yInput ∈ Reals) =!= True,
    Return[Indeterminate]]; (* Return NaR. *)

  {x, y} = {Abs[xInput], Abs[yInput]};
  If[y < x, {x, y} = {y, x}]; (* Insure x ≤ y *)

  If[y == maxPos || y == 0, Return[y]]; (* Exception cases *)

  t = y × √(1 + (x / y)²) ; (* t is within one ULP of r *)

  q = t × t - (x × x + y × y); (* q is a quire, so exact *)
  p = q̄; (* Rounding preserves the sign. *)
  If[p == 0, Return[t]];
  (* If root is exact, we're done. *)

  If[p < 0, (* Else find the bounding posits for t. *)
    {rLo, rHi} = {t, pToX[xToP[t] + 1]}, (* else *)
    {rLo, rHi} = {pToX[xToP[t] - 1], t}
  ];

  q = x × x + y × y; (* Compute the discriminant. *)
  q = q + q;
  q = q - rLo × rHi;
  q = q + q;
  q = q - rLo × rLo - rHi × rHi;
  p = q̄; (* Rounding preserves the sign. *)

  If[p > 0, Return[rHi]];
  If[p < 0, Return[rLo]]; (* else tie goes to even. *)
  If[BitAnd[xToP[rLo], 1] == 0, rLo, (* else *) rHi]
]
```

Run the test again, and confirm that there are no errors for all possible inputs, using posit⟨8,2⟩ as the test set. Notice that all $2^8 = 256$ values are tested for both inputs; the test includes NaR.

```
setPositEnv[{8, 2}];
set = Table[pToX[p], {p, NaR, -NaR - 1}];
hypotTable =
   Table[xToP[hypot[set[[i]], set[[j]]]] - xToP[√(set[[i]]² + set[[j]]²)],
   {i, 1, Length[set]}, {j, 1, Length[set]}];
ListPlot3D[hypotTable]
```

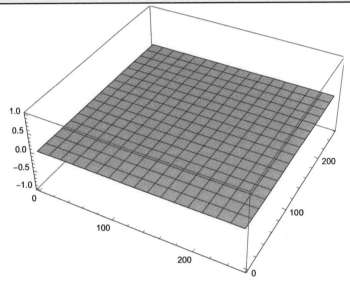

Error plot of quire-assisted computation of $\sqrt{x^2 + y^2}$ for standard 8-bit posits

The graphic output confirms that every value is rounded correctly.

What about the Twilight Zone? There, the midpoint is $\sqrt{r_{Lo} \cdot r_{Hi}}$. The test is even easier, just compute

$$q = x \times x + y \times y - r_{Lo} \times r_{Hi}$$

as the discriminant and test that for less than, greater than, or equal to zero as before. We did not need to do that for posit⟨8,2⟩ because luckily there were no cases where the two ways to compute the discriminant differed in sign. Had we used posit⟨9,0⟩ as the test case, say, there would have been several incorrect roundings without using a different test for the Twilight Zone. Including that test in **hypot** is left as an exercise for the reader.

Appendix D *L* Table for 8-Bit Posits

This Appendix shows how to construct a fast arithmetic *L* table for multiplications of standard 8-bit posits. That is, we find a set of integers where the sums have the same rankings as the products of the posits they correspond to, so that we can replace multiplication hardware with integer maps and integer adders that are quite a bit faster and take up less area for low precision.

D.1 First Solve the Densest Binade

The binade [1, 2) is always inside the fovea. If you can find the rank table for that, it can be used to find the rank tables for all the other binades, easily. That binade for ⟨8, 2⟩ standard posits has the following values:

$$\left\{1, \frac{9}{8}, \frac{5}{4}, \frac{11}{8}, \frac{3}{2}, \frac{13}{8}, \frac{7}{4}, \frac{15}{8}, 2\right\}$$

By including the open endpoint at 2, we will find out how to chain the binades together. Scaling that set does not change the ranking, so it is easier to read if we multiply it by 8:

$$x = \{8, 9, 10, 11, 12, 13, 14, 15, 16\}$$

There are 43 distinct products in the multiplication table for those nine values:

✕	8	9	10	11	12	13	14	15	16
8	64	72	80	88	96	104	112	120	128
9	72	81	90	99	108	117	126	135	144
10	80	90	100	110	120	130	140	150	160
11	88	99	110	121	132	143	154	165	176
12	96	108	120	132	144	156	168	180	192
13	104	117	130	143	156	169	182	195	208
14	112	126	140	154	168	182	196	210	224
15	120	135	150	165	180	195	210	225	240
16	128	144	160	176	192	208	224	240	256

Multiplication table for significands with three fraction bits

Therefore, the multiplication rank plot has 43 ranks, rank **0** to rank **42**:

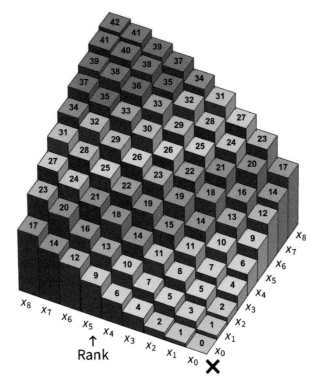

Rank table for products of significands with three fraction bits

This brute-force code finds the equivalent *L* table and prints it:

```
Clear[L];
For[L1 = 10, L1 < ∞, L1++,
  For[L2 = L1 + 1, L2 < 2 L1, L2++,
    For[L3 = 2 L1 + 1, L3 < L1 + L2, L3++,
      If[2 L2 > L1 + L3,
        For[L4 = L1 + L2 + 1, L4 < L1 + L3, L4++,
          If[2 L3 > L2 + L4 && L2 + L3 > L1 + L4,
            For[L5 = 2 L2 + 1, L5 < L1 + L4, L5++,
              If[2 L4 > L3 + L5,
                For[L6 = L2 + L3 + 1, L6 < L1 + L5, L6++,
                  If[2 L5 > L4 + L6,
                    L7 = L2 + L4;
                    L8 = 2 L4 - L1;
                    If[2 L6 > L5 + L7 && 2 L7 > L6 + L8,
                      L = {0, L1, L2, L3, L4, L5, L6, L7, L8};
                      Abort[]]]]]]]]]]]]
L
```

$Aborted

{0, 18, 33, 47, 60, 72, 83, 93, 102}

Despite the deep nesting, the code runs in a fraction of a second. Here is the addition table for that set of integers *L*:

+	0	18	33	47	60	72	83	93	102
0	0	18	33	47	60	72	83	93	102
18	18	36	51	65	78	90	101	111	120
33	33	51	66	80	93	105	116	126	135
47	47	65	80	94	107	119	130	140	149
60	60	78	93	107	120	132	143	153	162
72	72	90	105	119	132	144	155	165	174
83	83	101	116	130	143	155	166	176	185
93	93	111	126	140	153	165	176	186	195
102	102	120	135	149	162	174	185	195	204

Addition table ranks for *L* = {0, 18, 33, 47, 60, 72, 83, 93, 102} match the multiplication table ranks.

It produces a 3D rank plot identical to the one for multiplication.

D.2 Chain *L* Sets for Other Binades

Because these work like logarithms, the binade from 2 to 4 can be found by adding the last element of the *L* list to itself and chaining the original *L* list to the new one.

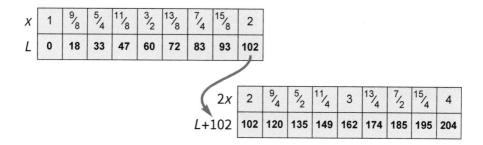

Multiplying *x* by 2 corresponds to adding 102 to *L*.

Or we can regard that as the two binades from $\frac{1}{2}$ to 2. The integers range from 0 to 204. We had to include the value 2 in the set *x* to find out that a binade is of size 102 in the *L* list.

Repeat the chaining process to cover four binades, and again to cover eight binades:

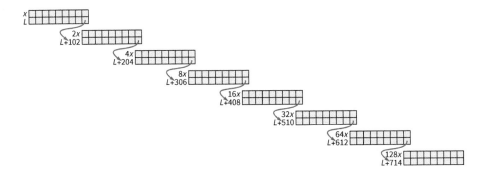

The fovea of standard posits has eight binades; repeat the chaining process to cover all eight.

The *L* list now ranges from 0 to 816 and covers the eight binades in the fovea of 8-bit standard posits, visible as eight notches in the wobble.

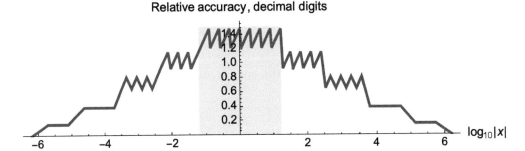

Accuracy plot for posit(8,2) with the fovea highlighted in light blue

D.3 Chain Binades with Lower Accuracy

Just outside the fovea, both left and right, are binades with only two fraction bits instead of 3. For example, the [16, 32) binade contains only the four posit values {16, 20, 24, 28}. The chaining idea still works, but we leave out every other integer in the *L* list for the fovea, the ones grayed out here:

$$\{\mathbf{0}, 18, \mathbf{33}, 47, \mathbf{60}, 72, \mathbf{83}, 93, \mathbf{102}\}$$

Chain that four times to the beginning and end of the *L* found so far. Then simplify the binade values to {**0**, **60**, **102**} for the binades with a single fraction bit. And so on. The next page shows *Mathematica* code that produces the *L* for all positive real values represented by 8-bit standard posits:

```
L = {0, 18, 33, 47, 60, 72, 83, 93, 102}; (* 3-bit fraction *)
L = Union[L, L + Last[L]]; (* double to 2 binades *)
L = Union[L, L + Last[L]]; (* double to 4 binades *)
L = Union[L, L + Last[L]]; (* fovea has 8 binades *)


Le = {0, 33, 60, 83, 102}; (* 2-bit fraction *)
Le = Union[Le, Le + Last[Le]]; (* double to 2 binades *)
Le = Union[Le, Le + Last[Le]]; (* prepend/append 4 binades *)
L = Union[Le, L + Last[Le], Le + Last[L] + Last[Le]];


Le = {0, 60, 102}; (* 1-bit fraction *)
Le = Union[Le, Le + Last[Le]]; (* double to 2 binades *)
Le = Union[Le, Le + Last[Le]]; (* prepend/append 4 binades *)
L = Union[Le, L + Last[Le], Le + Last[L] + Last[Le]];


Le = {0, 102}; (* fraction begins at first ghost bit *)
Le = Union[Le, Le + Last[Le]]; (* double to 2 binades *)
Le = Union[Le, Le + Last[Le]]; (* prepend/append 4 binades *)
L = Union[Le, L + Last[Le], Le + Last[L] + Last[Le]];


Le = {0, 204}; (* Twilight Zone, spacing is 4x apart*)
Le = Union[Le, Le + Last[Le]]; (* prepend/append 2 binades *)
L = Union[Le, L + Last[Le], Le + Last[L] + Last[Le]];


Le = {0, 408}; (* minPos and maxPos are 16x away *)
L = Union[Le, L + Last[Le], Le + Last[L] + Last[Le]];
```

Here is the set *L* as a paragraph:

L
{0, 408, 612, 816, 918, 1020, 1122, 1224, 1284, 1326, 1386, 1428, 1488, 1530, 1590, 1632, 1665, 1692, 1715, 1734, 1767, 1794, 1817, 1836, 1869, 1896, 1919, 1938, 1971, 1998, 2021, 2040, 2058, 2073, 2087, 2100, 2112, 2123, 2133, 2142, 2160, 2175, 2189, 2202, 2214, 2225, 2235, 2244, 2262, 2277, 2291, 2304, 2316, 2327, 2337, 2346, 2364, 2379, 2393, 2406, 2418, 2429, 2439, 2448, 2466, 2481, 2495, 2508, 2520, 2531, 2541, 2550, 2568, 2583, 2597, 2610, 2622, 2633, 2643, 2652, 2670, 2685, 2699, 2712, 2724, 2735, 2745, 2754, 2772, 2787, 2801, 2814, 2826, 2837, 2847, 2856, 2889, 2916, 2939, 2958, 2991, 3018, 3041, 3060, 3093, 3120, 3143, 3162, 3195, 3222, 3245, 3264, 3324, 3366, 3426, 3468, 3528, 3570, 3630, 3672, 3774, 3876, 3978, 4080, 4284, 4488, 4896}

On the next page is a table of the positive posit bit strings, the value they represent, and their corresponding *L* values.

Posit	Value	L	Posit	Value	L	Posit	Value	L	Posit	Value	L
00 000 001	$6. \times 10^{-8}$	0	00 100 001	0.07	2058	01 000 001	1.1	2466	01 100 001	20.	2889
00 000 010	$1. \times 10^{-6}$	408	00 100 010	0.08	2073	01 000 010	1.3	2481	01 100 010	24.	2916
00 000 011	$4. \times 10^{-6}$	612	00 100 011	0.086	2087	01 000 011	1.4	2495	01 100 011	28.	2939
00 000 100	0.00002	816	00 100 100	0.09	2100	01 000 100	1.5	2508	01 100 100	30.	2958
00 000 101	0.00003	918	00 100 101	0.1	2112	01 000 101	1.6	2520	01 100 101	40.	2991
00 000 110	0.00006	1020	00 100 110	0.11	2123	01 000 110	1.8	2531	01 100 110	50.	3018
00 000 111	0.0001	1122	00 100 111	0.12	2133	01 000 111	1.9	2541	01 100 111	56.	3041
00 001 000	0.0002	1224	00 101 000	0.125	2142	01 001 000	2.	2550	01 101 000	60.	3060
00 001 001	0.0004	1284	00 101 001	0.14	2160	01 001 001	2.3	2568	01 101 001	80.	3093
00 001 010	0.0005	1326	00 101 010	0.16	2175	01 001 010	2.5	2583	01 101 010	100.	3120
00 001 011	0.0007	1386	00 101 011	0.17	2189	01 001 011	2.8	2597	01 101 011	110.	3143
00 001 100	0.001	1428	00 101 100	0.19	2202	01 001 100	3.	2610	01 101 100	130.	3162
00 001 101	0.0015	1488	00 101 101	0.2	2214	01 001 101	3.3	2622	01 101 101	160.	3195
00 001 110	0.002	1530	00 101 110	0.22	2225	01 001 110	3.5	2633	01 101 110	200.	3222
00 001 111	0.003	1590	00 101 111	0.23	2235	01 001 111	3.8	2643	01 101 111	220.	3245
00 010 000	0.004	1632	00 110 000	0.25	2244	01 010 000	4.	2652	01 110 000	300.	3264
00 010 001	0.005	1665	00 110 001	0.28	2262	01 010 001	4.5	2670	01 110 001	400.	3324
00 010 010	0.006	1692	00 110 010	0.3	2277	01 010 010	5.	2685	01 110 010	500.	3366
00 010 011	0.007	1715	00 110 011	0.34	2291	01 010 011	5.5	2699	01 110 011	800.	3426
00 010 100	0.008	1734	00 110 100	0.38	2304	01 010 100	6.	2712	01 110 100	1000.	3468
00 010 101	0.01	1767	00 110 101	0.4	2316	01 010 101	6.5	2724	01 110 101	1500.	3528
00 010 110	0.012	1794	00 110 110	0.44	2327	01 010 110	7.	2735	01 110 110	2000.	3570
00 010 111	0.014	1817	00 110 111	0.47	2337	01 010 111	7.5	2745	01 110 111	3000.	3630
00 011 000	0.016	1836	00 111 000	0.5	2346	01 011 000	8.	2754	01 111 000	4000.	3672
00 011 001	0.02	1869	00 111 001	0.56	2364	01 011 001	9.	2772	01 111 001	8000.	3774
00 011 010	0.023	1896	00 111 010	0.6	2379	01 011 010	10.	2787	01 111 010	20000.	3876
00 011 011	0.027	1919	00 111 011	0.7	2393	01 011 011	11.	2801	01 111 011	30000.	3978
00 011 100	0.03	1938	00 111 100	0.75	2406	01 011 100	12.	2814	01 111 100	70000.	4080
00 011 101	0.04	1971	00 111 101	0.8	2418	01 011 101	13.	2826	01 111 101	$3. \times 10^5$	4284
00 011 110	0.05	1998	00 111 110	0.9	2429	01 011 110	14.	2837	01 111 110	$1. \times 10^6$	4488
00 011 111	0.055	2021	00 111 111	0.94	2439	01 011 111	15.	2847	01 111 111	$2. \times 10^7$	4896
00 100 000	0.06	2040	01 000 000	1.	2448	01 100 000	16.	2856	10 000 000	NaR	—

Color-coded posit bit strings for positive values and NaR, value represented, and *L* value

The **Value** columns show only as many significant decimals as needed such that they would convert back to the posit (see Section 2.8). For example, the posit `01111010` represents the exact value 16 384, but it is shown as 20 000, and `xToP[20000]` returns the correct bit string `01111010`. This helps make the table a little more readable.

D.4 Hardware Considerations

The sign of a product is the exclusive OR (XOR) of the signs of the inputs, so there is no need for the L values to cover both positive and negative inputs. Similarly, the cases of a 0 or NaR exception as an input are easily dealt with. We can therefore regard the inputs to the multiplier to be 7 bits by ignoring the sign.

We seek to map the 7-bit inputs to the corresponding L value as an unsigned integer. A 12-bit unsigned integer can represent values 0 to 4095, but the largest L value is 4896, slightly outside the range, so L will need to be 13 bits. Using a method similar to the "wired ROM" diagram in Section 9.4.4, the integer mapping of 7 bits to 13 bits can be done with 7 inverters, 127 AND gates, and 13 OR gates, a remarkably economical way to express every entry in the preceding table. Two such circuits should be constructed so they can operate simultaneously on the multiplier inputs.

Because there could be a carry, the adder of the two 13-bit L values needs to be 14 bits wide. Those 14-bit integer values then map to the output products. While the rounded posit values could be programmed into the wired ROM as output, that would be a missed opportunity if the reason for the multiplications is a dot product.

There are 1078 possible values for the sum of two L values, corresponding to the 1078 possible products of the input posit values. The exact products can all be written as $m \times 2^j$ where m is an *odd* integer, since any factors of 2 can be incorporated into the exponent j. In fact, once reduced to an odd m, there can only be 33 possible values for m:

$$m \in \{1, 3, 5, 7, 9, 11, 13, 15, 21, 25, 27, 33, 35, 39, 45, 49, 55, 63, 65,$$
$$75, 77, 81, 91, 99, 105, 117, 121, 135, 143, 165, 169, 195, 225\}$$

Because all the m values are odd, the last bit is always a 1 and need not be stored; it can be supplied by the hardware, like a "hidden bit." A 7-bit string then suffices to store every m value. The values of j range from -48 to 48, so again 7 bits suffice to store the amount of the shift into the quire, which is 128 bits in size for 8-bit standard posits.

The adder for the *L* values and for the accumulation of products into the quire must of course use techniques like carry lookahead, and there are well-known techniques for making integer addition as fast as possible.

For applications like AI, signal processing, image processing, and other applications that tolerate low relative accuracy but rely heavily on dot products, the wired ROM described here can be very fast, power-efficient, and chip area-efficient.

Appendix E Linear Solver Tests

E.1 The Four Solvers Tested

As a gold standard, we need a linear solver with no rounding. If the inputs are rational numbers (floats or posits), then the result will be an exact rational number unless the system is singular. Call it `linearSolve2By2`:

```
linearSolve2By2[{{a_, b_}, {c_, d_}}, {u_, v_}] := Module[{t, x, y},
  If[a × d - b × c == 0, Return["System is singular."]];
  t = c / a;
  y = (v - t × u) / (d - t × b);
  x = (u - b × y) / a;
  {x, y}
]
```

There is no pivoting needed here to protect against the case where **a** is zero, since there are no zero values in the **RRSet** test set. There are also no cases of a zero determinant, since the values are random 64-bit floats and the chances of a determinant being exactly zero are not impossible but extremely low (about 1 in $2^{106} \approx 10^{32}$). *Mathematica* uses extended-precision arithmetic to perform exact operations on rational numbers, and the result will not in general be expressible exactly as a float or a posit, that is, $m \times 2^j$ where m and j are integers. For example, solve $\begin{pmatrix} 1/2 & -3/8 \\ 80 & -5 \end{pmatrix} \cdot \begin{pmatrix} u \\ v \end{pmatrix} = \begin{pmatrix} 12 \\ -23 \end{pmatrix}$ as follows:

```
linearSolve2By2[{{1 / 2, -3 / 8}, {80, -5}}, {12, -23}]
```

$$\left\{ -\frac{549}{220}, -\frac{1943}{55} \right\}$$

The routine provides the exact answer in the Math Layer to compare against.

Solver 1 is **gauss,** a solver that uses Gaussian elimination without pivoting, rounding to the nearest posit.

```
gauss[{{a_, b_}, {c_, d_}}, {u_, v_}] := Module[{t, x, y},
   t = ‾c‾/‾a‾;
   y = ‾(v‾-‾t‾×‾u)‾ / ‾(d‾-‾t‾×‾b)‾;
   x = ‾(u‾-‾b‾×‾y)‾ / a;
   {x, y}
]
```

It fails this criterion:

■ If the problem is singular, that should be detected and reported.

There is also no protection from dividing by zero in the **t = c / a** step, which is another reason for the "not recommended" rating in the Ratings table in Section 12.3.5. The "price" of the routines here is their operation count, including rounding operations and any **If** tests. The price of **gauss** is 15 (three divides, three subtracts, three multiplies, and six roundings).

Solver 1 can easily be equipped with zero determinant protection and partial pivoting by putting a wrapper around it, so here is Solver 2:

```
gaussPivot[{{a_, b_}, {c_, d_}}, {u_, v_}] := Module[{det},
   det = ‾a‾×‾d‾-‾b‾×‾c‾;
   If[det == 0, Return["System is singular."]];
   If[Abs[c] > Abs[a], (* Swap rows? *)
     gauss[{{c, d}, {a, b}}, {v, u}], (* else *)
     gauss[{{a, b}, {c, d}}, {u, v}]
   ]
]
```

The determinant and **If[abs[c]>Abs[a]**...] tests take eight operations, raising the price from 15 to 23. We will see that pivoting markedly reduces the average rounding error from about 40 million ULPs to 100 000 ULPs, although that is still alarmingly high.

Solver 3 is Cramer's rule; it only requires five lines of code:

```
cramer[{{a_, b_}, {c_, d_}}, {u_, v_}] := Module[{det},
  det = a × d - b × c;
  If[det == 0, Return["System is singular."]];
  {(u × d - b × v) / det, (a × v - u × c) / det}
]
```

The price is 18 operations.

Lastly, Solver 4 is Cramer's rule with the possibility of refining the solution if the quire says that for the computed x, $A \cdot x - b \neq 0$.

```
cramerRefined[{{a_, b_}, {c_, d_}}, {u_, v_}] :=
  Module[{det, sx, sy, tx, ty},
  det = a × d - b × c;
  If[det == 0, Return["System is singular."]];
  {sx, sy} = {(u × d - b × v) / det, (a × v - u × c) / det};
  {tx, ty} = {u, v} - {{a, b}, {c, d}}.{sx, sy};   (*2 roundings*)
  If[{tx, ty} == {0, 0}, {sx, sy},
    {sx + tx × d - b × ty / det, sy + a × ty - tx × c / det}]
]
```

The price is 37 operations. Even though the work inside the second **If** clause only executes a fraction of the time, we still count it because it would have to be laid out in a dataflow circuit like those described in Chapter 12.

E.2 Can They Find Expressible Exact Answers?

Recall this criterion:

- If the exact answer is expressible in the number format, the solver should find it.

We need to generate test problems for which **A**, **x**, and **u** are all in the vocabulary of the numerical environment.

First, generate **RRSet** by the usual method:

```
len = 2^16; SeedRandom[31 416];
log10RRSet =
  Sort[Table[RandomVariate[NormalDistribution[0, 3]], {i, 1, len}]];
RRSet = Sort[10^log10RRSet *
    Table[(2 * RandomInteger[{0, 1}] - 1), {i, 1, len}]];
```

Set the environment to the standard posit⟨32,2⟩, and create a random number generator **f** that converts a number randomly chosen from **RRSet** to a posit, then zeros the rightmost 13 bits of the fraction by shifting right by 13 bits and back again. This makes it more likely that the problem **A·x = u** has an exact solution. The random number seed is reinitialized so that results will be repeatable. The number of trials **nTrials** is set to 10 000, and we initialize the number of cases where each method successfully finds the exact solution to 0.

```
setPositEnv[{32, 2}]
f := Module[{x = RRSet[[RandomInteger[{1,65 536}]]], p},
  p = xToP[x];
  p = BitShiftLeft[BitShiftRight[p, 13], 13];
  pToX[p]
]
SeedRandom[314 159];
nTrials = 10 000;
nGauss = nGaussPivot = nCramer = nCramerRefined = 0;
```

The testing loop creates a random **A** and a random **x** and computes **u** exactly as **A·x**, then tests If **u** = **ū**; if it is, then **u** is expressible without rounding, and it seems like a solver should be able to find that exact answer. Unfortunately, only one of the four solvers tested met that criterion. The Gaussian solvers found the solution for more than half of the test cases.

```
For[i = 1, i ≤ nTrials, i++,
  u = {π, π};
  While[u ≠ ū, (* Generate A, x until A·x is an exact posit. *)
   {A, x} = {{{f, f}, {f, f}}, {f, f}};
   u = A.x;
  ];
  If[gauss[A, u] == x, nGauss++];
  If[gaussPivot[A, u] == x, nGaussPivot++];
  If[cramer[A, u] == x, nCramer++];
  If[cramerRefined[A, u] == x, nCramerRefined++]
 ]
Print[Row[{"Gauss exact rate:           ",
    DecimalForm[nGauss * 100. / nTrials, 3], "%"}]];
Print[Row[{"GaussPivot exact rate:      ",
    DecimalForm[nGaussPivot * 100. / nTrials, 3], "%"}]];
Print[Row[{"Cramer exact rate:          ",
    DecimalForm[nCramer * 100. / nTrials, 3], "%"}]];
Print[Row[{"CramerRefined exact rate: ",
    DecimalForm[nCramerRefined * 100. / nTrials, 3], "%"}]];
```

```
Gauss exact rate:          53.8%
GaussPivot exact rate:     55.8%
Cramer exact rate:         40.1%
CramerRefined exact rate: 100.%
```

Based on those results, we can fill out one column of the Ratings table:

Ratings *2–by–2 linear solvers*

	Excellent	Very good	Good	Fair	Poor

Brand & model	Price	Overall score		Finds exact answer	Finds closest answer	Worst-case error	Average error
		0 100					
		P F G VG E					
32–BIT POSIT FORMAT *These cost less than legacy high–precision solvers.*							
Gauss No pivoting*	15			○			
Gauss Partial pivoting	23			○			
Cramer Base model	18			◒			
Cramer Refined	37			◉			

*We rate this brand *not recommended*. Use can result in severe injury to a calculation.

Partial ratings of linear solvers when the solution can be expressed without rounding error

E.3 Scoring the Inexact Cases

For the next three columns of the evaluation table, we make no restriction that the exact answer is expressible. Here we check how often it finds the closest representable answer, per this criterion:

■ If the exact answer is not expressible in the number format, the answer should be the pair of numbers closest to the exact answer.

```
f := RRSet[[RandomInteger[{1,Length[RRSet]}]]]
SeedRandom[314 159];
nTrials = 10 000;
{nGauss, gaussAvg, gaussWorst} = {0, 0, 0};
{nGaussPivot, gaussPivotAvg, gaussPivotWorst} = {0, 0, 0};
{nCramer, cramerAvg, cramerWorst} = {0, 0, 0};
{nCramerRefined,
    cramerRefinedAvg, cramerRefinedWorst} = {0, 0, 0};
```

The loop also measures the worst-case and average error in ULPs.

```
For[i = 1, i ≤ nTrials, i++,
 {A, u} = {{{f, f}, {f, f}}, {f, f}};
 xComp = LinearSolve[A, u];
 (* Gold standard answer, rounded *)
 {pX, pY} = xToP[xComp];

 xGauss = linearSolve2By2r[A, u];
 {pGX, pGY} = xToP[xGauss];
 ulpError = Abs[pGX - pX] + Abs[pGY - pY];
 If[ulpError == 0, nGauss++];
 gaussWorst = Max[gaussWorst, ulpError];
 gaussAvg += ulpError;

 xGaussPivot = If[Abs[A[[1,1]]] < Abs[A[[2,1]]], linearSolve2By2r[
    Reverse[A], Reverse[u]], linearSolve2By2r[A, u]];
 {pGPX, pGPY} = xToP[xGaussPivot];
 ulpError = Abs[pGPX - pX] + Abs[pGPY - pY];
 If[ulpError == 0, nGaussPivot++];
 gaussPivotWorst = Max[gaussPivotWorst, ulpError];
 gaussPivotAvg += ulpError;

 xCramer = cramer[A, u];
 {pCX, pCY} = xToP[xCramer];
 ulpError = Abs[pCX - pX] + Abs[pCY - pY];
 If[ulpError == 0, nCramer++];
 cramerWorst = Max[cramerWorst, ulpError];
 cramerAvg += ulpError;

 xCramerRefined = cramerRefined[A, u];
 {pCRX, pCRY} = xToP[xCramerRefined];
 ulpError = Abs[pCRX - pX] + Abs[pCRY - pY];
 If[ulpError == 0, nCramerRefined++];
 cramerRefinedWorst = Max[cramerRefinedWorst, ulpError];
 cramerRefinedAvg += ulpError;
]
```

Print the summarized test results:

```
Row[{"Gauss correct roundings:          ",
  N[100 * nGauss / nTrials, 3], "%"}]
Row[{"Gauss worst error:               ", gaussWorst, " ULPs"}]
Row[{"Gauss average error:             ",
  N[gaussAvg / nTrials, 3], " ULPs"}]

Row[{"GaussPivoting correct roundings:   ",
  N[100 * nGaussPivot / nTrials, 3], "%"}]
Row[{"Gauss with pivoting worst error:   ",
  gaussPivotWorst, " ULPs"}]
Row[{"Gauss with pivoting average error: ",
  N[gaussPivotAvg / nTrials, 3], " ULPs"}]

Row[{"Cramer correct roundings:          ",
  N[100 * nCramer / nTrials, 3], "%"}]
Row[{"Cramer worst error:                ",
  cramerWorst, " ULPs"}]
Row[{"Cramer average error:              ",
  N[cramerAvg / nTrials, 3], " ULPs"}]

Row[{"CramerRefined correct roundings:    ",
  N[100 * nCramerRefined / nTrials, 3], "%"}]
Row[{"CramerRefined worst error:         ",
  cramerRefinedWorst, " ULPs"}]
Row[{"CramerRefined average error:       ",
  N[cramerRefinedAvg / nTrials, 3], " ULPs"}]
```

```
Gauss correct roundings:            26.7%

Gauss worst error:                  4 225 314 317 ULPs

Gauss average error:                4.35 × 10^7 ULPs

GaussPivoting correct roundings:    39.6%

Gauss with pivoting worst error:    942 896 664 ULPs

Gauss with pivoting average error:  1.18 × 10^5 ULPs

Cramer correct roundings:           27.0%

Cramer worst error:                 1854 ULPs

Cramer average error:               5.50 ULPs

CramerRefined correct roundings:    100.%

CramerRefined worst error:          0 ULPs

CramerRefined average error:        0 ULPs
```

Based on those results, we can fill out the rest of the Ratings table. I gave a high weighting to the last column, the average error.

The Cramer "Base model" had a worst-case error of 1854 ULPs, which for 32-bit posits is actually a small relative error. The fovea of a posit⟨32,2⟩ has 28 bits, so even when an answer is off by 1854 ULPs, it likely has about five correct decimals in the answer.

Ratings *2–by–2 linear solvers*

		Excellent	Very good	Good	Fair	Poor
		◉	⊜	○	⊖	●

Brand & model	Price	Overall score		Finds exact answer	Finds closest answer	Worst-case error	Average error
		0 100	P F G VG E				
32–BIT POSIT FORMAT *These cost less than legacy high–precision solvers.*							
Gauss No pivoting*	15	3		○	●	●	●
Gauss Partial pivoting	23	21		○	○	●	●
Cramer Base model	18	63		⊖	●	○	⊖
Cramer Refined	37	99		◉	◉	◉	◉

*We rate this brand *not recommended*. Use can result in severe injury to a calculation.

Ratings based on the tests shown in this Appendix

Appendix F Linear Solvers

F.1 A Float Linear Solver

Here is a linear solver using elimination with pivoting, with rounding to the nearest float. Fused multiply-adds are used to halve the number of rounding errors:

```
linearSolveFloat[A_, u_] :=
 Module[{a = Transpose[Join[Transpose[A], {u}]],
   i, j, k, n = Length[u], piv, temp},
  For[i = 1, i ≤ n, i++, (* Forward elimination *)
   piv = i - 1 + (Position[a[[i;;n,i]], Max[a[[i;;n,i]]]])[[1,1]];
   For[j = 1, j ≤ n + 1, j++, (* Swap rows. *)
    temp = a[[piv,j]]; a[[piv,j]] = a[[i,j]]; a[[i,j]] = temp
   ];
   For[j = 2, j ≤ i, j++,
    a[[i,j-1]] = a[[i,j-1]] / a[[j-1,j-1]]; (* Note: no singularity check. *)
    For[k = 1, k ≤ j - 1, k++;
     a[[i,j]] = a[[i,j]] - a[[i,k]] × a[[k,j]] (* Main kernel. *)
    ]
   ];
   For[j = i + 1, j ≤ n + 1, j++,
    For[k = 1, k ≤ i - 1, k++,
     a[[i,j]] = a[[i,j]] - a[[i,k]] × a[[k,j]] (* Main kernel. *)
    ]
   ]
  ];
  For[i = n, i ≥ 1, i--, (* Backsolve *)
   For[k = i + 1, k ≤ n, k++, a[[i,n+1]] = a[[i,n+1]] - a[[i,k]] × a[[k,n+1]]];
   a[[i,n+1]] = a[[i,n+1]] / a[[i,i]]];
  a[[1;;n,n+1]] (* Solution is in column n+1. *)
 ]
```

For a large N-by-N linear system, the bulk of the work will be done in the lines labeled `(* Main kernel. *)`, which execute order N^3 times.

Protection against singular matrices can be optionally inserted as a test of whether $a_{[j-1,j-1]}$ is zero just before dividing by it. The "$;;$" notation in *Mathematica* means a range of indices, so $a_{[1;;n,n+1]}$ means the $n + 1^{\text{th}}$ column of elements in the **a** array.

F.2 A Posit Linear Solver

The posit version of the elimination solver is similar to the float version except for rounding. For *N* equations, the quire reduces the number of roundings to order N^2.

```
linearSolvePosit[A_, u_] :=
   Module[{a = Transpose[Join[Transpose[A], {u}]],
     i, j, k, n = Length[u], quire, piv, temp},
    For[i = 1, i ≤ n, i++, (* Forward elimination *)
     piv = i - 1 + (Position[a[i;;n,i], Max[a[i;;n,i]]])[1,1];
     For[j = 1, j ≤ n + 1, j++, (* Swap rows. *)
      temp = a[piv,j]; a[piv,j] = a[i,j]; a[i,j] = temp
     ];
     For[j = 2, j ≤ i, j++,
      a[i,j-1] = ‾a[i,j-1] / a[j-1,j-1]‾;
      quire = a[i,j]; (* Exact dot product: *)
      For[k = 1, k ≤ j - 1, k++, quire = quire - a[i,k] * a[k,j] ]
     ];
     a[i,j] = ‾quire‾
    ];
    For[j = i + 1, j ≤ n + 1, j++,
     quire = a[i,j]; (* Exact dot product: *)
     For[k = 1, k ≤ i - 1, k++, quire = quire - a[i,k] * a[k,j]];
     a[i,j] = ‾quire‾
    ]
   ];
   For[i = n, i ≥ 1, i--, (* Backsolve *)
    quire = a[i,n+1]; (* Exact dot product *)
    For[k = i + 1, k ≤ n, k++, quire = quire - a[i,k] * a[k,n+1]];
    a[i,n+1] = ‾quire‾;
    a[i,n+1] = ‾a[i,n+1] / a[i,i]‾
   ];
   a[1;;n,n+1] (* Solution is in column n+1. *)
  ]
```

Appendix G Takum Format

Because the only parameter in a takum is the number of bits of precision, and because takums fundamentally follow posit rules, the takum environment can be set using `setPositEnv`. The precision is global environment variable `nbitsP`. However, there are two additional parameters to set. The first is whether the system is logarithmic or linear. Set `logQ` to `True` for logarithmic takums and to `False` for linear takums before using takum format conversion routines. For example,

```
logQ = True;
```

Second, if we want to use logarithmic takums, then we also need to set the base of the logarithm to `tB` (short camelCase for takum base). The base `tB` is ignored if the `logQ` flag is `False`.

```
tB = 10^{1/16};
```

The `tToX` function does the usual checks for the validity of the input integer, the 0 and NaR exception cases, and also the case of `nBitsP` being as small as 2, which means the real value represented is just the sign of the input.

The `a` array is the binary digits of the input. It is a programming convenience to tack on enough ghost bits to make the total bit string at least 13 bits long so we are guaranteed to have at least one explicit fraction bit. For example, we can then extract the three regime bits (bits 3, 4, and 5 if numbered from left to right) with `r=Take[a,{3,5}]` and not have to worry about exceptions for when there are fewer than 5 bits in the bit string.

The `c` variable (for characteristic) is used to hold the integer value represented by the regime and the exponent bits after the regime. Remember that there is a hidden leading 1 bit, and then 1 is subtracted from the literal meaning of the exponent bits as an unsigned integer.

The fraction field is stored in **f**. The last line assembles the real value from the sign of the input, the scaling of 2^c, times either the base **tB** to the power **c+f** for logarithmic takums, or $1+$**f** for linear takums.

A technical detail specific to *Mathematica*: A linear takum is returned as a rational number, but a logarithmic takum is rounded to a machine-precision (16-decimal) float using **N**[**tB**$^{\text{c+f}}$, 16]. Otherwise, *Mathematica* keeps the expression in symbolic form, which after a few calculations leads to a very complicated expression that is slow to work with. Here is the whole routine for takum-to-real, called **tTox**:

```
tTox[t_ /; positQ[t]] := Module[{a, c, m, p, r, x},
  If[t == NaR, Return[Indeterminate]]; (* Exception values *)
  If[t == 0, Return[0]];
  If[nBitsP == 2, Return[Sign[t]]];
  a = IntegerDigits[Floor[t], 2, nBitsP];
  If[Length[a] < 13, a = Join[a, Table[0, 13 - Length[a]]]];
  r = Take[a, {3, 5}];
  If[a[[2]] == 0, r = {1, 1, 1} - r]; (* Direction bit check *)
  r = FromDigits[r, 2];
  c = If[r == 0, 0, FromDigits[Take[a, {6, 5 + r}], 2]];
  c = If[a[[2]] == 0, -2^(r+1) + 1 + c, 2^r - 1 + c];
  p = Length[a] - r - 5;
  f = FromDigits[Take[a, -p], 2] / 2^p;
  Sign[t] × If[logQ, N[tB^(c+f), 16], 2^c × (1 + f)]
]
```

As with **fTox** (float-to-real) and **pTox** (posit-to-real), this routine takes a quantity from the Compute Layer to the Math Layer. Now we need the inverse of that, a routine that can take any real number and put it in takum format, or **xToT**. This is a fairly long piece of code to explain, so I will rely on comments in the code.

```
xToT[x_] := Module[
  {a = Table[0, 4], b = If[logQ, tB, 2], e = 0, f = 0, i, r, t, y},
  Which[
   (x ∈ ℝ) =!= True, NaR, (* Make sure input is real. *)
   x == 0, 0, (* Dispense with the other exception case, 0. *)
   y = Abs[x]; (* Simplify coding by making it positive. *)
   y ≥ tToX[-NaR - 1], Sign[x] × (-NaR - 1), (* |x|≥maxReal? *)
   y ≤ tToX[1], Sign[x], (* |x|≤ minReal? *)

   True, (* Find regime bits *)
   If[y ≥ 1, {r, a⟦1⟧} = {0, 1}; While[y ≥ b^(2^(r+1)-1), r++];
    y = N[y / b^(2^r-1)], (* else y < 1. *)
    {r, a⟦1⟧} = {7, 0}; (* Count down from 7, not up to 7. *)
    While[y < b^(1-2^(8-r)), r--]; (* Now we know the regime. *)
    y = N[y / b^(1-2^(8-r))]]; (* Use numeric, not symbolic ops. *)
   a⟦2;;4⟧ = IntegerDigits[r, 2, 3]; (* r as base 2, 3 digits *)
   If[a⟦1⟧ == 0, r = 7 - r]; (* a⟦1⟧ = direction bit. *)

   While[y ≥ b, {e, y} = {e + 1, y / b}] (* Find exponent bits *);
   a = Join[a, IntegerDigits[e, 2, r]]; (* Bit string grows. *)

   i = 5 + r; (* i is a pointer to the next bit. *)

   If[logQ,
    While[i ≤ Max[13, nBitsP],
     {i, y} = {i + 1, y × y}; (* Loop to find log base b *)
     If [y ≥ b, {a, y} = {Append[a, 1], y / b}, a = Append[a, 0]]
    ], (* else *)
    y = y - 1; (* Remove hidden bit; find fraction bits *)
    While[i ≤ Max[13, nBitsP],
     {i, y} = {i + 1, y + y};
     If [y ≥ 1, {a, y} = {Append[a, 1], y - 1}, a = Append[a, 0]]]
   ];
   i--;
   (* Back off to point to last assigned bit. *)

   t = FromDigits[Take[a, nBitsP - 1], 2] (* Build integer *);

   If[a⟦nBitsP⟧ ≠ 0, (* Banker's rounding logic as usual *)
    If[a⟦nBitsP-1⟧ ≠ 0 || y > 0, t = t + 1]];
   If[x < 0, -t, t]
   (* apply sign to the takum as an integer. *)
  ]
]
```

While *Mathematica* allows the vector symbol over a variable to notate vector quantities, like $\vec{b}$, the notation is called **OverVector** and can actually be assigned a function "round to nearest takum" just as we use **UnderBar** to round to nearest float and **OverBar** to round to nearest posit. The technique is the same: Convert whatever is under the vector line to takum format, then convert the takum back to a real number:

```
x⃗ := tToX[xToT[x]];
SetAttributes[tToX, Listable];
SetAttributes[xTot, Listable];
SetAttributes[OverVector, Listable]
```

Graphically test the round trip from a takum to real and back:

```
setPositEnv[{16, 0}];
DiscretePlot[xToT[tToX[t]], {t, NaR, -NaR - 1}, ImageSize → 230]
```

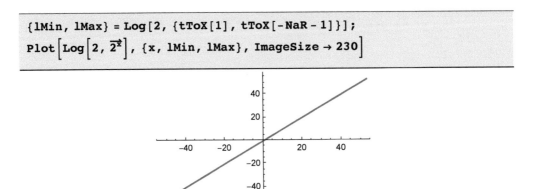

Takum-to-real-to-takum conversion appears perfect for all 65536 bit patterns.

Graphically test the round trip from a real to a takum and back:

```
{lMin, lMax} = Log[2, {tToX[1], tToX[-NaR - 1]}];
Plot[Log[2, 2^x⃗], {x, lMin, lMax}, ImageSize → 230]
```

Real-to-takum-to-real conversion appears perfect for reals spanning their dynamic range

A bug in the routine for either conversion direction usually shows up instantly in such plots as something other than a diagonal line.

Index